Megadrought and Collapse

Megadrought and Collapse

From Early Agriculture to Angkor

Edited by HARVEY WEISS

OXFORD
UNIVERSITY PRESS

Oxford University Press is a department of the University of Oxford. It furthers
the University's objective of excellence in research, scholarship, and education
by publishing worldwide. Oxford is a registered trade mark of Oxford University
Press in the UK and certain other countries.

Published in the United States of America by Oxford University Press
198 Madison Avenue, New York, NY 10016, United States of America.

© Oxford University Press 2017

CIP data is on file at the Library of Congress
ISBN 978–0–19–932919–9

9 8 7 6 5 4 3 2 1
Printed by Sheridan Books, Inc., United States of America

CONTENTS

CONTRIBUTORS

Avner Ayalon is Senior Researcher at the Department of Geochemistry, Geological Survey of Israel. His research focuses on paleoclimate reconstructions using the geochemistry and isotopic composition of cave deposits from Israel and their relations to human habitation and culture. His recent publications treat late Pleistocene climate change.

Miryam Bar-Matthews is Senior Researcher at the Department of Geochemistry, Geological Survey of Israel. Her major research interest is the reconstruction of well-dated, highly resolved continental paleoclimate records using speleothems, mainly in Israel and the Levant and focusing on critical changes in the regional hydrological cycle, and the data's implications towards an understanding of rapid aridification/desertification processes and their impact on human settlement and migration patterns.

Ofer Bar-Yosef is the MacCurdy Research Professor of Prehistoric Archaeology, Department of Anthropology, Harvard University. With an interest in human cultural evolution, he has conducted joint field research in a number of Palaeolithic and Neolithic sites in Western Asia and Europe (Israel, Egypt, Turkey, Georgia, Czech Republic) and is currently undertaking fieldwork in China. His early training in physical geography and archaeology has led him to study climate change and its potential impact on prehistoric forager cultures and early farmers in West and East Asia. His many published works include 19 co-edited volumes.

Juan Pablo Bernal-Uruchurtu is an investigator at the Centro de Geociéncias of the Universidad Autónoma de México. His research is dedicated to the reconstruction of past climates using analyses of geochemical proxies in cave stalagmites, with the objective of elucidating the consequences of abrupt climate change in Mexico and elsewhere. He utilizes elemental and isotopic techniques for high-precision analyses of climatic change.

Brendan Buckley holds the position of Lamont Research Professor at the Lamont-Doherty Earth Institute of Columbia University. He is one of the leading experts in tropical tree-ring studies with a focus on reconstruction of past hydroclimate variability in the Asian tropics. He is a strong advocate for interdisciplinary research and his work in Southeast Asia has led to a rethinking of the role of climate in shaping the course of human history for the region. One particular example was the discovery of Southeast Asian climate variability over the past millennium and the role of severe and/or extended periods of drought or pluvial in the demise of the Khmer kingdom at Angkor.

Dorian Burnette is Assistant Professor in the Department of Earth Sciences at the University of Memphis. His research interests include historical meteorology and climatology, dendroclimatology, societal impacts of weather and climate extremes, and computer programming applications. His current research focuses on the reconstruction of extreme weather and climate events using instrumental, documentary, and tree-ring data.

Lee Clare is Research Lecturer at the German Archaeological Institute (DAI), Berlin. His PhD, completed at the University of Cologne in 2013, was dedicated to the study of Early Holocene climate-culture interaction in the Eastern Mediterranean during Rapid Climate Change (8.6–8.0 ka calBP). His other areas of scientific interest include prehistoric warfare and conflict mitigation mechanisms, radiocarbon dating and absolute chronologies, and, more recently, the cognitive-evolutionary development of modern humans in relation to neolithization processes. He is currently research coordinator for the Early Neolithic site of Göbekli Tepe, Southeast Turkey.

Edward R. Cook is Ewing Lamont Research Professor and the Director of the Lamont-Doherty Earth Observatory Tree-Ring Laboratory. He has spent the past 40 years conducting tree-ring research in many parts of the world and developing statistical methods and software for tree-ring analyses. Over the past 15 years, he has produced drought atlases (gridded spatial reconstructions of past drought and wetness) from tree-ring networks for North America, Monsoon Asia, and Europe. Covering 1000 or more years, these drought atlases have identified the occurrence of past megadroughts far greater in duration than any known droughts in modern times.

Roland Fletcher is Professor of Theoretical and World Archaeology at the University of Sydney. His book *The Limits of Settlements Growth* (1995) reconciles long-term, cross-cultural, and short-term, contextually unique analyses of settlement trajectories. He is Director of the Greater Angkor Project, an international collaboration with Cambodian and French researchers that has shown Angkor to be a low-density city dependent on a massive water management infrastructure and vulnerable to extreme climate instability at the end of the Medieval Warm Period.

Daniel Griffin is Assistant Professor of Geography at the University of Minnesota. His research interests include climate science, environmental dynamics, and dendrochronology. He specializes in developing sub-annual tree-ring chronologies for the reconstruction of season-specific environmental history. His doctoral research focused on the recent paleoclimate history of the summer monsoon in the American Southwest.

David A. Hodell is Woodwardian Professor of Geology in the Department of Earth Sciences at the University of Cambridge. He is a Fellow of Clare College and Director of the Godwin Laboratory for Palaeoclimate Research. He has made fundamental contributions to the study of Earth's past climate, from the impact of climate change on ancient civilizations to the processes controlling the Pleistocene Ice Ages. Together with colleagues and students, he has uncovered critical evidence for understanding past relationships among ancient civilizations, climate, and environment, principally in Mesoamerica and India.

David Kaniewski is Associate Professor in the Department of Biology and Geosciences, University Paul Sabatier-Toulouse 3 (France). His research focuses on paleoenvironmental and paleoclimatic reconstructions, mainly in the Central and Eastern Mediterranean. His publications have focused upon Late Bronze Age/Iron Age climate change and cultural decline during the 3.2 ka BP (1200 BC) event.

Douglas Kennett is Professor of Environmental Archaeology in the Department of Anthropology at Penn State University. His current interests include the study of human sociopolitical dynamics under changing environmental conditions, human impacts on ancient environments, and behavioral response to abrupt climate change in the past. During the last 20 years, his work has included a comparative focus on the formation, consolidation, and breakdown of complex societies in North America, Mesoamerica, and South America within the context of global climate change.

Alan L. Kolata is Bernard E. and Ellen C. Sunny Distinguished Service Professor of Anthropology at the University of Chicago. He leads ongoing interdisciplinary research projects studying human-environment interactions over the past 3000 years in the Lake Titicaca basin of Bolivia, on the north coast of Peru, and most recently in Thailand and Cambodia. His recent research interests include comparative work on agroecological systems, human-environment interactions, food, energy, and water systems, agricultural and rural development, and archaeology and ethnohistory, particularly in the Andean region.

Matthew Lachniet is Professor of Geosciences at the University of Nevada Las Vegas. His research focuses on isotopic records of climate change in the tropics, arid lands, and permafrost regions derived from proxy material in caves, surface, and rain waters, and from permafrost. He has research experience in Latin America and North America. He is also director of the Las Vegas Isotope Science Laboratory for stable isotopic analyses of geologic, hydrologic, and biologic materials.

Christophe Pottier is Associate Professor at the École française d'Extrême-Orient (EFEO). He directed the EFEO research center in Siem Reap—Angkor (Cambodia) from its reopening in 1992 until 2009. Since 2012 he heads the EFEO in Bangkok at the Princess Maha Chakri Sirindhorn Anthropology Center. As an architect, he directed restoration and conservation works at Angkor, in particular at the Royal Terraces in the center of Angkor Thom. In parallel, he initiated various archaeological researches on ancient architecture, urbanism, and mapping. Since 2000 he has directed the Cambodian-French Archaeological Mission on the Angkor Region (Mafkata) and co-directed the Greater Angkor Project at the University of Sydney.

Shih-Yu Simon Wang is Associate Professor of Climate at Utah State University in Logan, Utah. With a PhD in Meteorology, Wang studies climate dynamics and weather extremes and has a particular interest in developing climate prediction. He teaches data mining that involves the analysis and application of climate data.

David W. Stahle is Distinguished Professor in the Department of Geosciences and Director of the Tree-Ring Laboratory at the University of Arkansas. His research concentrates on the development of long climate-sensitive tree-ring chronologies from the United States and Latin America, the reconstruction and analysis of climate variability and change, and the social and environmental impacts of past climatic extremes.

Lonnie Thompson is Distinguished University Professor and School of Earth Sciences Senior Research Scientist, Byrd Polar Research Center, Ohio State University. He has led 58 ice-core drilling expeditions in the Polar Regions, as well as on tropical and subtropical ice fields. The resulting ice-core-derived climate histories have resulted in major revisions in the field of paleoclimatology, in particular by demonstrating how tropical regions have undergone significant climate variability, countering an earlier view that higher latitudes dominate climate change.

Elise van Campo is Research Director at the Institute of Ecology and Environment, National Center for Scientific Research (France). Her research interests are Quaternary paleoecology and paleoclimatology in Mediterranean and tropical domains of the Old World. She now focuses her research on ecological responses to rapid climate changes and their effects on ancient societies.

Harvey Weiss is Professor of Near Eastern Archaeology and Environmental Studies at Yale University. For the past thirty years he has directed the Tell Leilan Project, Syria, with research focused upon late prehistoric and early historic Mesopotamian states and empires, their regional settlement dynamics, and their responses to climate changes. He previously edited *Seven Generations since the Fall of Akkad* (2012) synthesizing data from the Khabur Plains, northeastern Syria, on the societal effects of the 2200–1900 BC global megadrought.

Bernhard Weninger has directed the radiocarbon laboratory at the University of Cologne since 1993 and has specialized in climate-archaeological research in the Levant, Turkey, the Aegean, and Southeast Europe. After receiving his PhD in Prehistoric Archaeology at the University of Frankfurt am Main in 1992, he served as a researcher in the Troy Project led by Prof. Manfred Korfmann at the University of Tübingen. His main interests are in the Neolithic and Bronze Age Mediterranean and in palaeoclimatology.

| Megadrought, Collapse, and Causality

HARVEY WEISS

Recent discoveries of megadroughts, severe periods of drought lasting decades or centuries, during the course of the Holocene have revolutionized our understanding of modern climate history. Through advances in paleoclimatology, researchers have identified these periods of climate change by analyzing high-resolution proxy data derived from lake sediment cores, marine cores, glacial cores, speleothem cores, and tree rings. Evidence that megadroughts occurred with frequency and abruptly over the last 12,000 years, a timespan long assumed to be stable compared to earlier glacial periods, has also altered our understanding of societies' trajectories. The fact that severe, multi-decadal or century-scale droughts coincided with societal collapses well known to archaeologists has challenged established multi-causal analyses of these events. Megadroughts, impossible to predict and impossible to withstand, may have caused political collapse, regional abandonment, and habitat tracking to still-productive regions. The nine megadrought and societal collapse events presented in this volume extend from the foraging-to-agriculture transition at the dawn of the Holocene in West Asia to the fifteenth-century AD collapse of the Khmer Empire in Angkor (Cambodia). Inevitably, this collection of essays also raises challenges to causal analyses of societal collapse and for future paleoclimatic and archaeological research.

MEGADROUGHTS ARE REAL. Very difficult drought conditions in the western United States, apparent fifteen years ago, have approached megadrought (Cook et al. 2004; Kogan & Wei 2015). In the past, megadroughts spanned more than a decade to several centuries. They covered wide expanses of territory defined by the El Niño-Southern Oscillation (ENSO) and the Pacific Decadal Oscillation and monsoons (some globally teleconnected) that delivered regional precipitation, and they had characteristic precipitation reductions, with abrupt or gradual onsets and terminations (Meehl & Hu 2006; Sinha et al. 2011; Lake 2011). Megadroughts are now well documented in Africa (Verschuren, Laird, & Cumming 2000; Maley & Vernet 2015), India (Sinha et al. 2011), East Asia (Cook et al. 2010), the Andes (Ledru et al. 2013; Thompson et al. 2013), Europe and the Mediterranean (Cook et al. 2015), and North America (Cook et al. 2004; Stahle et al. 2007).

Over the past millennium alone, especially during the eleventh to fifteenth centuries in Europe and North America, megadroughts have been abrupt, severe, extensive, and prolonged (Asmeron et al. 2013; Cook et al. 2014, 2015.) Then and still earlier, megadroughts forced societal collapse: they presented insurmountable obstacles in rainfed-agriculture regions and reduced river flow conditions in irrigation-agriculture regions, thereby undermining the economy and political structure of agriculture dependent societies. The collapse of such societies, evident archaeologically and historically, manifested as political and social disintegration and the abandonment of drought-stricken regions accompanied by migration, or habitat tracking, to agriculture sustaining landscapes. However, the coincidence of megadrought and societal collapse is a recent discovery within archaeology and paleoclimatology and far from completely described and understood. As a contribution toward that understanding, this volume details nine past major megadroughts and societal collapses, from the origins of agriculture in the Near East 12,000 years ago to the demise of Angkor in Cambodia in the fifteenth century AD.

Studies of postglacial Holocene climate history and archaeology have been converging for some thirty years. Beginning in the 1970s and '80s, paleoclimate researchers' attention was drawn increasingly to the transition from the glacial Pleistocene to the post-Pleistocene—that is, the modern, or Holocene, period. The nature of that transition, thought to be gradual in the 1950s, was revealed through analyses in the early 1990s of the Greenland ice cores—especially GISP2 (Greenland Ice Sheet Project 2)—to be an abrupt climate change known as the Younger Dryas event (Alley et al. 1993). The rapidity of that event, and especially its abrupt decadal terminus and the onset of fully modern postglacial conditions, was a geological surprise that would set the stage for further research into the nature of global Holocene climates. Revelation of the dynamic qualities of the post-Younger Dryas climates of the modern or Holocene period and their global or regional punctuation by periods of megadrought, decadal- and century-scale abrupt climate changes previously unknown, has been a major achievement of the past few decades of research.

Paleoclimate Revolution

The major tools in the paleoclimate revolution are paleoclimate proxy data retrieved from natural long-term archives. The archives are the sedimentary deposits on ocean floors, lake floors, drip water created stalagmites in caves, annual snowfall laminations on glaciers, and annual growth rings on trees. Each of these five types of natural archives preserves sequences of isolable and datable intervals of deposition or growth that provide proxies for different kinds of climate conditions. Lake bottom sediments, for instance, preserve in oxygen-free conditions the annual pollen from lake-shore plant life, and thus record the datable and climate specific changes in lake-shore plant pollen (Zolitschka 2007; Ojala et al. 2012). Similarly, ocean cores preserve the microscopic diatoms that are specific to temperature conditions, or terrestrial dust derived from places subject to arid conditions, or tephra deposits windblown into the oceans from specific volcanoes (Hughen & Zolitschka 2007). Accumulations of ceiling drip on cave floors preserve in stalagmites, or speleothems, the mineral content of the water above the cave. That mineral content includes minute amounts of stable oxygen and carbon isotopes, which are indicators of both humid conditions and temperature above the cave at the specific time of the incremental cave drips (Lachniet 2009; Wong & Breecker 2015). The ring width of annual growth observed in trees, a variable determined by precipitation and temperatures levels, also provides a precisely dated climate record (Hughes et al. 2011; Salzer et al. 2013). Mapping the distribution through space and time of tree-ring sequences (with annual resolution dating) has proven to be a key identifier of megadroughts through the past two millennia across North America, Europe, and Asia (Cook et al. 2004; 2007; 2010; 2015). Lastly, glaciers preserve the annual laminations of snow at high latitudes and high tropical elevations. The isotopic composition of the oxygen in the ice identifies the sea surface temperature of the water's source, as well as precipitation, while major ions such as ammonium and nitrate unravel regional atmospheric circulation variability, and sulfate concentrations mark datable volcanic eruptions (Alley et al. 1993; Severinghaus, et. al. 1998; Vimeux et al. 2009; Thompson et al. 2013). Each of these five paleoclimate proxies has provided new and essentially revolutionary evidence based on high-resolution dating of megadroughts from the end of the Pleistocene (the last glacial ages) through the modern period.

Archaeological Studies of Societal Collapse

While this revolution in the earth's climate history has evolved, archaeologists have been moving from a post-World War II fascination with the social evolutionary processes that surrounded the genesis of civilizations to a concern for their periodic terminations or collapse. Societal collapses appear in the archaeological record as political devolution, urban abandonment and settlement dispersal, regional abandonment, and migration (or habitat tracking) to sustainable environments. Interest in societal collapse accelerated in the late 1970s and 1980s. One manifestation, prompted by C. S. Holling (1973), was Robert Adams's analysis of Mesopotamian collapses, which focused on the role of economic maximization policies in weakening societies' adaptive capacities (1978). A second early influence was research by Michael Moseley and his colleagues into the natural causes of Andean societal collapses (Moseley & Deeds 1982; Moseley et al. 1983). By the late 1980s, the evolving interest in collapse was synthesized in the publication of two volumes (Tainter 1988; Cowgill & Yoffee 1988) that broadly sought to select cogent examples of collapse and to synthesize the then current explanations for them. In the edited volume by George Cowgill and Norman Yoffee (1988), the nature of collapse was defined in uncertain terms, but where possible the explanatory forces were deemed mostly internal, endogenous processes surrounding resource exploitation, population pressure, warfare, and societal dysfunctions. In no case were climate changes addressed as causes of collapse, and "climate" did not appear in the volume's index. Joseph Tainter (1988) argued that climate change as a societal collapse force never occurred, and that in the cases he addressed the law of diminishing returns operated on societal scales to induce collapse. He dismissed the usual, countervailing roles territorial expansion, resource expansion, and technological innovation played against diminishing returns in the production process.

The study of collapse was redirected vigorously in the early 1990s with the publication of four forays into Holocene climate research. Izumi Shimada and Lonnie Thompson linked the collapse of the Moche state in the north Andes to annual lamination Quelccaya ice-core records of a "series of severe sixth-century droughts, including one of the severest droughts of the past 1,500 years, spanning AD 562 to 594" (Shimada et al. 1991). Charles Ortloff and Alan Kolata (1993) linked the collapse of the Tiwanaku state on the shores of Lake Titicaca (Bolivia) to a major twelfth-century AD drought event recorded nearby in the Quelccaya glacial core, as well as in the lake's sediment cores. Harvey Weiss and colleagues (1993) linked the 2200 BC Akkadian Empire collapse in Mesopotamia, and synchronous collapses from the Aegean to the Indus, with local evidence for a two- or three-century drought episode hypothesized to have been interregional. David Hoddell, Jason Curtis, and Mark Brenner (1995) retrieved and analyzed a lake sediment core in Guatemala and discovered a major drought event that was synchronous with the eighth century AD Maya collapse. The Akkadian and Maya collapse studies received significant public and scientific attention. The Hoddell, Curtis, and Brenner article was published with a cautionary prefatory note by a Mayan archaeologist questioning the role of climate change and drought in societal collapse. Cautionary comments from other archaeologists also prefaced the article by Weiss and colleagues. The worlds of archaeology and climatology had converged—uneasily—on abrupt, century-scale climate change, megadroughts, and societal collapses.

Twenty-first Century Studies of Megadrought and Collapse

The research has advanced and retreated since. Peter deMenocal (2001) and Weiss and Raymond Bradley (2001) drew attention in the journal *Science* to the new and numerous paleoclimate data synchronous with societal collapse. deMenocal's survey was the first to emphasize new megadrought-specific and El Niño data linked to records of societal collapse. Weiss and Bradley stressed the adaptive quality of societal collapse in the face of insurmountable climate degradation and extended their findings to modern climate-change challenges. A popular volume by Jared Diamond (2005) offered a synthetic treatment of causes for some societal collapses, though it failed to capture the essence of the new Holocene paleoclimatology. Diamond assigned collapse to multiple causes, including "unsustainable" societal decisions, such as exhausting resources, generating warfare, diminishing the agricultural base, and overpopulating.

New paleoclimatic and archaeological data for Holocene climate changes synchronous with societal collapse continued to generate strong reactions among some archaeologists, who understood such research to be an elective ideological stance, a backward step to the discredited "climate determinism" popularized in the early twentieth century by Ellsworth Huntington, a history lecturer at Yale University (Huntington 1907; 1915). Huntington had promoted, with scant archaeological and paleoclimate data, the notion that climate alone had determined the demise of several civilizations, just as he had determined the nature of their cultures, lazy and slow-thinking in the tropics, energetic and sharp-thinking in northern latitudes. Huntington's thesis was rejected as too simple and, in Owen Lattimore's mocking words, as "the romantic explanation of hordes of erratic nomads, ready to start for lost horizons at the joggle of a barometer in search of suddenly vanishing pastures" (Manley 1944).

Warnings about the dangers of climate determinism have since been one strand of response to the steady production of high-resolution paleoclimate data for abrupt and decadal to century-scale change coincident with, and apparently the cause of, region-wide agricultural failures and societal collapses. Such a focus on a purported ideological stance—perhaps at odds with the archaeologists' own deeply held stances—can lead to a myopic view of the testable paleoclimatic and archaeological data for regional settlement and agricultural production. For example, critics of what is deemed "climate determinism" often misinterpret paleoclimate data as indicating minimal climate change during the documented megadroughts (Wilkinson et al. 2007; Roberts et al. 2011; Butzer 2012), misinterpret the archaeological data as indicating minimal evidence for regional collapse, and confuse continuous adaptive occupations with habitat-tracking refugia (Rosen 2007; Schwartz 2007; Kuzucuoğlu 2007). Needless to say, the testing and measurement of all causal factors, including dynamic climatic conditions and societal responses, as well as related social processes, is certainly the goal of modern researchers.

The publication of *Questioning Collapse*, edited by Patricia McAnany and Norman Yoffee (2009) marked a major oppositional step in societal collapse and abrupt climate change studies. Here the new hard data for abrupt Holocene climate change were labeled equivocal and unconvincing, and even the archaeological data for societal collapses, including regional depopulations and abandonments, were denied in a politically correct sweep. In common with other works, paleoclimate data for drought and megadrought were dismissed as merely popular concerns (Tainter 2015). The data were deemed not real, or, if real, certainly not as precise as claimed and largely the product of a current fascination with "climate." In a similar vein, the continued existence of Maya and Ancestral Pueblo peoples—even Christians from northern Iraq, among others—was cited as proof that their ancient societies had not experienced collapse. Indeed, it was suggested that a kind of ethnocentric delusion informed arguments for the collapse of these ancient societies.

In search of explanatory paradigms that address collapse, some archaeologists have recently resurrected "resilience theory." A simple model of ecological systems' responses to stress factors, resilience theory has been applied to social systems and adopted as an explanatory framework for major transitions and transformation, including societal collapse (Holling 1973; Holling & Gunderson 2002; Redman & Kinzig 2003; Redman 2005; Costanza et al. 2007; Faulseit ed. 2016). Resilience, the ability of a social system to adapt to disequilibrating forces or, still more generally, "the ability to maintain, or quickly restore, in the face of a challenge, conditions considered highly desirable" (Cowgill 2012: 304), had been applied before to the long arc of Mesopotamian history, as noted above (Adams 1978). Recent archaeological applications, however, fit phases of social and environmental change into resilience theory's four-phase sequence of "exploitation," "conservation," "release" or "collapse," and "reorganization." The external forces that cause "release" or "collapse" are, of course, as variable as the social and environmental conditions at times of collapse and reorganization. Hence resilience theory "describes a process but fails to explain how the process is transformed" (Kidder et al. 2016: 73). Equally problematic is the incommensurability between resilience theory and the social sciences: agency, conflict, knowledge, and power are absent from resilience theory (Olsson et al. 2015). Similarly, of course, resilience theory lacks a megadrought phase.

One older line of societal collapse analysis exemplified within Karl Butzer's *Archaeology as Human Ecology* (1982), now resynthesized (Butzer 2012), takes coincident multicausality to be essential for societal collapse. This social science theme, that the interplay of multiple factors is almost always more critical than any single factor, is often reprised in societal collapse analyses (Diamond 2005; Costanza et al. 2007), but is now challenged by the new paleoclimate data for megadrought. Political analysts of effective causal forces had long understood, even in the social science of the early twentieth century, that "multiple causes of an event may be out of phase and only have an effect when they occur together" (Feyerabend 2010: 106, citing Lenin 1964). Butzer's claim, for instance, is that coincident multicausality generated the Old Kingdom collapse in Egypt at 2200 BC, or 4.2 ka BP (4200 years ago), which was synchronous with the 2200 BC abrupt megadrought in Mesopotamia (see Weiss, this volume). According to Butzer a "concatenation of triggering economic, subsistence, political, and social forces probably drove Egypt across a threshold of instability, setting in train a downward spiral of cascading feedbacks" (Butzer 2012: 3634). He imagines disruption of international trade, erroneously hypothesized and then dated 100 years too early, and supposes "contending elite groups may originally have used an impending crisis of succession to undermine royal legitimacy, in an effort to steer more power to the provinces" (ibid.). To arrive at this conclusion, Butzer rejects the dating of the palaeoclimate proxies for abrupt onset megadrought: "Considering the partial match of these rough dates, it is possible but unproven that Nile failures may have helped trigger collapse of the Old Kingdom" (ibid.: 3633).

However, the dating is robust for both the Nile flow abrupt reduction (Stanley et al. 2003; Marshall et al. 2011; Blanchet et al. 2013) and the disruptor of the sources of Nile flow, the Indian monsoon. The Indian Monsoon disruption is in fact documented by the highest resolution data available for the 4.2 ka BP (2200 BC) event, from the Mawmluh Cave speleothem (Berkelhammer et al. 2012), and supported by the lower resolution, but nevertheless compelling Kotla Dahar lake sediment core (Dixit, Hodell, & Petrie 2014). The chronology of the subsequent First Intermediate Period, the Egyptian Old Kingdom collapse, is also now robustly radiocarbon dated (Ramsey et al. 2010) and is virtually synchronous with the similarly high-resolution, radiocarbon dated Akkadian Leilan IIc collapse in northern Mesopotamia (Weiss et al. 2012). The megadrought disruption of Nile flow is a certainty at ca. 2200 BC, while multiple other causes hypothesized by Butzer remain undocumented.

A related analysis has relied upon the "correlation is not causality" argument (Coombes & Barber 2005). For example, while the ca. 4.2 ka BP (2200 BC) megadrought was coincident with Egyptian and Mesopotamian collapse, Coombes and Barber argue, there may have been other sufficient and necessary causes—although they simultaneously maintain that synchronous collapse across varied regions supports climate forcing at ca. 2200 BC. However, archaeological data shows that preindustrial human societies could not withstand megadrought, that circumstances rendered irrigation innovations physically impossible, and that they adapted through political collapse and migration to sustainable environments. Other ancillary causes are either not documented or seem minor. In most cases of megadrought, the stricken societies analyzed in this volume, including the Natufian, Anatolian Neolithic villages, Akkadian Empire, Late Bronze Age eastern Mediterranean, Teotihuacan, Ancestral Pueblo, and Angkor, were manifestly unsustainable when unpredicted megadrought occurred. Some pre-collapse environmental or social stressors have long been suggested for the Maya and Tiwanaku cases discussed here, but these forces have receded as causal explanations to ancillary or possibly superfluous amplifier status compared to megadrought, which brings unmitigated vegetation die-off and precludes nonlinear societal responses (Breshears et al. 2005).

Present and Future Studies of Megadrought and Collapse

In view of the recent history of research, discussions, and analysis, the function of this volume is to synthesize advances in the most prominent archaeological and historical situations where abrupt climate changes--specifically megadroughts—and societal collapses have coincided in a causal link. To be sure, varieties of evidence are required to identify abrupt megadrought as a causal force in societal collapse. Societal collapse must be observed, defined, and measured in the archaeological record, and coincident abrupt climate change must be observed and measured in the paleoclimate record. Abruptness, magnitude, and duration appear to be the important paleoclimate variables, since they co-occur at levels that made collapse inevitable due to the unavailability in some circumstances of natural or social tools and resources with which to counter the effects of megadrought upon local subsistence resources and technologies (Dillehay & Kolata 2004). In such situations the stress of reduced subsistence resources, cultivable landscapes, and agricultural production forces adaptive regional abandonment, migration, or habitat tracking, to accessible agricultural refugia. The archaeological and palaeoclimatic challenge is to define predictive levels of megadrought abruptness, magnitude, and duration relative to pre-abrupt climate change social and nautral resources.

In several cases discussed in this volume, megadrought is not the only disruptive causal force in evidence archaeologically or historically at the time of collapse. Or, megadrought itself generates other causes that promote societal collapse. Here we enter the delicate arena of weighing multiple causal factors, or factors elicited in the search for multiple causal factors. In a variety of situations, for example, short-term drought is understood to be a natural condition that can provoke famine, a social condition, and underscores the well-documented societal variables that mediate reduced agricultural production and distribution (Davis 2001). In some cases, a mode of subsistence alteration, from agriculture to pastoral nomadism, from millet to cereal farming, or multi-cropping, provides resilience and is also visible or suggested within the megadrought affected societies (see, for example, Petrie et al. 2016).

Also addressed in this volume are the termination of megadrought and variable societal responses, from resettlement to permanent relocation, to the return of pre-megadrought climatic and environmental conditions. The most salient examples of that variability are the essentially unexplained, opportunistic Amorite resettlement of northern Mesopotamia (Weiss 2014), the continued post-collapse desertion of the central Maya lowlands (Turner & Sabloff 2012), and the post-collapse population dispersal at Angkor (Lucero, Fletcher, & Coningham 2015). Adaptation and sustainability, alongside resilience or culturally perceived viability, remain, therefore, key processes in these societal collapses and post-collapse transformations.

These observations ultimately raise two questions, however. What is the weight of megadrought climate change in the causal nexus of societal collapse? And would collapse have occurred without megadrought? One path toward resolution of these questions might be the careful replacement of the loose archaeological usage "cause," with its array of descriptive and functional subcategories. As Nancy Cartwright has noted, "One factor can contribute to the production or prevention of another in a great variety of ways. There are standing conditions, auxiliary conditions, precipitating conditions, agents, interventions, contraventions, modifications, contributory factors, enhancements, inhibitions, factors that raise the number of effects, factors than only raise the level, etc." (1999: 119). The archaeological challenge, of course, would be tests of the verisimilitude of such categorizations. Another analytic path forward might be the quantified and modeled simulation of societal trajectories using archaeological chronologies, paleoclimate data, subsistence and settlement pattern data (Altaweel 2008, Axtell et al. 2002, d'Alpoim Guedes et al. 2016, Weiss & Booth 2014). Of course, such simulations would only be as useful as their data resolution and their models' verisimilitude. A third path might be the deployment of influence diagrams (Howard & Matherson 2005; Pearl 2000) with highly resolved archaeological data. These quantitative measures might allow us to begin to weigh causes beyond the unlikely multicausal democracy presumed within "concatenated events."

The nine examples presented in this volume extend from the terminal Pleistocene to the Little Ice Age, and from Southeast Asia to West Asia, the Mediterranean, the US Southwest, Mesoamerica, and the Andes. Each case has been the subject of prominent and numerous earlier analyses and publications, but is reexamined in detail here. In each case, the researchers explore the role of abrupt megadrought climate change in societal collapse and set the stage for further research, both within these examples and other cases yet untreated in detail.

The Collapse of Foraging and Introduction of Cultivation 12,800–11,500 Years Ago

The origins of agriculture occurred independently at several times and places during the Holocene, but first appeared in the West Asia with the early transition from foraging to cultivation and then to farming in the Levant region of Israel, Palestine, and western Syria. This West Asian trajectory from foraging to full-blown agriculture took at least 20,000 years, across the span from the Late Glacial Maximum to the Holocene. Not until the end of the Pre-Pottery Neolithic A period, ca. 11,000 years ago, is the evidence of genetic selection and domestication visible in archaeobotanical assemblages.

Archaeological and paleoclimate research has focused much attention on the hunter-gatherer-forager transition to early cultivation (the sowing and harvesting of wild plants in tilled soil)—the first step towards eventual plant domestication and farming within sedentary agricultural villages (Willcox, Nesbitt, & F. Bittman 2012). It was the discovery of the Younger Dryas Period event that marked a new phase in both paleoclimate and origins of agriculture research. Following upon the Late Glacial Maximum, about 23 thousand years ago, as the earth's surface was gradually warming, the Younger Dryas event was an abrupt climate reversal to colder and drier conditions between 12.8 and 11.5 thousand years ago. The archaeological concern since has been to identify in West Asia the cultural periods and hunter-gatherer-forager subsistence alterations that coincided with the abrupt alterations in environmental conditions during the Younger Dryas event. How did forager societies in West Asia adapt to the Younger Dryas event, its onset, and its abrupt terminus? Here, Bar-Yosef, Bar-Matthews and Ayalon review and synthesize the West Asian archaeological data for early plant cultivation and, as well, the synchronous data for the cold and possibly dry Younger Dryas event that is available from the speleothem record at Soreq Cave, Israel.

During the Late Glacial Maximum, ca. 23,000 to 18,000 years ago, forager groups spread across the Levant. This was a generally humid period with a short dry spell represented at the Ohalo II site, which was founded ca. 19,400 years ago when Sea of Galilee lake levels fell. The earliest plant cultivation at Ohalo II, well preserved due to the settlement's burning and subsequent submersion, was undertaken for a brief time (Weiss 2009). The termination of the dry period possibly marked the end of the cultivation experiment. In any case, subsequent attempts with plant cultivation are not documented until the late Younger Dryas event.

Bar-Yosef, Bar-Matthews and Ayalon understand the Younger Dryas event as a period of "changing and unstable environmental circumstances" during which "demographic pressure created by large numbers of forager groups caused regional upheaval, competition for resources, and local migrations from more arid areas . . . into the sown land"—that is, the western flank of the "big bend" in the central Euphrates River. Several early cultivation sites are now known from this region, making for an unusual density that is probably only in part due to the amount of archaeological research in that area. Adding interest and complication to the Younger Dryas event's adaptive effects are the now contradictory analyses of Younger Dryas period precipitation in the Levant. Soreq Cave speleothem analyses set the initial standard for understanding reduced precipitation during the Younger Dryas (Orland et al. 2012). The following year, however, analyses of Dead Sea rock varnish seemed to indicate a wetter Younger Dryas period (Liu, Broecker, & Stein 2013). New high-resolution proxy records from Lake Van, Turkey, definitive for Anatolia, the northern Levant, and northern Mesopotamia, indicate rapid increase of arid desert-steppe plants in the pollen assemblages and a dramatic drought signature peak in the stable oxygen isotope record for the Younger Dryas period (Pickarski et al. 2015). More recently, Hartman and colleagues (2016) suggest that gazelle tooth enamel indicates consumption of humid grasses during the cold Younger Dryas while Cheng and colleagues (2015) observe little Younger Dryas climate change at Jeita Cave. How these sets of contradictory data will shake out is unknown, but one discovery remains: at least twelve years of intense drought occurred at the end of the Younger Dryas event, as recorded in Soreq Cave (Orland et al. 2012).

The Bar-Yosef, Bar-Matthews, and Ayalon hypotheses are realistic assessments of both the Younger Dryas event's effects upon the fully occupied forager landscape and the archaeological data supporting an early shift to cereal cultivation in the Levant. Of course we need additional tests of the demographic pressure and landscape packing estimated by Munro (2009) and utilized here, along with still higher resolution synchronisms between the archaeological and paleoclimate data. Nevertheless, "regional upheaval, competition for resources, and local migrations" in the wake of the Younger Dryas event reside at the foundations of human behavioral ecology models for the origins of agriculture (Gremillion, Barton, & Piperno 2014).

Abrupt Climate Change and the Spread of Farming to Europe 8600-8000 Years Ago

The earliest Holocene abrupt climate change occurred between ca. 8.6 and 8.0 ka cal BP, a more than 500-year long period of extreme cold that is documented globally. Weninger and Clare explain that the onset of this event is synchronous with the long-distance movement of the earliest farming communities from West Asia into the Aegean. Further, they demonstrate that the end of this cold period is synchronous with the further spread of early farmers out of the Aegean at around 8.2 ka cal BP. Thereafter, it took only about a century for farming to be adopted throughout Southeast Europe. Weninger and Clare treat the chronology and distribution of this megadrought event in full and stress that it manifested as two superimposed events in West Asia and Europe. The first was the Siberian High, a period of intensely cold and fast-flowing air masses that intruded into West Asia beginning about 8600 years ago (8.6 ka BP) and within which a briefer and higher magnitude occurrence, the 8.2 ka BP event proper, appeared about 8200–8000 years ago (van der Pflicht et al. 2011).

The 8.2 ka BP was pan-global in extent, with a strong signature in Europe, the Mediterranean, West Asia, and China. In West Asia, particularly prominent 8.2 ka BP records are available from Soreq Cave (Bar Matthews et al. 1999) and the Dead Sea (Litt et al. 2012; Migowski et al. 2006) and present a remarkable depositional hiatus at about 8000 years ago. In central Anatolia, Nar Gölü (Dean et al. 2015) provides a new lake sediment record for the event. Although its magnitude in a variety of proxies is relatively minor compared to the Younger Dryas event in Europe and West Asia, the 8.2 ka BP event is the highest magnitude abrupt climate change of the Holocene in most globally distributed records (Walker et al. 2012).

Considering these drought events, and establishing the synchronism between archaeological events and processes and the abrupt climate changes, Weninger and Clare define two fundamental understandings for the spread of farming, or neolithization, of West Asia and Europe. The first process, coincident with the 8.6 cal BP event, was the arrival of the first settlements to the Turkish lakes district, the western coast of Turkey, the Turkish Aegean, and the coastal Levant. This was a migration to milder climate regions during the Siberian High cold spell. The second set of adaptations, coincident with the 8.2 cal BP event, was a regional site abandonment labeled the "Yarmoukian crisis." During this period the highlands of Jordan and the Jordan Valley were abandoned, with resettlements restricted to lower lying and moister areas, and was followed by swift neolithization movements from the Aegean coast to northeastern Hungary.

Two recent reviews of archaeological and paleoclimate data for the 8.2 cal BP event dismiss this synchrony of 8.2 cal BP abrupt climate change and adaptive societal collapse responses in West Asia. Researchers at Sabi Abyad, in the Balikh Valley of northern Syria, believe the site was occupied continuously during this period and consider the continuity to be major evidence for the adaptive capabilities and resilience of early Neolithic village farmers (van der Pflicht et al. 2011). In fact the village occupation on one flank of the site (A) was abandoned at 8.2 cal BP and, possibly 50 to 100 years later, a new village (B) grew on the opposite flank of the site—a scenario similar to that at Çatalhüyük discussed by Clare and Weninger. The hiatus between the Sabi Abyad occupations is accommodated within standard deviations in their radiocarbon ages, which are confused with occupational durations. Indeed, the entire radiocarbon date analysis from Sabi Abyad has been deemed useless (Bayliss 2015). There are, additionally, no data from Sabi Abyad for cumulative site-size alterations, nor for regional settlement and alterations, which would be the anticipated expression of regional adaptations to reduced precipitation, colder temperatures, and reduced agricultural production. Rather, much site data points to adaptive responses to altered subsistence conditions at its radically new village: the expansion of tholoi residences, the introduction of a large mobile pastoralist population, a decrease in pig herding, an increase in sheep and goat for milk and fiber, and a sudden increase in clay sealings as personal property markers (van der Pflicht et al. 2011: 231). These fundamental agricultural and social alterations are attributed, in part, to the coincidence of the 8.2 cal BP event and the changed climatic and environmental conditions. That is, they were adaptations to the 8.2 cal BP event. Other forces that may have conditioned these alterations and adaptations are not mentioned and, indeed, might be difficult to observe, if they existed. Hence it is not altogether clear if the demise of village A represents that society's collapse and if the rise of the new village B represents post-collapse adaptations by relocation to a new 8.2 cal BP environment. Or, are these distinctions only a matter of semantics?

Another group of researchers (Flohr et al. 2016), find little chronological synchronism between the 8.2 cal BP event and major disruptions of this period, the abandonment of some settlements, and the neolithization-migration or habitat tracking that is detailed by Weninger and Clare. However, Flohr and her colleagues focus only on the brief 200-year period of the 8.2 ka BP event, which is possibly difficult to isolate in many archaeological records, whereas Weninger and Clare analyze the superimposed abrupt climate changes 8.6–8.0 cal BP. Lastly, it should be noted that a careful recent study documents regional abandonment in Scotland as a response not only to the magnitude of the 8.2 cal BP event's impact on climate change but also its abruptness (Wicks & Mithen 2014).

Megadrought and Imperial Collapse in Mesopotamia: 2200–1900 BC

The pan-global Holocene abrupt climate change event ca. 4200 years ago (4.2 ka BP), a megadrought that extended from ca. 2200–1900 BC, has prompted controversy among archaeologists as its pattern of coincident and synchronous societal collapses rewrites the archaeology of regions extending from the western Mediterranean to the Aegean to the Near East, the Indus, and beyond to Africa, China, and even north America. This global abrupt climate change event occurred in historical time in Mesopotamia, coincident with the collapse of the Akkadian Empire, and therefore challenges the traditional, text-based historiography of Mesopotamia's earliest empire and its successor states.

The 4.2 ka BP megadrought brought a 30–50 percent reduction in precipitation and cooling across the Mediterranean, west to east (with coincident heavy rains in Alpine Europe: Zanchetta et al. 2016), and across West Asia, Central Asia, and eastern Asia, Africa, and the western hemisphere, as expressed in marine, lake, speleoethem, glacial core, and tree-ring records. Intriguingly, there is yet no consensus explanation for this geological event with global expression, although various region-wide explanations, such as solar insolation variability Magny et al. 2013; Staubwasser et al. 2003), onset of El Niño conditions (Fisher 2011), and Atlantic Oscillation displacement (Booth et al. 2005), have been hypothesized.

The event is linked temporally, spatially, and causally with the regional settlement abandonments and political collapses of the Akkadian Empire (Weiss et al. 2012), the Old Kingdom in Egypt (Marshall et al. 2011; Revel et al. 2014; Welc et al. 2014), Early Bronze Age IV Levantine (Harrison 2012), and Early Bronze Age III Anatolian (Massa & Şahoğlu 2015) and Aegean societies (Davis 2013), Languedoc Late Neolithic and Rhône Valley Bell-Beaker societies in southern France (Carozza et al. 2015), and the Harappan-period cities of the Indus Valley (Ponton 2012). The chronological linkages across these regions are now exceptionally tight. One important example is the AMS (accelerator mass spectrometry) radiocarbon-dating synchronism at 2200 BC for the Leilan IIc Akkadian imperial collapse in northern Mesopotamia and the beginning of the First Intermediate Period in Egypt (Ramsey et al. 2010; Weiss et al. 2012).

Before its global expression was known, and before its abruptness, magnitude, and duration had been defined, if only approximately, the case for the 4.2 ka BP megadrought and linked dry-farming collapses and habitat tracking was sometimes criticized as a "climate determinism" argument that erroneously imputed causal force to climate change while ignoring the social forces that must have been involved in, for example, the Akkadian or Old Kingdom collapses. These were extensions of an older argument noted above (Butzer 1982), that if Nile River flow failed it was not a determinant of Old or New Kingdom Egyptian collapse, but must have been only one of several synchronous negative forces (political, social, and economic concatenations). Presumed here is a still older argument that only collections of forces have large social effects—an earlier 20th century position that is maintained by some to this day (Costanza et al. 2007). For the most part, such arguments dissolve against the highly resolved dating of the megadrought, its abruptness, and magnitude, and its synchronism with transregional societal collapses, regional abandonments and habitat tracking. Butzer (2012), for example, misdates the Akkadian collapse by a century or more, misses its synchronism with the First Intermediate Period in Egypt, and rejects the paleoclimate proxies for both Mediterranean westerlies and Nile River flow diminution. Similarly illuminating is the counterfactual question, "We know Nile flow failure occurred; would Egyptian collapse have occurred without Nile flow failure?" The problem of weighing causes, and types of causes (Cartwright 1999: 119), is a large social science problem to which the 4.2–3.9 ka BP event and its multiple hemispheric collapses draw attention.

The Late Bronze Age Collapse: Thirteenth to Eleventh Centuries BC

One of the great mysteries of Near Eastern and Mediterranean archaeology, and the subject of countless articles and monographs, including bestsellers (Cline 2014), is the crisis and collapse of the Late Bronze Age societies of the eastern Mediterranean and the Near East between ca. 1250–1100 BC, and the "dark age" and birth of the Early Iron Age societies that immediately followed. The drama of this process is uniquely recorded in Egyptian bas-reliefs and inscriptions and in Ugaritic and Hittite inscriptions, and focused in part upon a series of famines in West Asia, the Levant, and Egypt and the invasions of the "Sea Peoples" from coastal southeastern Europe. Recorded as groups with strange non-Semitic, non-Egyptian names, some clearly derived from Aegean and Anatolian toponyms, they included the Peleshet (Philistines) of biblical record who settled in these times upon the coast of Palestine, ready to greet the Israelites already entering the same land through the interstices of the region-wide Late Bronze Age collapses. The invasions by sea and land, with small ships and ox-drawn carts, resulted in the archaeologically visible destruction of many large and wealthy Late Bronze Age urban centers of the Levant and the blanket replacement of their cultures, with few surviving features, in the Early Iron Age (Killebrew & Lehmann, 2013).

Why were small bands of "Sea Peoples" able to topple the well-established and militarily powerful Late Bronze Age urban kingdoms of the Aegean and West Asia? Episodes of famine were already well known from the textual record for this period, but Kaniewski, van Campo, and colleagues were the first to situate and measure the hypothesized droughts within their definition of the independent paleoclimate record for the Aegean and the Levant in the thirteenth to tenth centuries BC (2013). The subsequent publication of a similar and confirmatory drought record (Langgut, Finkelstein, & Litt 2013), derived from the pollen analysis of a Lake Tiberias sediment core, showed the abrupt megadrought ca. 1250–1100 BC followed by an abrupt return of pre-drought humid conditions. Additional confirmatory data for the megadrought in the Levant derive from two new Dead Sea cores (Neugebauer et al. 2015). A new speleothem record from the Peloponnese, Greece (Boyd 2015), adds to the previously retrieved and coincident records of Aegean drought.

A recent critique of the data for late Bronze Age megadrought focuses upon the limited number of lake sediment core radiocarbon dates deployed to prove sub-decadal coincidence with the Late Bronze Age collapse (Knapp & Manning 2016). The authors conclude, however, that cumulative evidence for the aridification period, including its radiocarbon dating, is now essentially incontrovertible.

The "Sea Peoples" invasions and successes, however, require further description and explanation, as all agree. Can we develop a high-resolution chronology of drought and habitat tracking from southeastern Europe to the Levant and the Hittite Empire's Anatolian domains that explains the Sea Peoples' migrations and invasions and the sequences of drought and famine (Kaniewski, Guiot, & van Campo 2015)? Synchronous disruption of the Mediterranean westerlies, the driver of precipitation for the eastern Mediterranean, is visible in the paleoclimate records for the North Atlantic Oscillation and confirms the Aegean, Levantine, and Anatolian proxy records (Geirsdóttir et al. 2013; Blair, Geirsdóttir, & Miller 2015; Olsen, Anderson, & Knudsen 2012). The variability within and across these varied proxy records will be essential for articulating and explaining the dramatic historical details of the Late Bronze Age collapses across the Mediterranean and West Asian worlds. Similarly, we can now require archaeological definitions and explanations for the cultural and environmental impacts and adaptations of the successor Iron Age civilizations, and the "new opportunities" created for post-collapse reorganization. Impressively, the examination of the Philistine impact on southern Levantine floral ecosystems has already been initiated (Frumin et al. 2015).

The Basin of Mexico Megadrought and Teotihuacan Collapse: Eighth Century AD

The largest city in the pre-Hispanic New World, covering twenty-one square kilometers, was Teotihuacan, in the Basin of Mexico, with a population estimated at 125,000 sustained by spring-fed irrigation agriculture. At its height, in the sixth century AD, Teotihuacan was the center of a civilization whose economic, political, and artistic influence extended to the Yucatan and to the US Southwest (Millon 1970; Sanders, Parsons, & Santley 1979; Cowgill 2015). The great city flourished for about seven centuries, which in itself poses an archeological challenge, until its mysterious demise ca. 600–800 AD. Numerous studies have been dedicated to the dates of abandonment of the city's public buildings, The Great Fire, signs of internal struggle, and the distribution and dates of various socially significant artifact categories. Virtually all manner of endogenous and exogenous forces have been invoked and explored as causes for the abandonment of the city, including regional deforestation and soil erosion (Manzanilla 2003; Sanders, Parsons, & Santley 1979), but abrupt megadrought was only documented by Lachniet and colleagues in a breakthrough paper in 2012.

In this volume, Lachniet and Bernal revisit the data from the 2400-year sta-lagmite record from Juxtlahuaca Cave, which they have dated ultra-precisely, with uncertainties of less than about ten years. The results of their analysis document Basin of Mexico abrupt rainfall variations switching from wet to dry within ten to thirty years, and 30–50 percent precipitation reductions during a century-long drought centered on AD 750 that was coincident with the collapse of the city. Specifically, the peak dry conditions in their high-resolution rainfall reconstruction coincide with earlier proposed dates for the city's population decrease in the seventh or eighth centuries AD (Cowgill 1997).

Lachniet and Bernal underscore the coincidence of the city's rise, as well, with a period of above average precipitation and likely, therefore, higher agri-cultural productivity. Hence the rise and fall of the city were coincident with variant and measurable natural conditions for social exploitation and a natu-ral force—megadrought—that proved insurmountable. The effects of mega-drought upon the city's agricultural base only amplified, Lachniet and Bernal observe, the variety of consequent internal and violent frictions that may have accelerated the urban collapse and which now require archaeological reanaly-sis. It must be noted, as well, that the approximately two-century rate of the city's abandonment is a function of lengthy ceramic periodizations. This Teotihuacan abandonment period may shorten considerably when it is rede-fined by high-resolution radiocarbon dates.

Drought and the Classic Maya Collapse, AD 800–1100

The Maya abandonment and collapse of the Guatemalan lowlands and Yucatan highlands in the ninth century AD has been the most discussed societal col-lapse in the western hemisphere, famously, with virtually every possible cause discussed at length. A few researchers in the 1980s dared to suggest that drought may have been a cause, but it was not until the retrieval and analysis of the sediment core at Lake Chichancanab (Hoddell, Curtis, & Brenner 1995) that hard data were available to support this hypothesis (Gill 2001). Since then impressive progress has been made in the retrieval of additional paleoclimate proxy data, the highest resolution and most accurately dated now being the Yok Balum cave, southern Belize, speleothem analyses (Kennett et al. 2012).

In this volume, Kennett and Hodell review the precise Yok Balum stalag-mite data and the archaeological data for the Maya collapse to derive several fundamental conclusions that revise previous attempts at Terminal Classic Maya climate and collapse synthesis. The Early Classic period (AD 440–640) was a period of anomalously high rainfall that was followed by a drying trend between AD 660 and 1000. Superimposed upon this drying trend were three severe and multi-decadal droughts. The first occurred between AD 820 and 870, the second around 930, and the third and most severe, between 1020 and 1100. According to Kennett and Hodell, the abrupt and severe megadroughts between 820 and 870 were "the unpredictable external shock" upon central and southern Maya lowlands already vulnerable to (1) agricultural failure from soil erosion caused by deforestation, and (2) political collapse due to high connectivity across regional sociopolitical nodes. In their view, the central and southern Maya lowlands settlements were more vulnerable to drought than "the less connected and therefore less vulnerable" northern lowlands settlements. The variability between the central region abrupt collapse and northern region gradual collapse, a subject of long-standing debate among Mayan archaeologists, was probably due to the differences in the regional political organization of settlements.

Additionally, it has been noted that the northern lowlands have over 6000 cenotes, steep-sided sinkholes fed by groundwater (Lucero 2006), while many settlements that survived the drought were also situated at fresh water refugia (Valdez & Scarborough 2014). Kennett and Hodell indicate, however, that the most severe drought had major consequences for the northern region settlements as well. To this day, indigenous Maya village agriculturalists continue to dwell in Mexico and Guatemala, of course, but the palace and temple dominated urban landscapes of pre-Hispanic Maya civilization never recovered from these megadrought episodes.

Here, as in other megadrought and collapse analyses, the effective causality is clouded by the absence of quantified or quantifiable relationships. Would Terminal Classic Mayan polities and agricultural production have been sustainable if they were not already, to some measure, soil-eroded field systems? Or was Terminal Classic megadrought, a 40 percent reduction in precipitation according to Medina-Elizade and Rohling (2012), insurmountable in all cases? The latter question remains open within other recent treatments of the Terminal Classic Maya megadroughts (Turner and Sabloff 2012; Webster 2014; Douglas et al. 2015, 2016).

The Tiwanaku Collapse at Lake Titicaca in the Thirteenth to Fourteenth Centuries AD

The collapse of the pre-Incan Tiwanaku Empire in the Andes is a classic example of megadrought-caused agricultural and political collapse, regional abandonment, and habitat tracking to refugia. For more than five hundred years, the Tiwanaku state prospered along the shores and the hinterland of Lake Titicaca with the products of intensive, flooded, raised-field agriculture. However, the onset of severe drought conditions in the thirteenth and fourteenth centuries AD reduced the level of Lake Titicaca by 12 meters, made raised field agriculture impossible, and diminished the subsistence of nucleated Tiwanaku settlement centers. The dissolution of Tiwanaku urbanism ensued, with dispersed populations moving into new, higher elevation refugia that afforded both water and defensive positions. In this volume, Kolata and Thompson note the tight synchronism between periods of Tiwanaku growth and collapse and a new higher resolution chronology for periods of high precipitation and drought, which is unambiguously expressed through archaeological settlement chronologies and the chronologies of two major adjacent paleoclimate proxy records, the Quelccaya glacial core and the Lake Titicaca sediment cores.

Without the intensive droughts of the thirteenth and fourteenth centuries, would the Tiwanaku state have survived? Kolata and Thompson note the gradual course of the Tiwanaku collapse beginning with earlier drought periods and continuing through to the mid-fifteenth century, as well as the existence of some early signs of social unrest, which suggest contributory collapse and resilience forces within the Tiwanaku social order. The "proximate cause" of the collapse, however, as noted early on, was located fundamentally within the megadrought's irremediable disruption of the highly adaptive Tiwanaku agricultural system (Ortloff & Kolata 1993; Dillehay & Kolata 2004). Tiwanaku social resilience to the megadrought was expressed as deurbanization, nomadization, and habitat tracking to sustainable high-elevation refugia.

Remarkably, within the framework created by Kolata and Thompson, a new genetic research project analyzing mitochondrial DNA samples is tracing population movements, migrations or habitat tracking, and the regional abandonments of megadrought stricken domains in the Andes at the time. The immigration of highland people to the coast at ca. AD 1150 generated rapidly increasing settlement density synchronous with the megadroughts in the southern Peruvian highlands. The pull force of increased precipitation in the lower valleys of the western Andean slopes was likely coincident with the push of Tiwanaku megadrought in the highlands (Fehren-Schmitz et al. 2014). The pan-Andean drought period, ca. AD 1250–1400, documented as well in recent lake sediment cores (Ledru et al. 2013), remains a key for understanding other Late Intermediate Period problems extending even to the causes of Wari collapse and Inka imperial expansion (Tung et al. 2016).

The Great Drought and Ancestral Pueblo Collapse in the late Thirteenth Century AD

Tree-ring studies focused upon chronology were developed by A. E. Douglass in the early twentieth century for archaeological purposes in the American Southwest and subsequently led to discovery of the Great Drought coincident with the Ancestral Pueblo (Anasazi) abandonments (Douglass 1929, 1935, 1936). Dendroclimatology defining long-term moisture balance now provides highly resolved and essential drought and megadrought data for the past two millennia across both hemispheres (Stahle et al. 2007; Cook et al. 2010; Cook et al. 2015). In the US southwest, the 400-year period of overall elevated aridity from AD 900 to 1300 was punctuated by four very dry epochs, the megadroughts centered at 936, 1034, 1150, and 1253 (Cook et al. 2004). The last epoch includes the Great Drought that extended from 1276 to 1297 and may have contributed to the Ancestral Pueblo collapse and abandonment of the San

Juan River, including settlements supported by dry-farming maize agriculture at Mesa Verde.

Carla Van West's GIS analysis of the Ancestral Pueblo settlement and resources first led to the hypothesis that drought alone did not account for the total abandonment of the region, that thousands could have remained, and that cultural forces had to have played a role as well (Van West 1994; Kohler 2010; Kohler, Varien, & Wright 2010). Research attention was thus refocused on the endogenous cultural forces that could have—but apparently did not—permit some remnant population to sustain themselves in the Colorado basin during the Great Drought. Some population apparently fled to agriculture sustaining springs, but only briefly (Kohler et al. 2008). By AD 1300 the entire population had emigrated to northern Arizona, western New Mexico, and the watershed of the Rio Grande (Varien 2010).

In this volume, David Stahle and his colleagues undertake remeasurement of tree-ring specimens from living trees and archaeological wood at Mesa Verde, Colorado, from AD 480 to 2008 to derive high-resolution earlywood and latewood width chronologies and, thereby, separately reconstruct a cool and early warm-season moisture history of the Ancestral Pueblo settlement region. They augment their results with seasonal reconstructions from the Mancos River to the east of Mesa Verde and from El Malpais National Monument in northwestern New Mexico. The conclusions derived from their study rein-force several salient facts. First, the Great Drought of AD 1276–1297 was one of the worst episodes of cool and early warm-season drought in a millennium. Second, dual-season drought occurred earlier in the thirteenth century, as well as its end. Third, early warm-season moisture conditions were generally below average over the San Juan region during the entire thirteenth century.

The Great Drought remains as the major determinant of regional collapse and Ancestral Pueblo migration in the late thirteenth century (Axtell et al. 2002), while other push and pull forces have proven extremely hard to identify in the archaeological and historical records. Kinship and refuge expectations, both difficult to discern or quantify against the material correlates of drought and migration, are perhaps the only partially non-climatic cultural forces that might explain why all fled the Great Drought while some thousands could have remained (Glowacki 2010; Berry & Benson 2010). A striking complement to the well-developed analyses of Ancestral Pueblo collapse and migration at the Great Drought is a new observation of immediate habitat tracking to refugia, such as the Pajarito Plateau, north central New Mexico (Bocinsky & Kohler 2014).

Flooding, Megadroughts, and the Collapse of Angkor in the Fourteenth Century AD

In the twelfth and thirteenth centuries AD, the Khmer capital at Angkor was a low-density megacity of about 750,000 persons spread over 1,000 square kilometers. Radiating from a center of massive public structures, temples, and water reservoirs (*baray*) was a vast city of residential structures grouped around their own temples, *baray*, and rice fields. Especially noteworthy (and described in this volume by Roland Fletcher et al.) was the city's elaborate water management system, an infrastructure that might have protected at least the elite from drought but that, ultimately, succumbed to monsoon flooding. By the fourteenth century most of Angkor was abandoned, and large areas of the surrounding plain had been converted to rice fields.

How and why this collapse occurred has largely been resolved through four closely coincident research projects. The Greater Angkor Project, a French, Australian, and Cambodian project from 2004–2009, alongside the Khmer Archaeology LiDAR Consortium, undertook GIS and LiDAR mapping of the city based upon earlier French ground mapping, along with selected excavations of crucial reservoir constructions and collapses (Evans et al. 2007, 2013). Edward Cook, meanwhile, directed the Monsoon Asia Drought Atlas collection of tree rings from more than 300 sites across the forested areas of Monsoon Asia and developed the annual dating and intensity of four historical droughts that had momentously affected the region (Cook et al. 2010). Brendan Buckley cored rare cypress trees in Vietnam, deriving a 759-year early monsoon drought index, that defined Angor Droughts I and II in the fourteenth and fifteenth centuries (Buckley 2010). Lastly, Mary Beth Day and colleagues (2012) retrieved lake sediment cores for the West Baray reservoir at Angkor that tracked reservoir drought, flooding, and channel failure in the fourteenth and fifteenth centuries.

The results from these teams of researchers are a detailed and complex history of intense megadrought during Angkor Drought I (AD 1345–1365) and II (AD 1401–1425), followed by intense flooding. The megadroughts reduced the megacity's agricultural rice production to unsustainable levels. The intense flooding destroyed essential reservoirs and channel structures. Essentially, the entire agro-production system was overwhelmed and failed. Political elites, alongside residential and agricultural worker populations, abandoned the dysfunctioning megacity and moved to Phnom Penh. Ancillary historical forces surrounded the droughts and floodings, but only reinforced the inevitable collapse.

Conclusions

Taken together, the nine essays in this volume challenge views that societal collapse is only a function of multiple social and environmental causes operating in concert, only a function of mistaken human decision-making, and only an underestimation of long-enduring civilizations and their traditions. The evidence suggests that abrupt megadroughts engendered agricultural and social conditions that could not be surmounted with agricultural or social innovations. Rather, societies adapted to these megadroughts with political collapse, regional abandonment, and population migration. Lesser climate changes—more gradual, lower in magnitude, shorter in duration—have not caused societal collapse, and modest, (if any) adaptive responses might well have defeated the instabilities caused by limited or short-term drought or colder temperatures. It is a challenging task, for instance, to locate adaptive societal responses to western Europe's gradual, half-a-degree centigrade temperature drop during the seventeenth century AD and the Little Ice Age (de Vries 1980; Parker 2014).

The analysis and study of the abrupt megadroughts of the Holocene provides several significant historical insights for the past, the present, and the future. First, the megadroughts and their societal impacts contribute to a dismantling of the historiographic view of conscious self-determinism, an intellectual tide that has been rising since the mid-nineteenth century in both the geological and social sciences. Second, data on the persistence and periodicity of Holocene megadroughts can yield geological insights into the current drought history of some regions, such as North America (Cook et al. 2014). Third, as already noted (Weiss & Bradley 2001), each of these preindustrial abrupt climate changes was a major and unanticipated natural event and thereby highlights, by contrast, contemporary climate change, which is anthropogenic, subject to societal decision-making, and an entirely new global phenomenon.

Acknowledgments

I join with my fellow authors and Oxford University Press in thanking Tulin Duda for the profound labors she invested in the editing of this volume. We are all very appreciative of the impact her wise literacy has had upon these essays.

References

Adams, R. McC. 1978. Strategies of Maximization, Stability, and Resilience in Mesopotamian Society, Settlement, and Agriculture. *Proceedings of the American Philosophical Society* 122: 329–335.

Alley, R. B., D. A. Meese, C. A. Shuman, A. J. Gow, K. C. Taylor, P. M. Grootes, J. W. C. White, M. Ram, E. D. Waddington, P. A. Mayewski, and G. A. Zielinski.1993. Abrupt increase in snow accumulation at the end of the Younger Dryas event. *Nature* 362: 527–529.

Altaweel, M. 2008. Investigating agricultural sustainability and strategies in northern Mesopotamia: results produced using a socio-ecological modeling approach. *Journal of Archaeological Science* 35: 821–835. doi:10.1016/j.jas.2007.06.012.

Asmeron, Y., V. J. Polyak, J. B. T. Rasmussen, S. J. Burns, and M. Lachniet. 2013. Multidecadal to multicentury scale collapses of Northern Hemisphere monsoons over the past millennium. *Proceedings of the National Academy of Sciences* 110: 9651–9656. doi: 10.1073/pnas.1214870110

Axtell, R. L., J. M. Epstein, J. S. Dean, G. J. Gumerman, A. C. Swedlund, J. Harburger, S. Chakravarty, R. Hammond, J. Parker, and M. Parker. 2002. Population growth and collapse in a multiagent model of the Kayenta Anasazi in Long House Valley. *Proceedings of the National Academy of Sciences* 99: 7275–7279.

Bar-Matthews, M., A. Ayalon, A. Kaufman, and G. J. Wasserburg. 1999. The Eastern Mediterranean paleoclimate as a reflection of regional events: Soreq cave, Israel. *Earth and Planetary Science Letters* 166: 85–95.

Bayliss, Alex. 2015. Quality in Bayesian chronological models in archaeology. *World Archaeology* 47: 677–700. doi 10.1080/00438243.2015.1067640

Berkelhammer, M., A. Sinha, L. Stott, H. Cheng, F. S. R. Pausata, and K. Yoshimura. 2012. An abrupt shift in the Indian Monsoon 4000 years ago. In L. Giosan, D. Q. Fuller, K. Nicoll, R. K. Flad, and P. D. Clift (eds.), *Climates, Landscapes, and Civilizations*. Washington, DC: American Geophysical Union. doi: 10.1029/2012GM001207

Berry, M. S., and L. V. Benson. 2010. Tree-ring dates and demographic change in the southern Colorado plateau and Rio Grande regions. In Kohler et al. (eds.), 2010: 53–74.

Blair, C., Á. Geirsdóttir, and G. H. Miller. 2015. A high-resolution multi-proxy lake record of Holocene environmental change in southern Iceland. *Journal of Quaternary Science* 30: 281–292.

Blanchet, C. L., R. Tjallingii, M. Frank, J. Lorenzen, A. Reitz, K. Brown, T. Feseker, and W. Brückmann. 2013. High- and low-latitude forcing of the Nile River regime during the Holocene inferred from laminated sediments of the Nile deep-sea fan. *Earth and Planetary Science Letters* 364: 98–110.

Bocinsky, R. K., and T. A. Kohler. 2014. A 2,000-year reconstruction of the rain-fed maize agricultural niche in the US Southwest. *Nature Communications*. 5:5618. doi: 10.1038/ncomms6618

Booth, R. K., S. T. Jackson, S. L. Forman, J. E. Kutzbach, E. A. Bettis III, J. Krieg, and D. K. Wright. 2005. A severe centennial-scale drought in mid-continental North America 4200 years ago and apparent global linkages. *The Holocene* 15: 321–328.

Boyd, M. 2015. Speleothems from Warm Climates: Holocene Records from the Caribbean and Mediterranean Regions. PhD thesis. Dept. Physical Geography, Univ. of Stockholm.

Breshears, D. D., N. S. Cobb, P. M. Rich, K. P. Price, C. D. Allen, R. G. Balice, W. H. Romme, J. H. Kastens, et al. 2005. Regional vegetation die-off in response to global change-type drought. *Proceedings of the National Academy of Sciences* 102: 15144–15148.

Buckley, B. M., K. J. Buckley, D. P. Anchukaitis, R. Fletcher, E. R. Cook, M. Sano, Le C. Nam, A. Wichienkeeo, T. That Minh, and T. Mai Hong. 2010. Climate as a contributing factor in the demise of Angkor, Cambodia. *Proceedings of the National Academy of Sciences USA* 107: 6748–6752. doi: 10.1073/pnas.0910827107

Butzer, K. W. 1982. *Archaeology as Human Ecology: Method and Theory for a Contextual Approach*. Cambridge: Cambridge University Press.

Butzer, K. W. 2012. Collapse, environment, and society. *Proceedings of the National Academy of Sciences* 109.10 (2012): 3632–3639.

Butzer, K. W., and G. H. Endfield. 2012. Critical perspectives on historical collapse. *Proceedings of the National Academy of Sciences* 109: 3628–3631.

Carozza, L., J.-F. Berger, C. Marcigny, and A. Burens. 2015. Society and environment in southern France from the third millennium BC to the beginning of the second millennium BC: 2200 BC as a tipping point? In H. Meller, H. W. Arz, R. Jung, and R. Risch (eds.), *2200 BC—A Climate Breakdown as a Cause for the Collapse of the Old World?* Halle: Tagungen des Landesmuseum für Vorgeschichte, 335–364.

Cartwright, N. 1999. *The Dappled World: A Study of the Boundaries of Science.* Cambridge: Cambridge University Press.

Cheng, H., A. Singha, S. Verheyden, F. H. Nader, X. L. Li, P. Z. Zhang, J. J. Yin, L. Yi, et al. 2015. The climate variability in northern Levant over the past 20,000 years. *Geophysical Research Letters* 42: 8641–8650. doi: 10.1002/2015GL065397

Cline, E. 2014. *1177 BC: The Year Civilization Collapsed.* Princeton, NJ: Princeton University Press.

Cook, B. I., J. E. Smerdon, R. Seager, and E. R. Cook. 2014. Pan-continental droughts in North America over the last millennium. *Journal of Climate* 27: 383–397.

Cook, E. R., C. A. Woodhouse, C. M. Eakin, D. M. Meko, and D. W. Stahle. 2004. Long-Term Aridity Changes in the Western United States. *Science* 306: 1015–1018. DOI: 10.1126/science.1102586

Cook, E. R., K. J. Anchukaitis, B. M. Buckley, R. D. D'Arrigo, G. C. Jacoby, W. E. Wright. 2010. Asian monsoon failure and megadrought during the last millennium. *Science* 328: 486– 489. DOI: 10.1126/science.1185188

Cook, E. R., R. Seager, M. A. Cane, D. W. Stahle.2007. North American drought: Reconstructions, causes, and consequences. *Earth-Science Reviews* 81: 93–134.

Cook, E. R., R. Seager, Y. Kushnir, K. R. Briffa, U. Büntgen, D. Frank, P. J. Krusic, W. Tegel, et al. 2015. Old World megadroughts and pluvials during the Common Era. *Science Advances* 1: e1500561.

Coombes, P., and K. Barber. 2005. Environmental determinism in Holocene research: causality or coincidence? *Area* 37 (3): 303–311.

Costanza, R., L. Graumlich, W. Steffen, C. Crumley, J. Dearing, K. Hibbard, R. Leemans, C. Redman, and D. Schimel. 2007. Sustainability or collapse: what can we learn from integrating the history of humans and the rest of nature? *AMBIO: A Journal of the Human Environment* 36: 522–527.

Cowgill, G. L. 1997. State and society at Teotihuacán, Mexico. *Annual Review of Anthropology* 26: 129–161.

Cowgill, G. 2012. Concepts of collapse or regeneration in human society. In D. L. Nichols and C. L. Pool (eds), *The Oxford Handbook of Mesoamerican Archaeology.* Oxford: Oxford University Press, 301–308.

Cowgill, G. L. 2015. *Ancient Teotihuacan: early urbanism in central Mexico.* Cambridge: Cambridge University Press.

Cowgill, G. L., and N. Yoffee. 1988. *The Collapse of Ancient States and Civilizations.* Tucson: University of Arizona Press.

d'Alpoim Guedes, J., S. Manning, and R. K. Bocinsky. 2016. A 5,500-Year Model of Changing Crop Niches on the Tibetan Plateau. *Current Anthropology* 57: 517–522. doi: 10.1086/687255

d'Alpoim Guedes, J., S. A. Crabtree, R. K. Bocinsky and T. A. Kohler. 2016. Twenty-first century approaches to ancient problems: Climate and society. *Proceedings of the National Academy of Science.* Doi: 10.1073/pnas.1616188113. [December 12, 2016 online before print].

Davis, J. L. 2013. Minding the gap: a problem in eastern Mediterranean chronology, then and now. *American Journal of Archaeology* 117: 527–533.

Davis, M. 2001. *Late Victorian Holocausts*. London: Verso.

Day, M. B., D. A. Hodell, M. Brenner, H. J. Chapman, J. H. Curtis, W. F. Kenney, A. L. Kolata, and L. C. Peterson. 2012. Paleoenvironmental history of the West Baray, Angkor (Cambodia). *Proceedings of the National Academy of Sciences* 109: 1046–1051. doi:10.1073/pnas.1111282109

Dean, J. R., M. D. Jones, M. J. Leng, S. R. Noble, S. E. Metcalfe, H. J. Sloane, D. Sahy, W. J. Eastwood, and C. N. Roberts. 2015. Eastern Mediterranean hydroclimate over the late glacial and Holocene, reconstructed from the sediments of Nar lake, central Turkey, using stable isotopes and carbonate mineralogy. *Quaternary Science Reviews* 124: 162–174.

deMenocal, P. B. 2001. Cultural responses to climate change during the Late Holocene. *Science* 292: 667–673.

deVries, J. 1980. Measuring the impact of climate on history: the search for appropriate methodologies. *Journal of Interdisciplinary History* 10: 599–630.

Diamond, J. 2005. Collapse: How societies choose to fail or succeed. New York: Penguin.

Dillehay, T. D. and A. L. Kolata. 2004. Long-term human response to uncertain environmental conditions in the Andes. *Proceedings of the National Academy of Sciences* 101: 4325–4330.

Dixit, Y., D. Hodell, and C. Petrie 2014. Abrupt weakening of the summer monsoon in northwest India ~4100 yr ago. *Geology* 42: 339–342.

Douglas, P., A. A. Demarest, M. Brenner, and M. Canuto. 2016. Drought impacts on the Lowland Maya civilization. *Annual Review of Earth and Planetary Sciences* 44: 613–645.

Douglas, P. M. J., M. Pagani, M. A. Canuto, M. Brenner, D. A. Hodell, T. I. Eglintone, and J. H. Curtis. 2015. Drought, agricultural adaptation, and sociopolitical collapse in the Maya Lowlands. *Proceedings of the National Academy of Sciences* 112: 5607–5612.

Douglass, A. E., 1929. The secret of the Southwest solved by talkative tree rings. *National Geographic Magazine* 56: 736–770.

Douglass, A. E., 1935. *Dating Pueblo Bonito and Other Ruins of the Southwest*. Contributed Technical Papers, Pueblo Bonito Series No. 1. Washington, DC: National Geographic Society.

Douglass, A. E., 1936. The central Pueblo chronology. *Tree-Ring Bulletin* 2: 29–34.

Evans, D., H. C. Pottier, R. Fletcher, S. Hensley, I. Tapley, A. Milne, and M. Barbetti. 2007. A comprehensive archaeological map of the world's largest pre-industrial settlement complex at Angkor, Cambodia. *Proceedings of the National Academy of Sciences* 104: 14277e–14282. doi: 10.1073/pnas.0702525104

Evans, D. H., R. J. Fletcher, C. Pottier, J.-B. Chevance, D. Soutif, B. S. Tan, S. Im, D. Ea, T. Tin, S. Kim, C. Cromarty, S. De Greef, K. Hanus, P. Bâty, R. Kuszinger, I. Shimoda, and G. Boornazian. 2013. Uncovering archaeological landscapes at Angkor using LiDAR. *Proceedings of the National Academy of Sciences* 110: 12595–12600. doi: 10.1073/pnas.1306539110

Faulseit, R. K. 2016. Collapse, resilience and transformation in complex societies: modeling trends and understanding diversity. In R. K. Faulseit (ed.), *Beyond Collapse: Archaeological Perspectives on Resilience, Revitalization . . . and*

Transformation in Complex Societies. Center for Archaeological Investigations, Occasional Paper No. 42. Southern Illinois University, 3–26.

Fehren-Schmitz, L., W. Haak, B. Mächtle, F. Masch, B. Llamas, E. Tomasto Cagigao, V. Sossna, K. Schittek, J. I. Cuadrado, B. Eiteld, and M. Reindelf. 2014. Climate change underlies global demographic, genetic, and cultural transitions in pre-Columbian southern Peru. *Proceedings of the National Academy of Sciences* 111: 9443–9448.

Feyerabend, P. 2010. *Against Method*. 4th ed. NY: Verso.

Fisher, D. A. 2011. Connecting the Atlantic-sector and the north Pacific (Mt Logan) ice core stable isotope records during the Holocene: The role of El Niño. *The Holocene* 21: 1117–1124.

Flohr, P., D. Fleitmann, R. Matthews, W. Matthews, and S. Black. 2016. Evidence of resilience to past climate change in Southwest Asia: early farming communities and the 9.2 and 8.2 ka events. *Quaternary Science Reviews* 136: 23–39. doi:10.1016/j.quascirev.2015.06.022

Frumin, S., A. M. Maeir, L. K. Horwitz, and E. Weiss. 2015. Studying ancient anthropogenic impacts on current floral biodiversity in the southern Levant as reflected by the Philistine migration. *Nature Scientific Reports* 5:13308 doi: 10.1038/srep13308

Geirsdóttir, Á., G. H. Miller, D. J. Larsen, and S. Ólafsdóttir.2013. Abrupt Holocene climate transitions in the northern North Atlantic region recorded by synchronized lacustrine records in Iceland. *Quaternary Science Review* 70: 48–62.

Gill, R. B. 2001. *The Great Maya Droughts: Water, Life, and Death*. Albuquerque: University of New Mexico Press.

Glowacki, D. M. 2010. The social and cultural contexts of the central Mesa Verde region during the Thirteenth Century migrations. In Kohler et al. eds. 2010: 200–221.

Gremillion, K.J., L. Barton, and D. R. Piperno. 2014. Particularism and the retreat from theory in the archaeology of agricultural origins. *Proceedings of the National Academy of Sciences* 111: 6171–6177.

Gunderson, L. H., and C. S. Holling. 2002. *Panarchy: understanding transformations in systems of humans and nature*. Washington, DC: Island.

Harrison, T. 2012. The southern Levant. In D. T. Potts (ed.), *A Companion to the Archaeology of the Ancient Near East*. New York: Wiley, 629–646.

Hartman, G., O. Bar-Yosef, A. Brittingham, L. Grosman, and N. D. Munro. 2016. Hunted gazelles evidence cooling, but not drying, during the Younger Dryas in the southern Levant. *Proceedings of the National Academy of Sciences* 113: 3997–4002. doi: 10.1073/pnas.1519862113

Hodell, D. A., J. H. Curtis, and M. Brenner. 1995. Possible role of climate in the collapse of Classic Maya civilization. *Nature* 375: 391–394.

Holling, C. S. 1973. Resilience and stability of ecological systems. *Annual Review of Ecology and Evolutionary Systematics* 4: 1–23.

Howard, R. A., and J. E. Matherson. 2005. Influence diagrams. *Decision Analysis* 2(3): 127–143.

Hughen, K.A., B. Zolitschka. 2007. Varved marine sediments. In S. A. Elias (ed.), *Encyclopedia of Quaternary Science*. Amsterdam: Elsevier, 3114–3123. http://dx.doi.org/10.1016/B0-44-452747-8/00064-8

Hughes, M. K., T. W. Swetnam, H. F. Diaz. 2011. *Dendroclimatology: Progress and Prospects*. New York: Springer.

Huntington, E. 1907. *Pulse of Asia*. Boston: Houghton Mifflin.

Huntington, E. 1915. *Civilization and Climate*. New Haven: Yale University Press.

Kaniewski, D., E. van Campo, J. Guiot, S. Le Burel, T. Otto, and C. Baeteman. 2013. Environmental Roots of the Late Bronze Age Crisis. *PlosOne 1* August 2013 8.8: 1–10.| e71004.

Kaniewski, D., J. Guiot, E. van Campo. 2015. Drought and societal collapse 3200 years ago in the Eastern Mediterranean: a review. *WIREs Clim Change* 2015. doi: 10.1002/wcc.345

Kennett, D. J., S. F. M. Breitenbach, J. Y. Aquino, J. Asmersom, J. U. L. Awe, B. J. P. Baldini, Bartlein, C. B. J. Culleton, C. C. Ebert, M. J. C. Jazwa, N. M. J. Macri, N. Marwan, V. Polyak, K. M. Prufer, H. E. Ridley, H. Sodemann, B. H. Winterhalder, and G. H. Haug. 2012. Development and disintegration of Maya political systems in response to climate change. *Science* 338: 788–791.

Kidder, T. R., L. Haiwang, M. J. Storuzum, and Q. Zhen. 2016. New perspectives on the collapse and regeneration of the Han Dynasty. In R. K. Faulseit (ed.), *Beyond Collapse: Archaeological Perspectivfes on Resilience, Revitalization and Transformation in Complex Societies*. Center for Archaeological Investigations, Southern Illinois University, Paper No. 42. Carbondale: Southern Illinois University Press, 70–98.

Killebrew, A. E., and G. Lehmann (eds). 2013.*The Philistines and Other "Sea Peoples" in Text and Archaeology*. Atlanta: Society of Biblical Literature.

Knapp, A. B., and S. W. Manning. 2016. Crisis in context: The end of the Late Bronze Age in the eastern Mediterranean. *American Journal of Archaeology* 120: 99–149.

Kogan, F., and W. Guo. 2015. 2006–2015 mega-drought in the western USA and its monitoring from space data. *Geomatics, Natural Hazards and Risk* 6: 651–658.

Kohler, T. A. 2010. A new productivity reconstruction for southwestern Colorado, and its implications for understanding Thirteenth Century depopulation. In T. A. Kohler, et al., (eds.), *Leaving Mesa Verde: Peril and Change in the Thirteenth-Century Southwest,* 102–127.

Kohler, T. A., M. D. Varien, A. M. Wright, and K. A. Kuckelman. 2008. Mesa Verde migrations. *American Scientist* 96: 146–153.

Kohler, T. A., M. D. Varien, and A. M. Wright (eds.). 2010. *Leaving Mesa Verde: Peril and Change in the Thirteenth Century Southwest*. Tucson: University of Arizona Press.

Kuzucuoğlu, C. 2007. Climatic and environmental trends during the third millennium B.C. in Upper Mesopotamia. In C. Kuzucuoğlu and C. Marro (eds.), *Sociétés humaines et changements climatiques à la fin du troisième millénaire: une crise a-t-elle eu lieu en Haute Mésopotamie?* Istanbul: Institut français d'études anatoliennes Georges-Dumézil, 459–80.

Lachniet, M. 2009. Climatic and environmental controls on speleothem oxygen-isotope values. *Quaternary Science Reviews* 28: 412–432.

Lachniet, M. S., J. P. Bernal, Y. Asmerom, V. Polyak, and D. Piperno. 2012. A 2400-yr rainfall history links climate and cultural change in Mexico. *Geology* 40: 259–262.

Lake, P. Sam. 2011. *Drought and Aquatic Ecosystems: Effects and Responses*. New York: Wiley-Blackwell.

Langgut, D., I. Finkelstein, and T. Litt. 2013. Climate and the Late Bronze collapse: new evidence from the southern Levant. *Tel Aviv* 40: 149–175.

Ledru, M.-P., V. Jomelli, P. Samaniego, M. Vuille, S. Hidalgo, M. Herrera, and C. Ceron. 2013. The Medieval Climate Anomaly and the Little Ice Age in the eastern Ecuadorian Andes. *Climate of the Past* 9: 307–321. doi:10.5194/cp-9-307-2013

Lenin, V.I. 1964. Left Wing Communism, an infantile disorder. *Collected Works*, Vol. 31. Orig. published 1920. Moscow: Progress Publishers, 1964, 12–118.

Litt, T., C. Ohlwein, F. H. Neumann, A. Hense, and M. Stein. 2012. Holocene climate variability in the Levant from the Dead Sea pollen record. *Quaternary Science Reviews* 49: 95–105.

Liu, T., W. S. Broecker, and M. Stein. 2013, Rock varnish evidence for a Younger Dryas wet period in the Dead Sea basin. *Geophys. Res. Lett.* 40: 2229–2235. doi:10.1002/grl.50492

Lucero, L. J. 2006. *Water and Ritual: The Rise and Fall of Classic Maya Rulers.* Austin: University of Texas.

Lucero, L. J., R. Fletcher, and R. Coningham. 2015. From "collapse" to urban diaspora: the transformation of low-density, dispersed agrarian urbanism. *Antiquity* 89: 1139–1154.

Magny, M., N. Combourieu-Nebout, J. L. de Beaulieu, V. Bout-Roumazeilles, D. Colombaroli, S. Desprat, A. Francke, S. Joannin, et al. 2013. North-south palaeohydrological contrasts in the central Mediterranean during the Holocene: tentative synthesis and working hypothese. *Climate of the Past* 9: 2043-2071. doi: 10.5194/cp-9-2043-2013

Maley, J., and R. Vernet. 2015. Populations and climatic evolution in north tropical Africa from the end of the Neolithic to the dawn of the Modern Era. *African Archaeological Review* 32: 179–232. DOI 10.1007/s10437-015-9190-y

Manley, G. 1944. Some recent contributions to the study of climatic change. *Quarterly Journal of the Royal Meteorological Society* 70: 197–220.

Manzanilla, L. 2003. The abandonment of Teotihuacán. In T. Inomata and R. W. Webb (eds.), *The Archaeology of Settlement Abandonment in Middle America.* Salt Lake City: The University of Utah Press, 91–101.

Marshall, M. H., H. F. Lamb, D. Huws, S. J. Davies, R. Bates, J. Bloemendal, J. Boyle, M. J. Leng, M. Umer, C. Bryant. 2011. Late Pleistocene and Holocene drought events at Lake Tana, the source of the Blue Nile. *Global and Planetary Change* 78: 147–161.

Massa, M., and V. Şahoğlu. 2015. The 4.2 ka BP climatic event in west and central Anatolia: combining paleoclimatic proxies and archaeological data. In H. Meller, H. W. Arz, R. Jung, and R. Frisch (eds.), *2200 bc—A Climatic Breakdown as a Cause for the Collapse of the Old World?* Halle: Museum für Vorgeschichte.

McAnany, P., and N. Yoffee (eds.). 2009. *Questioning Collapse: Human Resilience, Ecological Vulnerability, and the Aftermath of Empire.* Cambridge: Cambridge University Press.

Medina-Elizalde, M., and E. J. Rohling. 2012. Collapse of Classic Maya civilization related to modest reduction in precipitation. *Science* 335: 956–959.

Meehl, G., and A. Hu. 2006. Megadroughts in the Indian Monsoon region and southwest North America and a mechanism for associated multidecadal Pacific sea surface temperature anomalies. *Journal of Climate* 19: 1605–1623.

Migowski, C., M. Stein, S. Prasad, J. F. W. Negendank, and A. Agnon. 2006. Holocene climate variability and cultural evolution in the Near East from the Dead Sea sedimentary record. *Quaternary Research* 66: 421–431.

Millon, R. 1970, Teotihuacán: Completion of map of giant ancient city in the Valley of Mexico: *Science* 170: 1077–1082.

Moseley, M. E., and E. E. Deeds. 1982. The land in front of Chan Chan: agrarian expansion, reform, and collapse in the Moche Valley. In M. E. Moseley and K. C. Days (eds.), *Chan Chan, Andean desert city*. Albuquerque: University of New Mexico Press, 25–54.

Moseley, M. E., R. A. Feldman, C. A. Ortloff, and A. Narvaez. 1983. Principles of agrarian collapse in the Cordillera Negra, Peru. *Annals of Carnegie Museum* 52: 299–327.

Munro, N. 2009. Epipaleolithic subsistence intensification in the southern Levant: the faunal evidence. In J.-J. Hublin and M. P. Richards (eds.), *The Evolution of Hominin Diets: Integrating Approaches to the Study of Palaeolithic Subsistence*. Rotterdam: Springer, 141–155.

Neugebauer, I., A. Brauer, M. J. Schwab, P. Dulski, U. Frank, E. Hadzhiivanova, H. Kitagawa, T. Litt, V. Schiebel, N. Taha, N. D. Waldmann, and DSDDP Scientific Party. 2015. Evidences for centennial dry periods at ~3300 and ~2800 cal. yr BP from micro-facies analyses of the Dead Sea sediments. *The Holocene* 25: 1358–1372.

Ojala, A. E. K., P. Francus, B. Zolitschka, M. Besonen, and S.F. Lamoureux. 2012. Characteristics of sedimentary varve chronologies: a review. *Quaternary Science Reviews* 43: 45–60.

Olsen, J., N. J. Anderson, and M. F. Knudsen. 2012. Variability of the North Atlantic Oscillation over the past 5,200 years. *Nature Geoscience Letters* 5: 808-812. 5. doi: 10.1038/NGEO1589

Olsson, L., A. Jerneck, H. Thoren, J. Persson, and David O'Byrne. 2015. Why resilience is unappealing to social science: theoretical and empirical investigations of the scientific use of resilience. *Science Advances* 1: e1400217 22 May 2015

Orland, I. J., M. Bar-Matthews, A. Ayalon, A. Matthew, R. Kozdona, T. Ushikuboa, and J. W. Valleya. 2012. Seasonal resolution of eastern Mediterranean climate change since 34 ka from a Soreq Cave speleothem. *Geochimica et Cosmochimica Acta* 89: 240–255. doi:10.1016/j.gca.2012.04.035

Ortloff, C. R., and A. L. Kolata. 1993. Climate and collapse: agro-ecological perspectives on the decline of the Tiwanaku state. *Journal of Archaeological Science* 20: 195–221.

Parker, G. 2014. *Global Crisis: War, Climate Change and Catastrophe in the Seventeenth Century*. New Haven: Yale University Press.

Pearl, J. 2000. *Causality: Models, Reasoning and Inference*. Cambridge: Cambridge University Press.

Petrie, C. A., J. Bates, T. Higham, and R. N. Singh. 2016. Feeding ancient cities in South Asia: dating the adoption of rice, millet and tropical pulses in the Indus civilization. *Antiquity* 90: 1489–1504. Doi:10.15184/aqy.2016.210

Pickarski, N. O. Kwiecien, D. Langgut, and T. Litt. 2015. Abrupt climate and vegetation variability of eastern Anatolia during the last glacial. *Climate of the Past* 11: 1491–1505. doi: 10.5194/cp-11-1491-2015

Ponton, C., L. Giosan, T. I. Eglinton, D. Q. Fuller, J. E. Johnson, P. Kumar, and T. S. Collett. 2012. Holocene aridification of India. *Geophysical Research Letters* 39: L03704. doi: 10.1029/2011GL050722

Ramsey, C. B., M. W. Dee, J. M. Rowland, T. F. G. Higham, S. A. Harris, F. Brock, A. Quiles, E. M. Wild, E. S. Marcus, and A. J. Shortland. 2010. Radiocarbon-based chronology for Dynastic Egypt. *Science* 328: 1554–1557.

Redman, C. L., and A. P. Kinzig. 2003. Resilience of past landscapes: resilience theory, society, and the Longue Durée. *Conservation Ecology* 7: 14. http://www.consecol.org/vol7/iss1/art14

Redman, C. L. 2005. Resilience theory in archaeology. *American Anthropologist* 107: 70–77.

Revel, M., C. Colin, S. Bernasconi, N. Combourieu-Nebout, E. Ducassou, F. E. Grousset, Y. Rolland, S. Migeon, et al. 2014. 21,000 years of Ethiopian African monsoon variability recorded in sediments of the western Nile deep-sea fan. *Reg Environ Change*. DOI 10.1007/s10113-014-0588-x

Robert, N., D. Brayshaw, C. Kuzucuoğlu, R. Perez, and L. Sadori. 2011. The mid-Holocene climatic transition in the Mediterranean: causes and consequences. *The Holocene* 21: 3–13.

Rosen, A. M. 2007. *Civilizing Climate: Social Responses to Climate Change in the Ancient Near East*. Lanham, MD: Altamira Press.

Salzer, M. W., A. G. Bunn, N. E. Graham, and M. K. Hughes. 2013. Five millennia of paleotemperature from tree-rings in the Great Basin, USA. *Climate Dynamics* 42: 1517–1526. DOI: 10.1007/s00382-013-1911-9

Sanders, W. T., J. R. Parsons, and R. S. Santley. 1979. *The Basin of Mexico: Ecological Processes in the Evolution of a Civilization*. New York: Academic Press.

Schwartz, G. M. 2007 . Taking the long view on collapse: a Syrian perspective. In C. Kuzucuoğlu and C. Marro (eds.), Societés humaines et changement climatique à la fin du troisième millénaire: une crise a-t-elle eu lieu en haute Mésopotamie? Istanbul: Institut français d'études anatoliennes Georges-Dumézil, 45–67.

Severinghaus, J. P., T. Sowers, E. J. Brook, R. B. Alley, and M. L. Bender. 1998. Timing of abrupt climate change at the end of the Younger Dryas interval from thermally fractionated gases in polar ice. *Nature* 391: 141–146.

Shimada, I., C. B. Schaaf, L. G. Thompson, E. Mosley-Thompson. 1991. Cultural impacts of severe droughts in the prehistoric Andes: application of a 1,500-year ice core precipitation record. *World Archaeology* 22: 247–270.

Sinha, A., L. Stott, M. Berkelhammer, H. Cheng, R. Lawrence Edwards, B. Buckley, M. Aldenderfer, and M. Mudel. 2011. A global context for megadroughts in monsoon Asia during the past millennium. *Quaternary Science Reviews* 30: 47–62.

Stahle, D. W., K. Falko, F. K. Fye, E. R. Cook, and R. D. Griffin. 2007. Tree-ring reconstructed megadroughts over North America since A.D. 1300. *Climatic Change* 83: 133–149.

Stanley, J.-D., M. D. Krom, R. A. Cliff, and J. C. Woodward. 2003. Nile flow failure at the end of the Old Kingdom, Egypt: strontium isotopic and petrologic evidence. *Geoarchaeology* 18: 395–402.

Staubwasser, M., F. Sirocko, P. M. Grootes, and M. Segl. 2003. Climate change at the 4.2 ka BP termination of the Indus valley civilization and Holocene south Asian monsoon variability. *Geophysical Research Letter* 30.8.1425. DOI:10/1029/ 2002GL016822

Tainter, J. A. 1988. *The Collapse of Complex Societies.* Cambridge: Cambridge University Press.

Tainter, J. A. 2015. Why collapse is so difficult to understand. In R. K. Faulseit (ed.), *Beyond Collapse: Archaeological Perspectives on Resilience, Revitalization and Transformation in Complex Societies.* Carbondale, IL: Southern Illinois University Center for Archaeological Investigations. Occasional Paper 42.

Thompson, L. G., E. Mosley-Thompson, M. E. Davis, V. S. Zagorodnov, I. M. Howat, V. N. Mikhalenko, and P.-N. Lin. 2013. Annually resolved ice core records of tropical climate variability over the past ~1800 years. *Science* 340: 945–950. DOI: 10.1126/ science.1234210

Tung, T. A., M. Miller, L. DeSantis, E. A. Sharp, and J. Kelly. 2016. Patterns of violence and diet among children during a time of imperial decline and climate change in the ancient Peruvian Andes. In A. M. VanDerwarker and G. D. Wilson (eds.), *The Archaeology of Food and Warfare.* Cham, Switzerland: Springer International Publishing, 193–227.

Turner II, B. L., and J. A. Sabloff. 2012. Classic Period collapse of the central Maya lowlands: insights about human-environment relationships for sustainability. *Proceedings of the National Academy of Sciences* 109: 13908–13914.

Valdez, F., and V. Scarborough. 2014. The prehistoric Maya of northern Belize. In G. Iannone (ed.), *The Great Maya Droughts in Cultural Context: Case Studies in Resilience and Vulnerability.* Boulder: University of Colorado Press, 255–269.

van der Pflicht, J., P. M. M. G. Akkermans, O. Nieuwenhuyse, A. Kaneda, and A. Russell. 2011. Tell Sabi Abyad, Syria: radiocarbon chronology, cultural change, and the 8.2 ka event. *Radiocarbon* 53: 229–243.

Van West, C. R. 1994. *Modeling Prehistoric Agricultural Productivity in Southwestern Colorado: A GIS Approach.* Washington State University Department of Anthropology, Reports of Investigations 67. Cortez, CO: Crow Canyon Archaeological Center.

Varien, M. D. 2010. Depopulation of the northern San Juan region: historical review and archaeological context. In M. D. Varien, T. A. Kohler, and A. M. Wright (eds.), *Leaving Mesa Verde: Peril and Change in the Thirteenth-Century Southwest.* Tucson: University of Arizona Press, 1–33.

Verschuren, D., K. R. Laird, and B. F. Cumming. 2000. Rainfall and drought in equatorial east Africa during the past 1,100 years. *Nature* 403: 410–414.

Vimeux, F., P. Ginot, M. Schwikowski, M. Vuille, G. Hoffmann, L. G. Thompson, and U. Schotterer. 2009. Climate variability during the last 1000 years inferred from Andean ice cores: A review of methodology and recent results. *Palaeogeography, Palaeoclimatology, Palaeoecology* 281: 229–241.

Walker, M. J. C., M. Berkelhammer, S. Björck, L. C. Cwynar, D. A. Fisher, A. J. Long, J. J. Lowe, R. M. Newnham, S. O. Rasmussen, and H. Weiss. 2012. Formal subdivision of the Holocene Series/Epoch: a discussion paper by a working group of INTIMATE (integration of ice-core, marine and terrestrial records) and the

Subcommission on Quaternary Stratigraphy (International Commission on Stratigraphy). *Journal of Quaternary Science* 27: 649–659.

Webster, D. 2014. Maya drought and niche inheritance. In G. Iannone (ed.), *The Great Maya Droughts in Cultural Context.* Boulder: University of Colorado, 333–358.

Weiss, E. 2009. Glimpsing into a hut: the economy and society of Ohalo II's inhabitants. In A. S. Fairbairn and E. Weiss (eds.), *From Foragers to Farmers.* Oxford: Oxbow Books. 153–160.

Weiss, H., M.-A. Courty, W. Wetterstrom, F. Guichard, L. Senior, R. Meadow, and A. Curnow. 1993. The genesis and collapse of third millennium North Mesopotamian civilization. *Science* 261: 995–1004.

Weiss, H., and R. S. Bradley. 2001. What drives societal collapse? *Science* 291: 609–610.

Weiss, H., S. Manning, L. Ristvet, L. Mori, M. Besonen, A. McCarthy, P. Quenet, A. Smith, and Z. Bahrani. 2012. Tell Leilan Akkadian imperialization, collapse and short-lived reoccupation defined by high resolution radiocarbon dating. In H. Weiss (eds.), *Seven Generation Since the Fall of Akkad.* Wiesbaden: Harrassowitz, 163–192.

Weiss, H. 2014. The northern Levant during the Intermediate Bronze Age: altered trajectories. In A. E. Killebrew and M. Steiner (eds.), *The Oxford Handbook of the Archaeology of the Levant: c. 8000–332 bce.* Oxford: Oxford University Press, 367–387.

Weiss, H., and G. Booth. 2014. ArcheoSim 4.0. http://leilan.yale.edu/resources/archaeosim

Welc, F., and L. Marks. 2014. Climate change at the end of the Old Kingdom in Egypt around 4200 BP: new geoarchaeological evidence. *Quaternary International* 324: 124–133.

Wicks, K., and S. Mithen. 2014. The impact of the abrupt 8.2 ka cold event on the Mesolithic population of western Scotland: a Bayesian chronological analysis using "activity events" as a population proxy. *Journal of Archaeological Science* 45: 240–269.

Wilkinson, T., J. H. Christiansen, J. Ur, M. Widell, and M. Altaweel. 2007. Urbanization within a dynamic environment: modeling Bronze Age communities in Upper Mesopotamia. *American Anthropologist* 109: 52–68.

Willcox, G., M. Nesbitt, and F. Bittman. 2012. From collecting to cultivation: transitions to a production economy in the Near East. *Vegetation History and Archaeobotany* 21: 81–83.

Wong, C. I., and D. O. Breecker. 2015. Advancements in the use of speleothems as climate archives. *Quaternary Science Reviews* 127: 1–18.

Zanchetta, G., E. Regattier, I. Isola, R. N. Drysdale, M. Bini, I. Baneschi, and J. C. Hellstrom. 2016. The so-called "4.2 event" in the central Mediterranean and its climatic teleconnections. *Alpine and Mediterranean Quaternary* 29: 5–17.

Zolitschka, B. 2007. Varved lake sediments. In S. A. Elias (ed.), *Encyclopedia of Quaternary Science.* Amsterdam: Elsevier, 3105–3114.

A NOTE ON CHRONOLOGY

The convergence of paleoclimate and archaeological researches forces attention to the varied chronological terminologies used by their respective scientific communities. These usages have proven impossible to homogenize
across chapters in this volume. The usage "cal BP" represents calendar years
before the present, assumed to be 1950, as determined from the calibration of
radiocarbon dates, and the usage "cal BC" represents calendar years before the
birth of Christ, also determined from the calibration of radiocarbon dates. The
"ka BP" terminology used by paleoclimatologists represents thousands of calendar years before the present, measured by convention to 1950 as determined
from the calibration of radiocarbon dates or Uranium-Thorium dates. The
Mesopotamian historical chronology used by archaeologists is the probable
"revised Middle Chronology" calendar dates, with an uncertainty of about sixty
years for the third millennium BC. The "AD" dates used here by western hemisphere and East Asian paleoclimatologists and archaeologists are also derived
from calibrated radiocarbon dates alongside annual glacial ice lamination or
tree-ring counts, or both.

CHAPTER 1

12,000–11,700 cal BP

The Collapse of Foraging and Origins of Cultivation in Western Asia

OFER BAR-YOSEF, MIRYAM BAR-MATTHEWS, AND AVNER AYALON

We take up the question of "why" cultivation was adopted by the end of the Younger Dryas by reviewing evidence in the Levant, a subregion of southwestern Asia, from the Late Glacial Maximum through the first millennium of the Holocene. Based on the evidence, we argue that the demographic increase of foraging societies in the Levant at the Terminal Pleistocene formed the backdrop for the collapse of foraging adaptations, compelling several groups within a particular "core area" of the Fertile Crescent to become fully sedentary and introduce cultivation alongside intensified gathering in the Late Glacial Maximum, ca. 12,000–11,700 cal BP. In addition to traditional hunting and gathering, the adoption of stable food sources became the norm. The systematic cultivation of wild cereals begun in the northern Levant resulted in the emergence of complex societies across the entire Fertile Crescent within several millennia. Results of archaeobotanical and archaeozoological investigations provide a basis for reconstructing economic strategies, spatial organization of sites, labor division, and demographic shifts over the first millennium of the Holocene. We draw our conclusion from two kinds of data from the Levant during the Terminal Pleistocene and early Holocene: climatic fluctuations and the variable human reactions to natural and social calamities. The evidence in the Levant for the Younger Dryas, a widely recognized cold period across the northern hemisphere, is recorded in speleothems and other climatic proxies, such as Dead Sea levels and marine pollen records.

Introduction

Paleolithic archaeologists dealing with the deep past consider climate to be one determinant in the distribution of foraging societies and their adaptations. However, interactions between environmental changes and prehistoric lifeways have often been examined from the perspective of the hunters and rarely that of the gatherers. Mobility, expressed as an annual or seasonal strategy, would have allowed groups of foragers to move between ecological terrains in the savannah, along river valleys, or through topographic altitudes. Whether an entire group constantly changed location (residential movement) or teams of hunters went out to bring back food supplies (logistical strategy) there have been no claims made for "over-populated regions" (Binford 1980). Demographic pressure was certainly not a cause of prehistoric migrations, such as the four or five waves of dispersals from Africa (the "Out of Africa" theory). Generally, migrations within Africa or into Eurasia were characterized by acculturation and interbreeding, not physical conflict.

Environmental conditions are another factor that either must have facilitated or retarded movements of people to unknown empty or sparsely inhabited lands during the Pleistocene. Due to rapid demographic growth, the successful transformation after more than 2.6 million years of human evolution to a mixed economy heralded a new social organization (Bouquet-Appel 2011). By the end of the Younger Dryas (ca. 12,900 to 11,700 cal BP), the transition from foraging to farming is visible both archaeologically and archaeobotanically in the Levant. There remains a need to investigate why, how, and when this "tipping point" occurred.

Since the 1950s many scholars have tried to respond to these questions. Although views vary on whether cultivation or intensified gathering was conducted by the earliest Neolithic villagers during the first one and half millennia of the Holocene, most researchers support the idea that the assemblages of carbonized plant remains from the first millennium of the Holocene demonstrate a state of "pre-domestication" cultivation and that the "domestication syndrome" was achieved gradually (Simmons 2007; Rosen & Rivera 2012; Bar-Yosef, Mayer & Zohar 2010; Goring-Morris & Belfer-Cohen 2011; Belfer-Cohen & Goring-Morris 2011; Asouti & Fuller 2013; Bar-Yosef 2011, Bar-Yosef & Valla 2013; Willcox 2012). Others suggest that the mutational changes took place in a short timespan of perhaps a couple centuries (Hillman & Davies 1990a, b; Abbo, Lev-Yadaun, & Gopher 2011, 2012; Böhner & Schule 2008; Clare & Weninger 2013).

Levantine Climate and the Main Paleoclimatic Sources

To answer the question of "why" cultivation was adopted by the end of the Younger Dryas, we need to look at climatic evidence from the Late Glacial Maximum (hereafter, LGM) through the first millennium of the Holocene and consider the ways humans reacted when facing environmental challenges. We propose that humans used ephemeral adaptive techniques that often failed owing to social reasons. Some of these methods were embedded in the social memory of particular groups and eventually led to new successful trials.

Winter precipitation and centennial fluctuations over the last 15,000 years are highly relevant to any study of the origins of wild-plant cultivation in the Levant or the relative success of its first farming communities. The climate of the Levant is determined largely by its location at the eastern edge of the Mediterranean basin, a region affected by several atmospheric systems, chief among them the prevailing winds of the westerlies, an accumulation of cold air called the Siberian High, and the northern reach of the Intertropical Convergence Zone—literally, where the northeast and southeast trade winds convene and which impacts the arid belt of the Sahara and extends into the Arabian Peninsula (see fig. 1.1).

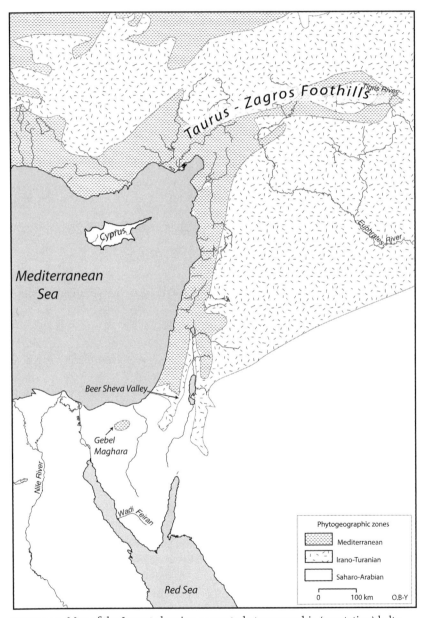

FIGURE 1.1 Map of the Levant showing current phytogeographic (vegetation) belts.

This combination of systems manifests in snowy winter precipitation in the higher mountains and dry summers. Temporary fluctuations may increase or decrease the amount of precipitation and bring very dry or wet summers. Rain is more abundant in the northern Levant, from the Taurus foothills through the upper Euphrates Valley, and decreases in the southern Levant, from the middle Euphrates Valley and Damascus basin to the southern tip of the Sinai Peninsula. Thus, the southern Levant suffers from low mean annual precipitation, shows higher inter-annual variability, and is more affected by droughts.

Paleoclimatic proxies suggest somewhat different patterns of rainfall from the LGM and Marine Isotope Stage 2 (MIS2) that lasted from ca. 24/23 to 19/18 ka BP through to the end of the Younger Dryas and the first millennium of the Holocene (ca. 11,700 cal BP). By combining results of the following different investigations, we can interpret the climate of various time slices from the LGM through the Younger Dryas and then pair it with the archaeological record:

- Soreq Cave speleothems (Bar-Matthews et al. 1996, 1997, 2003; Ayalon et al. 1998, 1999, 2004, 2013; Kaufman et al. 1998; Matthews et al. 2000; Kolodny et al. 2003; McGarry et al. 2004; Orland et al. 2009, 2012, 2014; Bar-Matthews & Ayalon 2011; Bar-Matthews 2014)
- Lake Lisan levels (Bartov et al. 2003; Stein et al. 2010; Kushnir & Stein 2010; Stein 2014; Torfstein et al. 2013a,b)
- Marine and terrestrial pollen and isotope records (Rossignol-Strick 1995; Almogi-Labin et al. 2009; van Zeist et al. 2009; Develle et al. 2010; Langgut et al. 2011).

In addition, we refer to climatic models (Enzel et al. 2008) and comment on the northern hemisphere paleoclimatic parallel time slices (Broecker et al. 2010; Denton et al. 2010; Wolff et al. 2010).

Each of these climatic proxies yields specific results and degrees of chronological resolution. Several studies of the Soreq Cave speleothems have been discussed elsewhere, but here we pay particular attention to the work of Orland et al. (2009, 2012, 2014), which reveals the critical issue of seasonality and, in particular, rainfall. Regarding seasonality, Rossignol-Strick's (1995) studies of marine pollen cores suggest conditions in the eastern Mediterranean during the Younger Dryas and the early Holocene that differs from the modern period. The revision of results from the terrestrial Hula Valley pollen core (van Zeist, Baruch, & Bottema 2009) supports further discussion.

Orland et al. (2012), following earlier work in the Soreq Cave, indicate that the $\delta^{18}O$ values of its speleothems reflect isotopic equilibrium with drip water that precipitated into the cave from rainfall on the hill above (Ayalon, Bar-Matthews, & Kaufman 1998). The $\delta^{18}O$ changes in Soreq Cave speleothems record the combined effect of vapor (sea source $\delta^{18}O$) and average annual precipitation through the Late Pleistocene and Holocene (Bar-Matthews et al. 2003; fig. 1.2). Kolodny, Stein & Machlus (2005) critiqued this conclusion and suggested that the long-term trends of the $\delta^{18}O$ values represent the changes in the water of the Eastern Mediterranean Sea. However, Almogi-Labin et al. (2009) compared the $\delta^{18}O$ of planktonic foraminifera from the Eastern Mediterranean Sea to the Soreq Cave $\delta^{18}O$ record to demonstrate that the first-order oscillation in the speleothem $\delta^{18}O$ is associated with sea sources, but the second-order effects are associated with rainfall, the relative elevation of the cave, and its distance from the sea. Since the sea level was up to 130 meters lower at that time, changing precipitation ratios are of major importance to speleothem $\delta^{18}O$ evidence.

Orland et al. (2009, 2012, 2014) employed a new technique combining high-resolution (Secondary Ion Mass Spectrometer, or SIMS) $\delta^{18}O$ with confocal laser fluorescence microscopy (CLFM) imaging (figs. 1.3–1.4). The combination of 10 μm-diameter spot analyses of ^{18}O revealed a saw-tooth pattern of fluorescence intensity across concentric growth bands for recent and mid-late Holocene-type climatic conditions. The beginning of each growth band was marked by a sharp onset of bright-fluorescent calcite ("bright calcite"), indicating spring laminae, followed by a gradual reduction in fluorescence intensity to dark-fluorescent calcite ("dark calcite"), indicating the end of summer-autumn laminae. The systematic recording of these fluorescent intensities allows a detailed study of individual annual growth bands. Moreover, the large data set assembled by Orland et al. allows new interpretations of climate change at the sub-annual scale. This high-resolution analytical method is ideal for examining both rapid climate

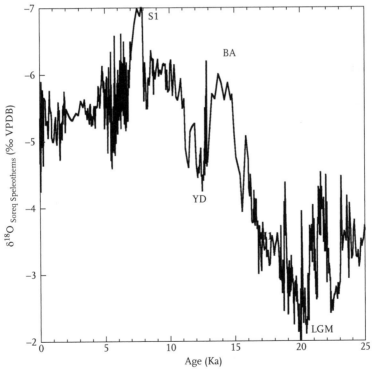

FIGURE 1.2 The sequence of Soreq Cave speleothem $\delta^{18}O$ from 25 ka BP through present day. (Reprinted with permission from Bar-Matthews et al. 2003.)

change events and seasonal climate differences between broad timespans—for example, between the Younger Dryas and the Holocene (figs. 1.3–1.4).

Rossignol-Strick (1995), who first studied marine pollen cores from bore-holes in the Eastern Mediterranean, noted a discrepancy between the dates of terrestrial pollen and marine cores, which she attributed to the contamination of samples by allochthonous substances (old carbonates) and hard-water effects. Yet a recent study by Langgut et al. (2011) found a better correlation with the climatic interpretations of the foraminifera record by Almogi-Labin et al. (2009). (Holocene information for the Aegean is from Rohling et al. 2002.)

For the first millennium of the Holocene, the terrestrial sequence from the Hula Valley produced aberrant dates (Rossignol-Strick 1995). This led van Zeist, Baruch, and Bottema (2009) to revise Baruch and Bottema's (1999) pollen sequence of the Holocene. A fuller sequence of an earlier study, by H. Tsukada, which unfortunately was not published in detail, provides complementary information from MIS 3 and later in the Hula Valley.

Additional isotopic sequences for terrestrial correlations are available from the karstic Lake Yammoûneh in the Lebanese mountains, ca. 2200 meters above sea level (Develle et al. 2010). The paleoclimatic record here is derived from a composite isotopic record obtained from three different ostracod taxa. The oxygen isotope signature of the most widespread species (*Ilyocypris inermis*) of ostracods ($\delta^{18}O_{ost}$) was converted to $\delta^{18}O$ values coeval with calcite. The calcite precipitated in equilibrium ($\delta^{18}O_C$) with the lake water ($\delta^{18}O_L$). Pollen studies from boreholes further enhanced the climatic interpretations.

Climate Models

Two models of southern Levant climate have been suggested (Robinson et al. 2006; Enzel et al. 2008). Both are based on several sources, as well as on atmospheric considerations. The Robinson et al. model suggests a dryer Younger Dryas. The Enzel et al. model, while focusing on the southern Levant, attributes a wetter Younger Dryas to the northern Levant. Using archaeobotanical evidence from numerous sites, the Enzel et al. hypothesis supports the view that the origins of cultivation took place in a "core area" in the northern Levant at a time when it probably benefited from better precipitation (Lev-Yadun et al. 2000; Bar-Yosef 2002; Kozlowski & Aurenche 2005; Gopher et al. 2001, 2013; Heun et al. 2012; but for an opposing view see Asouti & Fuller 2013; Weiss et al. 2004, 2006). These exclusive models are contradictory and need further testing.

Climatic Proxies in the Levant

Late Glacial Maximum (LGM)

The suggested dates for this period in Soreq Cave are 25–17/16 ka BP, based on the age model of Almogi-Labin et al. (2009). A speleothem from Jerusalem

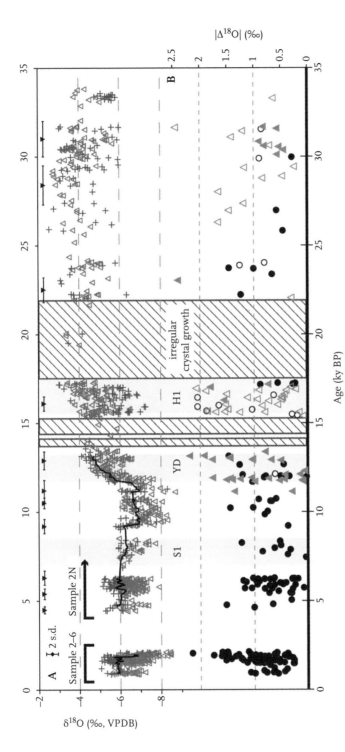

FIGURE 1.3 Soreq Cave speleothem sequence of the last 35,000 years showing fluorescence patterns reflecting seasonality and correlations, with known climatic fluctuations: Heinrich Event 1, the Younger Dryas, and Sapropel 1. (Reprinted with permission from Orland et al. 2012a.)

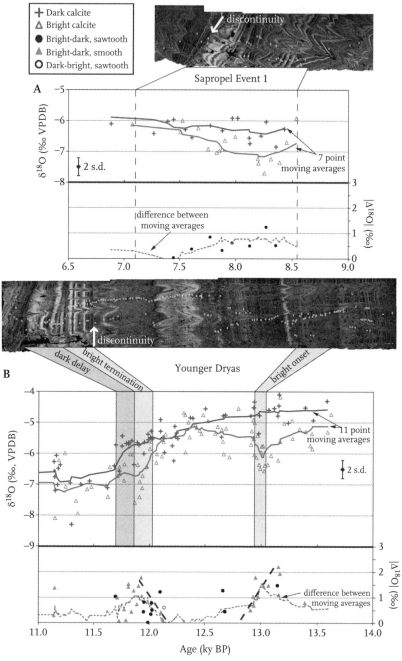

FIGURE 1.4 Detailed Soreq Cave speleothem sequences for Sapropel 1 and the Younger Dryas. (Reprinted with permission from Orland et al. 2012a.)

yielded additional data (Frumkin et al. 2000). The marine pollen record suggests the dates 26.6 to 22.4 ka BP or ca. 18 ka BP following an extreme cold episode ca. 19 ka BP (Langgut et al. 2011, p. 3967). From Torfstein et al. (2013a; see fig. 1.5), the MIS 2 dates are ca. 27–21 ka BP, incorporating H3 (Denton et al. 2005, 2010) and compared to NGRIP and Chinese cave speleothems. One difficulty with these dates is that if the cold episode is considered to be ca. 23/22–18 ka BP, then rapid dryness in the Dead Sea Basin is correlated in time with the Greenland Ice Sheet Project GISP2 and the refill of Lake Lisan took place in what is often considered the end of Marine Isotope Stage 3 (MIS 3).

The fluorescent bandings and $\delta^{18}O$ values in the Soreq speleothem during LGM are highly similar to those of the Heinrich 1 Stadial (H1), suggesting comparable climates that were unlike that of the Holocene. Yet an inferred empirical relationship between annual rainfall $\delta^{18}O$ values and rainfall amount during LGM and H1 based on the banding patterns and $\delta^{18}O$ seems uncertain. The regular "reversed" fluorescent banding pattern—that is, dark before bright banding—suggests a decreased seasonal rainfall gradient. The trends of the $\delta^{18}O$ observed in the LGM and H1 (fig. 1.3) probably indicate reduced differences in seasonal precipitation, occasional snowfall, and possibly the associated presence of vegetation that differed from the dominant pattern in the Judean Hills during the Holocene (Orland et al. 2012, 2014). In addition, with a lower sea level during the LGM, the cave was further inland than today, which would certainly have affected the $\delta^{18}O$ precipitation values above the cave (Almogi-Labin et al. 2009).

The arboreal pollen record declines considerably in LGMs. The coldest period in the sequence is at ca.19.0 ka, when tree frequencies were lowest (Langgut et al. 2011). This abrupt event correlates with similar conditions in Lake Yammoûneh (Develle et al. 2011) and is supported by a probably major change of ≈10°C (McGarry et al. 2004; Affek et al. 2008; Almogi-Labin et al. 2009), which is also indicated by the Mt. Hermon speleothems (Ayalon et al. 2013). Moreover, the Yammoûneh depositional sequence shows that from ca. 21 to 16 ka BP, frequent periods of desiccation existed, while pollen data indicate an open vegetation cover dominated by steppic and desert plants.

The upper sections of the Lake Lisan formation represent MIS 2 (fig. 1.5; Torfstein et al. 2013a). After a drop in high-water level at ca. 33 ka BP, an increase in precipitation fills the basin to a level of ca. 200 meters below mean sea level (bmsl), followed by a short drop before a return to a higher level. Finally, a small drop in the water level occurs before a major drop in H1 (starting at 17.1±0.5 ka BP).

Post-Late Glacial Maximum and Heinrich 1

The dates for this period depend on various sources. The archaeological record suggests that a slow rise in precipitation began around 18/17.5 ka BP and lasted, including a drier H1, until ca.14.5 ka BP. H1 is marked as 16.5 ka BP and was succeeded immediately by the onset of the Bølling-Allerød (B/A) period. According to Stein et al. (2010), the dates of H1 in the Dead Sea Basin are 17.4-16 ka BP. Torfstein et al. (2013b) suggest a basal U-Th date of H1 at 17.1±0.5 ka BP.

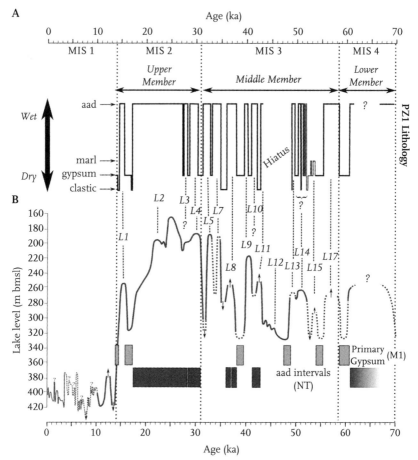

FIGURE 1.5 Lisan Lake-level fluctuations in the Dead Sea basin correlated with known climatic events (Reprinted with permission from Torfstein et al. 2013a.)

The H1 event in the Soreq record, characterized by a regular reversal in the fluorescent banding pattern, is dated to ca.16.5 ka BP (fig. 1.3). The banding pattern (fig. 1.4) differs from that of the Holocene. Furthermore, the mean of the $\delta^{18}O$ values during this period is higher than during both the Younger Dryas and the Holocene. Decreased seasonal rainfall gradients, regular snow cover, and different vegetation over the Judean hills may have been possible causes.

Considering marine pollen that reflects terrestrial vegetation, Langgut et al. (2011) suggest that lower tree cover and the dominance of steppic vegetation preceded the change that began with the deglaciation period. Similarly, the Dead Sea basin record also reflects a sequence of fluctuations indicating a constant decrease in precipitation with a maximum drop during H1 starting at ca. 17±0.5 ka BP (Torfstein et al. 2013a). An increase in precipitation level followed and is correlated with Bølling-Allerød.

Bølling-Allerød

Radiometric dates for this period are ca. 14.5/14 ka BP to 13/12.8 ka BP. Based on Soreq Cave U-Th dates (Almogi-Labin et al. 2009), the marine pollen record (Langgut et al. 2011) indicates a range from 14.6 to 12.3 ka BP. The vegetation indicates high arboreal pollen levels, mainly *Quercus calliprinos* type (evergreen oak) and a pronounced reduction in *Artemisia*. All records agree concerning the increase in precipitation. In the Dead Sea basin, Lisan Lake rises to ca. 240 meters bmsl from ca. 310 meters bmsl during the H1 event. Stein et al. (2010) argue that the Bølling-Allerød was the driest period and the Younger Dryas was the wettest. The Dead Sea curve shows a drop after the Bølling-Allerød and then rises again to reflect the onset of the Younger Dryas, when precipitation increased tremendously (fig. 1.5). The wet Younger Dryas in the Dead Sea basin differs markedly from the dry Younger Dryas in Lebanon (Verheyden 2008) and central Israel.

Younger Dryas

The 12.8–11.7 ka BP dates for the Younger Dryas are derived from global correlations based on marine and ice cores (Broecker et al. 2010; Denton et al. 2010). Modifications by local proxies include the marine pollen record (Langgut et al. 2011) with dates of 11.9-11.2 ka BP. A considerable rise in *Artemisia* values indicates a drier and somewhat colder period than the H1 and H2 events. Haase-Schramm et al. (2004), suggest that Lake Lisan recedes at 14.6 ka BP.

Orland et al. note: "markedly different fluorescent banding patterns exist, however, in growth prior to 10.5 ka BP. From 13.5 to 11 ka BP, the abrupt onset of fluorescence at the beginning of each band is commonly smoothed so that the majority of bands have a gradual gradient from bright to dark calcite as well as from dark to bright in the subsequent fluorescent band. . . . This sinusoidal fluorescence intensity describes the majority of bands during this period" (2012: p. 245).

Fluorescent banding patterns that are consistently different than in the Holocene indicate a decrease in seasonality and reduced but consistent supply of water to the cave during winter and summer. Since the width of these bands become both brighter and wider, the termination of the Younger Dryas was certainly marked by a significant climatic change that lasted at least 12 years (Orland et al. 2012).

First Millennium of the Holocene

In the Holocene, the ubiquitous pattern of minor fluctuations of both $\delta^{18}O$ and variable fluorescence (Orland et al. 2012) indicates that the modern seasonal regime of wet winters and dry summers has been consistent since 10.5 ka BP and even more so since 7.5 ka BP (Orland et al. 2012, figs. 2–5). This pattern occurred after the extreme wet climate indicated by the sapropel deposition in the Eastern Mediterranean (Bar-Matthews et al. 2000; 2003; Bar-Matthews 2014).

Analysis of the Hula pollen sequence shows that the early Holocene (11,700–10,500 cal BP) had low arboreal pollen, mostly represented by Tabor oak, and high frequencies of herbaceous pollen. This evidence is indicative of steppic vegetation. Van Zeist et al. (2009) suggest that these valley landscapes had open woodland with an understory rich in grasses. Post 7.5 ka BP a trend toward decreasing rainfall is apparent (Bar-Matthews et al. 2003). Superimposed on this trend is a 1500-year cyclical pattern of annual changes in rainfall (Bar-Matthews & Ayalon 2011). Information from the Dead Sea basin (Stein et al. 2010) supports this scenario and suggests that during the Holocene, the Dead Sea region and central and northern Israel responded similarly to climate change.

Archaeological Considerations

Although the cumulative paleoclimatic data demonstrate general trends, differences in dating caused by varying degrees of resolution make the correlation of climatic events with human cultural changes challenging. Human events are difficult to date accurately, since they often occur within short time intervals of between one to three centuries. Our approach, therefore, is first to examine demographic trends that began with the role of LGM as a genetic bottleneck and then to study changes in the spatial distribution of resources (Bar-Yosef 2002, 2011).

Experiments in plant cultivation were probably more common during earlier periods than has generally been assumed. Because of long- or short-term fluctuations in the predictability and accessibility of vegetal and animal sources and competition with other bands or tribes, some group members may well have been compelled to try new subsistence strategies. A recent study of cereal remains from Ohalo II, dated to ca. 23,000 cal BP, demonstrated that 36 percent of the 320 barley rachises and 25 percent of the 148 wheat rachises collected showed the rough disarticulation scars that mark a domesticated cereal. (The presence of more than 10 percent of the domestic-type in a given archaeobotanical assemblage is accepted as an indication for domestication.) Combined with evidence of "proto weeds," these findings suggest that some, or all, of the band practiced cultivation (Snir et al. 2015). Such evidence also demonstrates that our conceptions of Late Pleistocene foragers in the Levant are biased by the availability of well-preserved bones as opposed to poorly preserved plant remains. Thus, at most Late Pleistocene sites, we tend to identify hunting and trapping as the main activities.

Evidence of the manipulation of wild animal species has been noted during the terminal Pleistocene. One example is the import of boars to Cyprus (Vigne 2013, Bar-Yosef 2014). With a systematic phytolith and dynamic starch analysis, in spite of a lack of well-preserved plant remains, our meager knowledge of cultivation is far from balanced with what we know about hunting and herding. Hunting was intensified towards the end of the Epi-Paleolithic (Munro 2009). Since cultivation resulted in the "domestication syndrome," the genetic changes of plants facilitated the massive population growth that we attribute to the "Neolithic Revolution."

Archaeobotanical evidence reveals that intensified gathering of cereals led to the building of storage facilities among the newly established villages. During the early Holocene, plants and herded animals became practically domesticated and were transported long distances. To identify where and when this occurred and correlate this with climatic fluctuations, we must first discuss how humans react to environmental calamities. Next, we will examine how the paleoclimatic record compares with archaeological data.

Human Responses to Natural Calamities

Although the literature on resilience is increasing exponentially, most studies present a rather holistic view of human society at large, while only a few scholars address cultural variability. Historical records and relief workers' reports are our best sources for understanding how discrete human societies have reacted to natural disasters, be they earthquakes, floods, cold winters, or droughts.

The archaeological record of the Levant from the LGM to the mid-Holocene presents a challenging geographic puzzle of populations, intertwined with periods of competition for resources and conflicts (Rosen 2007; Rosen & Rivera 2012; Goring-Morris & Belfer-Cohen, 2010, 2011; Bar-Yosef 2011).

A focus purely on the evolutionary sequence from foraging to farming ignores relevant evidence for replacement, for failures due to lack of social flexibility, and for the detrimental effects of abrupt climatic change on several groups. Two centuries of reasonably frequent droughts could have been the trigger for cultural change, even as such a period appears as a minor blimp on a climatic graph. The current challenge is to understand the capacity of societies of foragers and farmers to be resilient in the face of an abrupt climate crisis (see Barnes et al. 2013). Ecologists originally defined the conceptual framework of "resilience" as "the magnitude of disturbance that can be tolerated before a socio-ecological system (SES) moves to a different region of state space controlled by a different set of processes" (Carpenter et al. 2001, p. 765; Holling, Gunderson, & Ludwig 2002). We believe that the better term is, rather, "adaptive capacity," which connotes the ability of societies to overcome, as well as survive, the impact of climatic calamities.

The overarching social factor that distinguishes Paleolithic, recent hunter-gatherer, and agro-pastoral societies from other forms of society is the quest for food. Finding food for themselves and their descendants was the main determinant of human reactions. To ensure the biological survival of the tribe, the right decisions needed to be made in spite of depleted environmental resources. During the MIS 3 stage of the Upper Pleistocene, before increasing cold conditions led to a new Ice Age (MIS 2), relative demographic densities in Eurasia were generally correlated with economic potentials or the carrying capacities of the variable habitats (Bouquet-Appel 2011). Optimal foraging was the common strategy for supplying meat and vegetal diet. Under favorable conditions, such societies have imposed additional limitations, such as taboos, or tensions with neighboring populations caused by competition and linguistic barriers. However, when environmental conditions worsened, basic issues of survival arose, and difficulties ensued for all, it appears that—as psychologist

Daniel Kahneman suggests—most judgments and choices were made intuitively and that "the rules that govern intuition [were] generally similar to the rules of perception" (2003: 1450). Foragers could expand their "homeland" or move into neighbors' territories, whether peacefully or by force. These actions would determine their future social interactions. To be sure, taking over a new territory or guarding an old one against raiders entailed costs in time, energy, and social relationships (see Dyson-Hudson & Smith 1978). When migrating to other territories, hunter-gatherers had three options:

1. Amicably join the locals;
2. Ignore the presence of locals;
3. Face the unpredictable consequences of fighting.

In terms of minimizing risk, as Kahneman points out, "natural and intuitive in a given situation is not the same for everyone: different cultural experiences favor different intuitions about the meaning of situations, and new behaviors become intuitive as skills are acquired" (2003: 1469). Therefore, we should not expect that the archaeological record of the Levant during the end of MIS 3 and the course of MIS 2 will reveal a uniform success story resulting from prevailing high precipitation, as recorded in the Dead Sea Basin. Rather, we can expect a geographic jigsaw puzzle of economically winning or losing social entities beginning with early Epi-Paleolithic period groups, such as the Kebaran or Nebekian.

The cognitive mechanism that enables foragers to make intuitive decisions is the long-term social memory preserved by band and tribal elders. Passed on across generations, diachronic observations relay information about environmental changes and past solutions. Conservationists working with extant populations of hunter-gatherers consider this knowledge of ecosystems and biodiversity to be highly important (Gadgil, Berkes, & Folke 1993). Such information may provide one component needed "to identify and encourage 'virtuous' tipping points in human systems and their coupling to environmental systems, to achieve sustainability" (Lenton 2013: 20).

The various strategies of foragers facing environmental degradation rarely included investment in the propagation of plant resources. Intensification of traditional exploitation techniques, migrations, and even physical conflicts would have been the best, though temporary, solutions. Poor choices ended with extinction.

Interpreting Climatic and Archaeological Records

The available proxies for the paleoclimate of the Levant presented above exhibit several difficulties. Water always existed in the system, and even during glacial periods ground water was available. The main debate concerns whether there was more or less rain during peak glacial periods. The second issue, for both environmental and archaeological sequences, concerns the achieved chronological resolution. Archaeologists combine archaeological data with climatic trends derived from a particular paleoclimatic source in order to draw conclusions regarding the degree of human resilience in the face of environmental challenges. The paleoclimate source generally employed has been the North Atlantic sequence as recorded in ice cores. Yet recent suggestions that geographic variation may make North Atlantic ice-core data less than ideal for the Levant has led to attempts to use local paleoclimatic source data.

Directly dated paleoclimatic evidence, which would have allowed a better correlation with the proxies of nonhuman sources, is missing from archaeological sites. On-site pollen is often poorly preserved and only partially represents the local environment. The analysis of non-pollen palynomorphs (NPPs), which yield more information about local environments and human activities, provides a partial remedy (Kvavadze et al. 2009; Carrión et al. 2010).

Additionally, the ^{14}C dates crucial for deciphering the impact of climatic calamities should derive from contexts representing the timing of the abandonment of the site and the earliest reoccupation in the same locale or nearby. Correlation with the Holocene, particularly for Neolithic farming communities, is the best available (Highman et al. 2007; Weninger & Joris 2008, Reimer et al. 2009; Weninger et al. 2009), since the climatic impacts of the Holocene are more clearly registered through village archaeology. Farmers, whose investments in houses, storage facilities, wells, and shrines increase their physical and psychological attachment to the particular location, react in a more active way to environmental changes than foragers who move away or die. For farmers, abandonment is often the last resort in face of a disaster, whereas hunter-gatherers would more certainly demonstrate an inherent social resilience (Holling, Gunderson, & Ludwig 2002; Rosen 2007; Rosen & Rivera 2012).

Given these caveats, we begin with the Late Glacial Maximum (ca. 24–19/18 ka BP), which was a relatively wet period with a few fluctuations, according to the Dead Sea levels (Torfstein et al. 2013a). The high levels of Lake Lisan could have been the result of the favorable climatic conditions of MIS 3 and a decrease in evaporation during the LGM. According to marine pollen data, this period was drier, with a decrease in arboreal pollen frequencies. Speleothem deposition was continuous in the western Judean Hills (Bar-Matthews et al. 2003; Bar-Matthews, 2014) during the LGM, suggesting the Levantine region sustained open woodlands along the coastal plain and the hilly belt from the Taurus foothills to the Judean hills. This area was clearly suitable for the continuous existence of hunter-gatherers. In the Jordan Valley, the eastern plateau (in Jordan), and Sinai, habitable localities were to be found mostly in oases or along wadi courses descending to the Jordan Valley.

Archaeologically, the lithic assemblages from numerous sites are usually the only preserved remains. These are assigned to the Masraqan, Qalkan, Nebekian and Kebaran complex cultural entities (ca. 23–18 ka BP) (Goring-Morris 1995, 2009 and references therein; Maher, Richter, & Stock 2012a). The Kebaran sites are rare in the semi-arid zone except for the mountains of southern Jordan and the oases of Kharaneh IV, Azraq, and Jilat, all within a

FIGURE 1.6 Distribution map of most of the Late Paleolithic/Early Epi-Paleolithic entities (ca. 24/23–18 ka BP).

radius of 20 kilometers (fig. 1.6; Maher et al. 2012b). Since the coastal plain was wider by about 10–15 kilometers due to the low sea levels, the geographic distribution has no data about the sites lost when the Mediterranean Sea rose to its Holocene levels. The locations of the known LGM sites correspond to the habitable conditions within the Mediterranean vegetation zone and the marginal Irano-Turanian steppic areas.

The best-recorded site from this period is Ohalo II, a camp of foragers on the shores of Lake Kinneret exposed when the lake shrank to 212 mbsl (Nadel et al. 2004). The site is dated to 22.3–22.8 ka BP, based on barley seed dates (Nadel et al. 1995, 2001) and numerous other charcoal samples and correlates with a relatively drier time within the LGM. The date is perhaps closer to H2, when the sharp drop in the lake level indicates a much drier climate than during the following millennia. The rich archaeobotanical assemblage from the site (with over 100,000 plant remains) contains many species, including fruits and seeds (Kislev, Nadel, & Carmi 1992; Weiss et al. 2004, 2008). These demonstrate the exploitation of an altitudinal cross-section of plants from the Lake Kinneret level (212 mbsl) to the top of the nearby eastern Galilee hills (600–800 asl). Cereals make up a small portion of the retrieved plant remains and probably represent a failed trial in cultivation rather than the results of gathering. Local foragers, presumably responding to improved gathering and hunting conditions in their environment, discontinued an experiment that could have heralded the earlier appearance of farming. Based on the Ohalo II example, other experiments in cereal cultivation likely occurred in the following millennia, but due to the poor preservation of plant remains in all excavated Epi-Paleolithic sites to date, there is yet no evidence for them.

In the northern Levant, similar ecological conditions provided adequate survival conditions for local foragers, but the distribution of the microlithic industries (also referred to as early Epi-Paleolithic) is poorly known except in the basin of el-Kowm in Syria. Due to the lack of targeted surveys, the vast area of southeast Turkey, from the Taurus foothills through the middle Euphrates area at the Syrian border, is still a *terra incognita* for Epi-Paleolithic archaeological entities.

Improved post-LGM environmental conditions, expressed as a slow temperature rise and increased winter precipitation, allowed the spread of the Geometric Kebaran Complex in the Mediterranean vegetation belt (ca. 18.5–15 ka BP) and in former semi-arid areas (at ca.18.5–16.5 ka BP). These Geometric Kebaran sites are located within three geographically parallel phytogeographic belts stretching through the Eastern Mediterranean. These include the Mediterranean forests, the Irano-Turanian semi-steppic belt, and the Saharo-Arabian desert region (Zohary 1973). The successful expansion of the Geometric Kebaran into the previously arid and semi-arid environments is exemplified by several sites (fig. 1.7). Among these, El-Kowm is a recently discovered major site (Cauvin & Coqueugniot 1988) similar to one in the Palmyra oasis (Fujimoto 1979). In the central Levant, Kharaneh IV is certainly the largest aggregation site known, covering an area of ca. 20,000 square meters (Maher, Richter, & Stock 2012a). Somewhat smaller is the Uyun al-Hammam site in Wadi Ziqlab (Maher 2007). Jebel Hamra J201 in southern Jordan has been excavated (Henry 1995), and some 200 kilometers further south in Wadi Feiran, in southern Sinai, the site of Wadi Sayakh marks the southernmost Geometric Kebaran location (Bar-Yosef & Killbrew 1984).

All these sites, spread over a 1000 kilometer-long axis, share a distinctive similarity in their flint core reduction sequence and their shaping of microlithic geometric forms (trapeze-rectangles) from blade/bladelet blanks, a

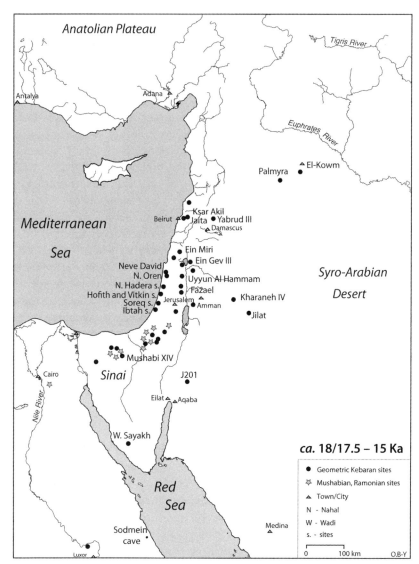

FIGURE 1.7 Distribution map of most of the Geometric Kebaran sites and invading southern entities, the Mushabian/Ramonian (ca. 18/17.5–15 ka BP). The map reflects the intensive research conducted in the southern Levant, but not yet in the north.

uniformity that demonstrates a shared Geometric Kebaran tool-making tradition and suggests that all the groups were closely related.

The same improved environmental conditions that existed while groups of Geometric Kebaran increased their populations and migrated further south into previously semi-arid areas also coincided with the expansion of northeast African foragers into the Levant. During the three millennia from 18/17.5–15.5 ka BP, a series of new groups entered Sinai from northeast Africa and from the Jordanian plateau (Garrad & Byrd 2013).

These new groups of hunter-gatherers, originally named Early and Late Mushabian and later renamed Mushabian and Ramonian (Goring-Morris 1995), occupied several sites that had been set up previously by members of the Geometric Kebaran as temporary camps. Typical Kebaran stone tools have been identified in the stratigraphy of at least two such sites, Mushabi XIV in Gebel Maghara (Bar-Yosef & Phillips 1977). The Mushabian tool-making operational sequence resembles North African characteristics: systematic and continuous use of the microburin technique and the exploitation of *piquant trièdre* techniques for shaping the La Mouillah point, their typical microlith. The descendant groups we recognize as Mushabian and Ramonian eventually invaded northern Sinai and the Negev, as indicated by the geographic distribution of their sites reaching the Judean foothills at the northern edge of the Beer-Sheva valley (fig. 1.7). The presence of Helwan lunates in the Ramonian entity is also recorded in the assemblages of Early Natufian. Indeed, by the time of the cold H1 event (ca.16.5–15.5 ka BP), the Levant accommodated numerous foraging groups, an unprecedented situation in the Paleolithic period. This relative demographic pressure has an archaeological basis and is fully supported by zooarchaeological investigations (Munro 2009).

The geographic location of the major landmass of the Levant between the sea and the desert is about 80–120 kilometers wide (Syro-Arabian Desert and Sinai; see fig. 1.7), while Jezirah, across the northern Levant, stretches from the sea to the southern Zagros foothills and reaches a width of about 300–400 kilometers. It is not surprising that the northern foraging societies that availed themselves of the river valleys of the Euphrates and the Tigris could continue to survive as they did during previous centuries, whereas the impact of demographic pressure was felt further south.

The emergence of the Early Natufian culture, previously defined as a "point of no return" (Bar-Yosef & Belfer-Cohen 1989; Belfer-Cohen & Bar-Yosef 2000) represents a major organizational change from the traditional mobile way of life practiced for many millennia (fig. 1.8). Natufian culture was characterized by the formation of sedentary and semi-sedentary hamlets, with simple dwellings of reeds and straw superstructures over stone foundations; changes in stone-tool making techniques, extensive mortar-and-pestle making and use; and orderly graveyards with ceremonial burials, some with body decorations, and evidence for feasting (Belfer-Cohen 1995; Byrd & Monahan 1995; Dubreuil & Grosman 2009; Edwards 2007; Goring-Morris et al. 1999; Grosman et al. 2008; Henry 1995; Janetski & Chazan 2004; Bocquetin & Bar-Yosef 2004; Le Dosseur 2008; Bar-Yosef Mayer & Porat 2008; Munro & Grosman 2010; Tchernov & Valla 1997; Weinstein-Evron 2009; Valla 1996,1998,1999, 2009; Bar-Yosef & Valla 1991, 2013; Edwards 2013).

There are two alternative scenarios that explain the emergence of the Natufian culture and the onset of the Neolithic Revolution. One suggests that improved conditions for stable subsistence led mobile foragers to stay put and establish hamlets. In the second scenario, mounting demographic pressures expressed in the need to control tribal territories caused by the H1 event saw the

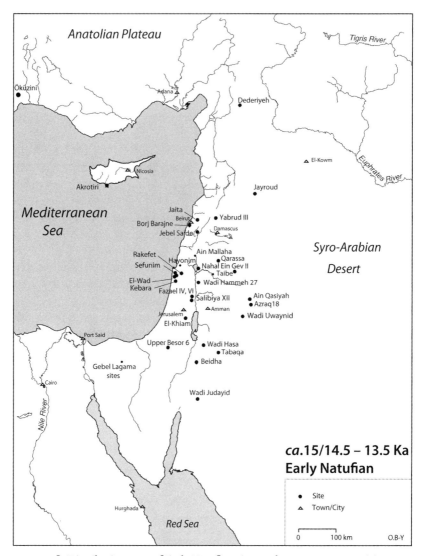

FIGURE 1.8 Distribution map of Early Natufian sites and contemporary entities (15–13/12.8 ka BP).

emergence of semi-sedentary or fully sedentary communities as an organizational agglomeration. Hamlets with in-site cemeteries signaled such territorial ownership. Munro (2009) describes the slow intensification of hunting, which supports cultural creation during rising site densities. The Bølling-Allerød, the warm and moist period ca. 14,500 to 12,700 cal BP, improved environmental conditions temporarily. This is indicated by a slight rise in the Dead Sea level (Torfstein et al. 2013a, b: figs. 5, 8), also indicated by Stein et al. (2010: fig. 2). The improvement of environmental conditions enabled the Early Natufian to survive for nearly two millennia. Several sites within the Mediterranean belt were temporarily abandoned and later resettled by the Late Natufian, as shown by the complex stratigraphy of Eynan, el-Wad Terrace, and Hayonim cave (Valla et al. 2007, 2010; Weinstein-Evron et al. 2012; Bar-Yosef 2002). The ephemeral nature of Late Natufian sites in this vegetation belt stands in contrast to a few sites in the Jordan Valley (Grosman et al. 2016).

Cultivation began either during the Younger Dryas or during the first millennium of the Holocene (cf. Willcox et al. 2009; Willcox 2012; Willcox & Savard 2011; Asouti & Fuller 2013) and was considered to be either "low level food production" (Smith 2001) or full-scale cultivation of the main food supply (Willcox & Stordeur 2012). In previously published models, the Younger Dryas was thought to be the tipping point for the emergence of systematic cultivation either during its last few centuries or immediately after the Final Natufian (Bar-Yosef & Belfer-Cohen 1989, 1992; Bar-Yosef 1998, 2002; Moore & Hillman 1992; Munro 2004). While the issue is still contested, the new evidence from Ohalo II clearly indicates active cultivation of wild cereals and the emergence of wild weeds many millennia earlier (Snir et al. 2015). This prehistoric experiment did not take hold, demonstrating that inventions and innovations may emerge from time to time only to be abandoned.

Until recently, the poor preservation of Natufian charcoal, seeds, and other plants has prevented the creation of a detailed radiometric chronology beyond stratified sites. Trials in dating animal and human bones were often unsuccessful. A series of dates derived from the renewed excavations of el-Wad Terrace and additional, isolated dates from other sites (Weinstein-Evron et al. 2012; Caracuta et al. 2016) improved the chronological control, which has allowed us to examine socioeconomic changes that took place during or towards the end of the Younger Dryas.

The conditions of the YD recorded in Soreq Cave have been supplemented recently by the analysis of gazelle tooth enamel carbonates from Hayonim and Hilazon caves indicating that the mean $\delta^{18}O$ and $\delta^{13}C$ from the Early to the Late Natufian, varied (Hartman et al. in press). This analysis demonstrates that there is no evidence for aridification during the YD as noted elsewhere (Robinson et al. 2006). The temperature of seawater dropped as indicated by the $\delta^{18}O$ value, but not annual precipitation, causing a modest negative shift in $\delta^{13}C$ values between the Early to Late Natufian phases (Hartman, 2012). These observations are supplemented by a detailed study of wood charcoal from el-Wad Terrace that indicates a decrease in precipitation during the Late Natufian (Caracuta et al. 2016). Therefore, the rapid onset of the seasonally unstable YD may have caused an initial drop in grain production because of the death of seeds from winter frost events and, more importantly, delayed germination and maturation of cereals—thus affecting the timing of the grain harvest.

During the period between 13.5–11.7 ka BP, major changes in the settlement pattern of the Natufian culture occurred across the Levant. Late and Final Natufian sites in the southern Levant gradually became ephemeral with flimsy dwellings, as recorded in Eynan (Ain Mallaha) by Valla et al. (2007, 2010). Only in special localities did a few Late Natufian sites continue to survive as permanent sedentary communities with pit-houses, such as in Nahal Ein Gen II on the western flanks of the Transjordanian plateau (Grosman et al. 2016).

Towards the end of the YD, the possibilities for Natufian groups and other complex foragers across the entire Levant were as follows:

1. Increased mobility led to regional adaptations, such as the Harifian culture, which invented the Harif Point, a new arrowhead that enhanced hunting success (Goring-Morris 1991; Goring-Morris & Belfer-Cohen 2011).
2. In the north, during the early centuries of the first millennium of the Holocene, there is evidence of increased sedentism of foragers along tributaries of the Tigris River at the Taurus foothills at sites such as Hallan Çemi and Körtik Tepe (Rosenberg & Redding 2000; Rosenberg 2011; Strakovich & Stiner 2009; Özdoğan et al. 2011a, b; Coşkun et al. 2012). No evidence for cultivation of cereals was found in those sites (Willcox & Savard et al. 2011), although einkorn wheat reached this area at that time from its main habitat in Karadag (Heun et al. 1997).
3. Intensified hunting, gathering, and part-time cultivation, possibly reflecting increased sedentism, began in the foothills west of the Euphrates valley (Tell Qaramel; Mazurowski & Kanjou 2012) and continued along the Euphrates River Valley (Willcox & Stordeur 2012; Willcox 2012).

Due to meager information concerning Epi-Paleolithic sites in SE Turkey, the bearers of the Late Natufian culture are represented only in Mureybet (Cauvin 2000; Ibañez 2008), Abu Huraira (Moore et al. 2000) in the Euphrates Valley, and Dederiyeh Cave (Tanno et al. 2013). (See fig. 1.9.) The extent to which Late Natufian culture was spread through the entire northern Levant is unknown.

Inside Dederiyeh Cave, several Natufian dwellings were erected successively. Good plant preservation (Tanno et al. 2013), dated by a long series of AMS dates, provides 13–12.5 ka BP (Yoneda et al. 2006). Willcox et al. (2008) suggested that in Late Natufian contexts cereals were cultivated, but that their state remained pre-domesticated, as deduced in other sites (Willcox 2012), including Yeni Mahalle in Şanliurfa (Çelik 2011). This data supports the notion of a "core area" in the northern Levant where cultivation began.

This core area where foragers became farmers was geographically identified on the basis of the earliest Neolithic communities, plant assemblages that demonstrate the first stage of cultivation, as well as the initiation of animal husbandry (fig. 1.10; Aurenche & Kozlowski 1999; Lev-Yadun et al. 2000; Gopher et al. 2001, Bar-Yosef 2001, 2002; Vigne 2013; Vigne et al. 2011). Kozlwoski & Aurenche (2005) refer to this area as "The Golden Triangle."

Under changing and unstable seasonal circumstances, the demographic pressure created by large numbers of forager groups caused regional upheaval, competition for resources, and local migrations from more arid areas, such as the Negev and the margins of the Syro-Arabian Desert, to arable land. Several human groups were the first to initiate cultivation. Once begun, the practice evolved in the core area and traditional foragers lost

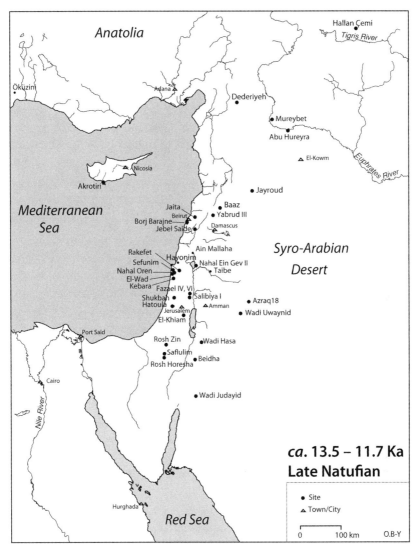

FIGURE 1.9 Distribution map of Late Natufian sites and contemporary entities (ca. 13/12.8–11.7 ka BP).

territories. They could survive only in marginal areas of the steppe and the Saharo-Arabian belt.

Conclusions

While correlations between culture and climate certainly exist, in the Levant, when comparing archaeological records and paleoclimate proxies, cultural shifts are richer in data and their chronological resolution is quite solid. Although a certain "plateau" in the ^{14}C calibration curve around the time of

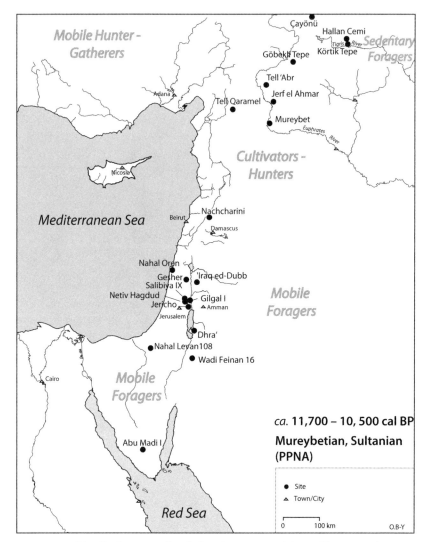

FIGURE 1.10 Distribution map of early farming communities and hunter-gatherer societies in the Levant (ca. 11.7–10.5 cal BP). Sites are generally dated to the Pre-Pottery Neolithic A.

the Younger Dryas complicates the resolution of dating, prehistoric cultures provide a clearer diachronic sequence than reconstructions of paleoclimate.

The cultural sequence is easier to examine for two reasons. First, there is a large body of botanical evidence, and second, archaeologists have uncovered new types of stone artifacts, building techniques, house plans, and site sizes. From the first centuries of the Neolithic period, there exists a rapid proliferation of axes-adzes, aerodynamically shaped symmetrical arrowheads, new types of sickle blades, and new shapes of grinding stones (11,700–11,200 cal BP).

Houses are still rounded, yet brick or adobe walls and generally flat roofs exist atop the stone foundations. Semi-underground "kiva–type" ceremonial buildings have been uncovered in several sites. Evidence of rituals, human figurines, and the somewhat later building of "temples" in villages and many more in cultic centers, such as Göbekli Tepe, reflect a major change in social organization with fast population growth. These cultural markers clearly demonstrate a change in lifestyle.

This demographic change could never have taken place without the successful cultivation of cereals. Foraging societies collapsed only in the Mediterranean vegetation belt and its shared ecozone with the Irano-Turanian steppe within the Fertile Crescent. Foragers in forested areas and especially in the semi-arid zone continued to survive for millennia, side by side with farming communities.

Why this happened is a highly debated topic, while the when and where issues are steadily being resolved. One favored response is that under "relative demographic pressures," climatic fluctuations such as the H1 event or the Younger Dryas (or both) of the Terminal Pleistocene affected clans or tribes of foragers in the northern Levant basin of the Euphrates River. Systematic cultivation of cereals certainly led to an increase in territorial control. Transfer of seeds and the sharing of cultivation knowledge and techniques in return for other commodities certainly boosted the very concept of farming. The rapid spread of cultivation and the ensuing herding and penning of goat, sheep, cattle, and pigs during the first two millennia of the Holocene enabled those who adopted the new subsistence system to become successful.

Acknowledgments

Research in Soreq Cave was supported by Israel Science Foundation (grants 20/01–13.0 and 910/05) and by the United States-Israel Binational Science Foundation (2010316). We thank the Geological Survey of Israel for its support of our work and the members of the Soreq Cave and Israeli Nature and Parks Authority for their cooperation. We are also grateful to D. Milson for editing the text.

References

Abbo, S., S. Lev-Yadun, and A. Gopher. 2011. Origins of Near Eastern plant domestication: hommage to Claude Levi-Strauss and "la pensée sauvage." *Genetic Resources and Crop Evolution* 58: 175–179.

Abbo, S., S. Lev-Yadun, and A. Gopher. 2012. Plant domestication and crop evolution in the Near East: on events and processes. *Critical Reviews in Plant Sciences* 31: 241–247.

Affek, H. P., M. Bar-Matthews, A. Ayalon, A. Matthews, and J. M. Eiler. 2008. Glacial/interglacial temperature variations in Soreq cave speleothems as recorded by "clumped isotope" thermometry. *Geochimica et Cosmochimica Acta*, 72: 5351–5360.

Almogi-Labin, A., M. Bar-Matthews, D. Shrike, M. Kolosovsky, B. Patente, B. Schulman, A. Ayalon, Z. Aizenshtat, and A. Matthews. 2009. Climatic variability during the last ca. 90 ka of the southern and northern Levantine basin as evident from marine records and speleothems. *Quaternary Science Reviews* 28: 2882–2896.

Asouti, E., and D. Q. Fuller. 2013. A contextual approach to the emergence of agriculture in southwest Asia: reconstructing early Neolithic plant-food production. *Current Anthropology* 54: 299–345.

Aurenche, O., and S. K. Kozlowski. 1999. *La Naissance du Néolithique au Proche Orient ou le Paradis Perdu*. Paris: Editions Errance.

Ayalon, A., M. Bar-Matthews, and E. Sass. 1998. Rainfall recharge relationship within a karstic terrain in the Eastern Mediterranean semi-arid region, Israel: $\delta^{18}O$ and δD characteristics. *Journal of Hydrology* 207: 18–31.

Ayalon, A., M. Bar-Matthews, and A. Kaufman. 1999. Petrography, strontium, barium and uranium concentrations, and strontium and uranium isotope ratios in speleothems as palaeoclimate proxies: Soreq Cave, Israel. *The Holocene* 9: 715–722.

Ayalon, A., M. Bar-Matthews, and B. Schilman. 2004. Rainfall isotope characteristics at various sites in Israel and the relationships with unsaturated zone water. *Geological Survey of Israel*. Report GSI/16/04.

Ayalon, A., M. Bar-Matthews, A. Frumkin, and A. Matthews. 2013. Last glacial warm events on Mount Hermon: the southern extension of the Alpine karst range of the east Mediterranean. *Quaternary Science Reviews* 59: 43–56.

Bar-Matthews, M. 2014. History of water in the Middle East and North Africa. In *Treatise in Geochemistry*, ed. H. D. Holland and K. K. Turekian. Oxford: Elsevier, 109–128.

Bar-Matthews, M., and A. Ayalon. 2011. Mid-Holocene climate variations revealed by high-resolution speleothem records from Soreq Cave, Israel and their correlation with cultural changes. *The Holocene* 21: 163–171.

Bar-Matthews, M., A. Ayalon, A. Matthews, E. Sass, and L. Halicz. 1996. Carbon and oxygen isotope study of the active water-carbonate system in a karstic Mediterranean cave: implications for paleoclimate research in semi-arid regions. *Geochimica et Cosmochimica Acta* 60: 337–347.

Bar-Matthews, M., A. Ayalon, and A. Kaufman. 1997. Late Quaternary paleoclimate in the eastern Mediterranean region from stable isotope analysis of speleothems at Soreq Cave, Israel. *Quaternary Research* 47: 155–168.

Bar-Matthews, M., A. Ayalon, and A. Kaufman. 2000. Timing and hydrological conditions of sapropel events in the Eastern Mediterranean, as evident from speleothems, Soreq cave, Israel. *Chemical Geology* 169: 145–156.

Bar-Matthews, M., A. Ayalon, M. Gilmour, A. Matthews, and C. J. Hawkesworth. 2003. Sea-land oxygen isotopic relationships from planktonic foraminifera and speleothems in the Eastern Mediterranean region and their implication for paleorainfall during interglacial intervals. *Geochimica et Cosmochimica Acta* 67: 3181–3199.

Barnes, J., M. Dove, M. Lahsen, A. Mathews, P. McElwee, R. McIntosh, F. Moore, J. O'Reilly, B. Orlove, R. Puri, H. Weiss, and K. Yager. 2013. Contribution of anthropology to the study of climate change. *Nature Climate Change* 3: 541–544.

Bartov, Y., S. L. Goldstein, M. Stein, and Y. Enzel. 2003. Catastrophic arid episodes in the Eastern Mediterranean linked with the North Atlantic Heinrich events. *Geology* 31: 439–442.

Bar-Yosef Mayer, D. E., and N. Porat. 2008. Green stone beads at the dawn of agriculture. *PNAS* 105: 8548–8551.

Bar-Yosef Mayer, D. E., and I. Zohar. 2010. The role of aquatic resources in the Natufian culture. *Eurasian Prehistory* 7: 31–45.

Bar-Yosef, O. 1998. The Natufian culture in the Levant, threshold to the origins of agriculture. *Evolutionary Anthropology* 6:150–177.

Bar-Yosef, O. 2001. From sedentary foragers to village hierarchies: the emergence of social institutions. In G. Runciman (ed.), The Origin of Human Social Institutions. London: The British Academy, 1–38.

Bar-Yosef, O. 2002. Natufian: a complex society of foragers. In *Beyond Foraging and Collecting: Evolutionary Change in Hunter-Gatherer Settlement Systems,* ed. B. Fitzhugh and J. Habu. New York: Kluwer Academic/Plenum, 91–148.

Bar-Yosef, O. 2011. Climatic fluctuations and early farming in West and East Asia. *Current Anthropology* 52 (Supplement 4): S175–S193.

Bar-Yosef, O. 2014. The homelands of the Cyprus colonizers. *Eurasian Prehistory* 10: 67–82.

Bar-Yosef, O., and A. Belfer-Cohen. 1989. The origins of sedentism and farming communities in the Levant. *Journal of World Prehistory* 3: 447–498.

Bar-Yosef, O., and A. Belfer-Cohen. 1992. From foraging to farming in the Mediterranean Levant. In *Transitions to Agriculture in Prehistory,* ed. A. B. Gebauer and T. D. Price. Madison, Wisconsin: Prehistory Press, 21–48.

Bar-Yosef, O., and A. Killebrew. 1984. Wadi Sayakh—A geometric Kebaran site in southern Sinai. *Paléorient* 10: 95–102.

Bar-Yosef, O., and J. L. Phillips, Eds. 1977. *Prehistoric Investigations in Jebel Meghara, Northern Sinai.* Jerusalem: Hebrew University.

Bar-Yosef, O., and F. R. Valla, Eds. 1991. *The Natufian Culture in the Levant.* Ann Arbor, Michigan: International Monographs in Prehistory.

Bar-Yosef, O., and F. R. Valla, eds. 2013. *Natufian foragers in the Levant: Terminal Pleistocene social changes in western Asia.* Ann Arbor, Michigan: International monographs in Prehistory.

Baruch, U., and S. Bottema. 1999. A new pollen diagram from Lake Hula: vegetational, climatic, and anthropogenic implications. In *Ancient Lakes: Their Cultural and Biological Diversity,* ed. H. Kawanabe, G. W. Coulter, and A. C. Roosevelt. Ghent, Belgium: Kenobe Productions, 75–86.

Belfer-Cohen, A. 1995. Rethinking social stratification in the Natufian culture: the evidence from burials. In *The Archaeology of Death in the Ancient Near East,* ed. S. Campbell and A. Green. Edinburgh: Oxbow Monographs, 9–16.

Belfer-Cohen, A., and O. Bar-Yosef. 2000. Early sedentism in the Near East: a bumpy ride to village life. In *Life in Neolithic Farming Communities: Social Organization, Identity, and Differentiation,* ed. I. Kuijt. New York: Plenum Press, 19–37.

Belfer-Cohen, A., and N. Goring-Morris. 2011. Becoming farmers: the inside story. *Current Anthropology* 52 (Supplement 4): S209–S220.

Binford, L. R. 1980. Willow smoke and dogs' tails: hunter-gatherer settlement systems and archaeological site formation. *American Antiquity* 45: 4–20.

Bocquentin, F., and O. Bar-Yosef. 2004. Early Natufian remains: evidence for physical conflict from Mt. Carmel, Israel. *Journal of Human Evolution* 47: 19–23.

Böhner, U., and D. Schyle. 2008. Radiocarbon CONTEXT Database 2002–2006. 10.1594/GFZCONTEXT.Ed1

Bouquet-Appel, J.-P. 2011. The agricultural demographic transition and after agriculture inventions. *Current Anthropology* 52 (Supplement 4): S497–S510.

Broecker, W. S., G. H. Denton, R. L. Edwards, H. Cheng, R. B. Alley, and A. E. Putnam. 2010. Putting the Younger Dryas cold event into context. *Quaternary Science Reviews* 29: 1078–1081.

Byrd, B., and C. M. Monahan. 1995. Death, mortuary ritual, and Natufian social structure. *Journal of Anthropological Archaeology* 14: 251–287.

Caracuta, V., M. Weinstein –Evron, R. Yeshurun, D. Kaufman, A. Tsatskin, and E. Boaretto. 2016. Charred wood remains in the Natufian sequence of el-Wad terrace (Israel): new insights into the climatic, environmental and cultural changes at the end of the Pleistocene. *Quaternary Science Reviews* 131: 20–32.

Carrión, Y. J., J. Kaal, J. A. López-Sáez, L. López-Merino, and A. Martínez Cortizas. 2010. Holocene vegetation changes in NW Iberia revealed by anthrapological and palynological records from a colluvial soil. *The Holocene* 20 (1): 1–14.

Cauvin, J. 2000. *The Birth of the Gods and the Origins of Agriculture*. Cambridge: University Press.

Cauvin, M.-C. and E. Coquegniot. 1988. L'oasis d'El Kowm et le Kébarien géométrique. *Paléorient* 14 (2): 270–282.

Carpenter, S., B. F. Walker, J. M. Andereis, and N. Abel. 2001. From metaphor to measurement: resilience of what to what? *Ecosystems* 4: 765–781.

Çelik, B. 2011. Şanlıurfa—yeni mahalle. In *The Neolithic of Turkey 2: The Euphrates Basin*, ed. M. Özdoğan, N. Başgelen, and P. Kuniholm. Istanbul: Archaeology and Art Publications, 139–164.

Clare, L., and B. P. Weninger. 2013. The dispersal of Neolithic lifeways: absolute chronology and rapid climate change in central and west Anatolia. In *10500–5200 BC: Environment, Settlement, Flora, Fauna, Dating, Symbols of Belief, with Views from North, South, East, and West*, ed. M. Özdoğan, N. Başgelen, and P. Kuniholm. The Neolithic in Turkey, vol. 6. Istanbul: Archaeology & Art Publications, 1–65.

Coşkun, A., M. Benz, C. Rössner, K. Deckers, S. Riehl, K.W. Alt, and V. Özkaya. 2012. New results on the Younger Dryas occupation at Körtik Tepe. *Neo-Lithics* 1/2: 25–32.

Denton, G. H., R. B. Alley, G. C. Comer, and W. S. Broecker. 2005. The role of seasonality in abrupt climate change. *Quaternary Science Reviews* 24: 1159–1182.

Denton, G.H., R. F. Anderson, J. R. Toggweiler, and R. L. Edwards, J. M. Schaefer, and A. E. Putnam. 2010. The last glacial termination. *Science* 328: 1652–1656.

Develle, A.-L., J. Herreros, L. Vidal, A. Sursock, and F. Gasse. 2010. Controlling factors on a paleo-lake oxygen isotope record (Yamoouneh, Lebanon) since Last Glacial Maximum. *Quaternary Science Reviews* 29 (7–8): 865–886.

Develle, A.-L., F. Gasse, L. Vidal, et al. 2011. A 250 ka sedimentary record from a small karstic lake in the Northern Levant (Yammoûneh, Lebanon): paleoclimaticimplications. *Palaeogeography, Palaeoclimatology, Palaeoecology*. 305: 10–27.

Dubreuil, L., and L. Grosman. 2009. Ochre and hide-working at a Natufian burial place. *Antiquity* 83: 935–954.

Dyson-Hudson, R., and E. Smith. 1978. Human territoriality: an ecological reassessment. *American Anthropologist* 80: 21–41.

Edwards, P. C. 2007. A 14000 year-old hunter-gatherer's toolkit. *Antiquity* 81: 865–876.

Edwards, P. C. 2013. *Wadi Hammeh 27, an early Natufian settlement at Pella, Jordan.* Leiden & Boston: Brill.

Enzel, Y., R. Amit, U. Dayan, O. Crouvi, R. Kahana, B. Ziv, and D. Sharon. 2008. The climatic and physiographic controls of the eastern Mediterranean over the late Pleistocene climates in the southern Levant and its neighboring deserts. *Global and Planetary Change* 60: 165–192.

Frumkin, A., D. C. Ford, and H. P. Schwarcz. 2000. Paleoclimate and vegetation of the last glacial cycles in Jerusalem from a speleothem record. *Global Biogeochemical Cycles* 14: 863–870.

Fujimoto, T. 1979. The Epi-Palaeolithic assemblages of Douara Cave. *Bulletin of the University Museum (Tokyo)* 16: 47–75.

Gadgil, M., F. Berkes, and C. Folke. 1993. Indigenous knowledge for biodiversity conservation. *Ambio* 22 (2–3): 151–156.

Garrard, A. H., and B. Byrd. 2013. *Beyond the Fertile Crescent: Late Paleolithic and Neolithic communities of the Jordanian steppe.* The Azraq basin project. Vol.1: project background and the Late Paleolithic (geological context and technology). Levant Supplementary Series, Vol.13. Oxford & Oakville: Oxbow.

Gopher, A., S. Abbo, and S. Lev-Yadun. 2001. The "when," the "where" and the "why" of the Neolithic Revolution. *Documenta Praehistorica* 28: 49–62.

Gopher A., S. Lev-Yadun, and S. Abbo. 2013. A response to "Arguments against the Core Area Hypothesis." *Tel-Aviv* 40: 185–198.

Goring-Morris, A. N. 1991. The Harifian of the Southern Levant. In *The Natufian Culture in the Levant,* ed. O. Bar-Yosef and F. R. Valla. Ann Arbor, Michigan: International Monographs in Prehistory, 173–234.

Goring-Morris, A. N. 1995. Complex hunter-gatherers at the end of the Paleolithic (20,000–10,000 BP). In *The Archaeology of Society in the Holy Land,* ed. T. E. Levy. London: Leicester University Press, 141–168.

Goring-Morris, A. N. 2009. Two Kebaran occupations near Nahal Soreq, and the reconstruction of group ranges in the early Epi-Paleolithic in the Israeli littoral. *Eurasian Prehistory* 6 (1–2): 75–94.

Goring-Morris, A. N., and A. Belfer-Cohen. 2010. Different ways of being, different ways of seeing . . . changing worldviews in the Near East. In *Landscapes in Transition: Understanding Hunter-Gatherer and Farming Landscapes in the Early Holocene of Europe and the Levant,* ed. B. Finlayson and G. Warren. London: Levant Supplementary Series & CBRL, 9–22.

Goring-Morris, A. N., and A. Belfer-Cohen. 2011. Neolithization processes in the Levant: the outer envelope. *Current Anthropology* 52 (Supplement 4): S195–S208.

Goring-Morris, A. N., P. Goldberg, Y. Goren, U. Baruch, and D. E. Bar-Yosef. 1999. Saflulim: a Late Natufian base camp in the central Negev Highlands, Israel. *Palestine Exploration Quarterly* 131(1): 1–29.

Grosman, L., N. D. Munro, and A. Belfer-Cohen. 2008. A 12,000-year-old Shaman burial from the southern Levant (Israel). *Proceedings of the National Academy of Sciences* 105 (46): 17665–17669.

Grosman, L., N. Munro, I. Abadi, E. Boaretto, D. Shaham, A. Belfer-Cohen, and O. Bar-Yosef. 2016. Nahal Ein Gev II, a Late Natufian Community at the Sea of Galilee. *PLoS ONE* 11(1): 1–32. doi: 10.1371/journal.pone.0146647.

Haase-Schramm A., S. L. Goldstein, and M. Stein. 2004. U-Th dating of Lake Lisan (late Pleistocene Dead Dea) aragonite and implications for glacial east Mediterranean climate change. *Geochim Cosmochim Acta* 68: 985–1005.

Hartman, G. O., A. Bar-Yosef, L. Brittingham, L. Grosman, and N. Monroe. In press. Hunted gazelles evidence cooling, but not drying, during the Younger Dryas in the southern Levant. *Proceedings of the National Academy of Science of the United States.*

Henry, D. O. 1995. Cultural evolution and interaction during the Epipaleolithic. In *Prehistoric Cultural Ecology and Evolution: Insights from Southern Jordan*, ed. D. O. Henry. New York: Plenum Press, 337–343.

Heun, M., R. Schafer-Pregl, D. Klawan, R. Castagna, M. Accerbi, B. Borghi, and F. Salamini. 1997. Site of einkorn wheat domestication identified by DNA fingerprinting. *Science* 278: 1312–1314.

Heun, M., S. Abbo, S. Lev-Yadun, and A. Gopher. 2012. A critical review of the protracted domestication model for Near-Eastern founder crops: linear regression, long-distance gene flow, archaeological, and archaeobotanical evidence. *Journal of Experimental Botany* 63: 4333–4341.

Higham, T. F. G., C. Bronk Ramsey, F. Brock, D. Baker, and P. Ditchfield. 2007. Radiocarbon dates from the Oxford AMS system: Archaeometry Datelist 32. *Archaeometry* 49: S1–S60.

Hillman, G. C., and M. S. Davies. 1990a. Domestication rates in wild-type wheats and barley under primitive cultivation. *Biological Journal of the Linnaean Society* 39: 39–78.

Hillman, G. C., and M. S. Davies. 1990b. Measured domestication rates in wild wheats and barley under primitive implications. *Journal of World Prehistory* 4: 157–222.

Holling, C. S., L. H. Gunderson, and D. Ludwig. 2002. In quest of a theory of adaptive change. In *Panarchy: Understanding Transformations in Human and Natural Systems*, ed. L. H. Gunderson and C. S. Holling. Washington, DC: Island Press, 3–22.

Ibáñez, J. J., ed. 2008. *Le site néolithique de Tell Mureybet (Syrie du Nord): En hommage à Jacques Cauvin*. Oxford: Archaeopress (BAR International Series 1843).

Janetski, J. C., and M. Chazan. 2004. Shifts in Natufian strategies and the Younger Dryas: evidence from Wadi Mataha, Southern Jordan. In *The Last Hunter-Gatherer Societies in the Near East*, ed. C. Delage. Oxford: John & Erica Hedges (BAR International Series 1320), 160–168.

Kahneman, D. 2003. Maps of bounded rationality: psychology for behavioral economics. *The American Economic Review* 93 (5): 1449–1475.

Kaufman, A., G. J. Wasserburg, D. Porcelli, M. Bar-Matthews, A. Ayalon, and L. Halicz. 1998. U-Th isotope systematics from the Soreq cave, Israel and climatic correlations. *Earth and Planetary Science Letters* 156: 141–155.

Kislev, M. E., D. Nadel, and I. Carmi. 1992. Epi-Palaeolithic (19,000 B.P.) cereal and fruit diet at Ohalo II, Sea of Galilee, Israel. *Review of Palaeobotany and Palynology* 71: 161–166.

Kolodny, Y., M. Bar-Matthews, A. Ayalon, and K. D. McKeegan. 2003. A high spatial resolution δ^{18}O profile of a speleothem using an ion-microprobe. *Chemical Geology* 197: 21–28.

Kolodny, Y., M. Stein, and M. Machlus. 2005. Sea-rain-lake relation in the Last Glacial East Mediterranean revealed by δ^{18}O-δ^{13}C in Lake Lisan aragonites. *Geochimica et Cosmochimica Acta* 69: 4045–4060.

Kozlowski, S. K., and O. Aurenche. 2005. *Territories, Boundaries, and Culture in the Neolithic Near East*. Oxford-Lyon: Archaeopress MOM (BAR International Series 1362).

Kvavadze E., O. Bar-Yosef, A. Belfer-Cohen, E. Boaretto, N. Jakeli, Z. Matskevich, and T. Meshveliani. 2009. 30,000-year-old wild flax fibers. *Science* 325: 1359.

Kushnir, Y., and M. Stein. 2010. North Atlantic influence on 19th–20th century rainfall in the Dead Sea watershed, teleconnections with the Sahel, and implication for Holocene climate fluctuations. *Quaternary Science Reviews* 29: 3843–3860.

Langgut D., A. Almogi-Labin, M. Bar-Matthews, and M. Weinstein-Evron. 2011. Vegetation and climate changes in the South Eastern Mediterranean during the Last Glacial-Interglacial cycle (86 ka): new marine pollen record. *Quaternary Science Reviews* 30: 3960–3972.

Le Dosseur, G. 2008. La place de l'industrie osseuse dans la Neolithisation au Levant Sud. *Paleorient* 34 (1): 59–89.

Lenton, T.M. 2013. Environmental tipping points. *Annual Review of Environment and Resources* 38: 1–29.

Lespez, L., Z. Tsirtsoni, P. Darcque, H. Koukouli-Chryssanthaki, D. Malamidou, R. Treuil, R. Davidson, G. Kourtessi-Philippakis, and C. Oberlin. 2013. The lowest levels at Dikili Tash, Northern Greece: a missing link in the Early Neolithic of Europe. *Antiquity* 87: 30–45.

Lev-Yadun, S., A. Gopher, and S. Abbo. 2000. The cradle of agriculture. *Science* 288: 1602–1603.

Maher, L. A. 2007. 2005 Excavations at the Geometric Kebaran site of 'Uyun al-Hammam, al-Koura District, northern Jordan. *ADAJ* 51: 263–272.

Maher, L. A., T. Richter, and J. T. Stock. 2012a. The Pre-Natufian Epipaleolithic: long-term behavioral trends in the Levant. *Evolutionary Anthropology* 21: 69–81.

Maher, L. A., T. Richter, D. Macdonald, M. D. Jones, L. Martin, and J. T. Stock. 2012b. Twenty thousand-year-old huts at a hunter-gatherer settlement in eastern Jordan. PLoS ONE 7(2). doi: 10.1371/journal.pone.0031447.

Marino, G., E. J. Rohling, F. Sangiorgi, A. Hayes, J. L. Casford, A. F. Lotter, M. Kucera, and H. Brinkhuis. 2009. Early and middle Holocene in the Aegean Sea: interplay between high and low latitude climate variability. *Quaternary Science Reviews* 28: 3246–3262.

Matthews, A., A. Ayalon, and M. Bar-Matthews. 2000. D/H ratios of fluid inclusions of Soreq Cave (Israel) speleothems as a guide to the Eastern Mediterranean Meteoric Line relationships in the last 120 ky. *Chemical Geology* 166: 183–191.

Mazurowski, R. F., and Y. Kanjou. 2012. *Tel Qaramel 1999–2007*. Polish Centre of Mediterranean Archaeology. Warsaw: University of Warsaw.

McGarry S., M. Bar-Matthews, A. Matthews, A. Vaks, B. Schilman, and A. Ayalon. 2004. Constraints on hydrological and paleotemperature variations in the Eastern Mediterranean region in the last 140 ka given by the δD values of speleothem fluid inclusions. *Quaternary Science Reviews* 23: 919–934.

Moore, A. M. T., and G. C. Hillman. 1992. The Pleistocene to Holocene transition and human economy in Southwest Asia: the impact of the Younger Dryas. *American Antiquity* 57: 482–494.

Moore, A. M. T., G. C. Hillman, and A. J. Legge (eds.). 2000. *Village on the Euphrates: From Foraging to Farming at Abu Hureyra*. Oxford: Oxford University Press.

Munro, N. D. 2004. Zooarchaeological measures of hunting pressure and occupation intensity in the Natufian: implications for agricultural origins. *Current Anthropology* 45 (Supplement): S5–S33.

Munro, N. 2009. Epipaleolithic subsistence intensification in the Southern Levant: the faunal evidence. In *The Evolution of Hominin Diets: Integrating Approaches to the Study of Palaeolithic Subsistence*, ed. J.-J. Hublin and M. P. Richards. Rotterdam: Springer, 141–155.

Munro, N. D., and L. Grosman. 2010. Early evidence (ca. 12,000 B.P.) for feasting at a burial cave in Israel. *Proceedings of the National Academy of Sciences* 107 (35): 15362–15366.

Nadel, D., E. Weiss, O. Simchoni, A. Tsatskin, A. Danin, and M. Kislev. 2004. Stone Age hut in Israel yields world's oldest evidence of bedding. *PNAS* 101: 6821–6826.

Nadel, D., I. Carmi, and D. Segal. 1995. Radiocarbon dating of Ohalo II: archaeological and methodological implications. *Journal of Archaeological Science* 22: 811–822.

Nadel, D., S. Belitzky, E. Boaretto, I. Carmi, J. Heinemeier, E. Werker, and S. Marco. 2001. New dates from the submerged Late Pleistocene sediment in the southern Sea of Galilee, Israel. *Radiocarbon* 43: 1167–1178.

Nadal, D., E. Weiss, O. Simchoni, A. Tsatskin, A. Danin, and M. E. Kislev. 2004. Stone Age hut in Israel yields world's oldest evidence of bedding. *PNAS* 101: 6821–6826.

Orland, I. J., M. Bar-Matthews, N. T. Kita, A. Ayalon, A. Matthews, and J. W. Valle. 2009. Climate deterioration in the eastern Mediterranean as revealed by ion microprobe analysis of a speleothem that grew from 2.2 to 0.9 ka in Soreq Cave, Israel. *Quaternary Research* 71: 27–35.

Orland, I. J., M. Bar-Matthews, A. Ayalon, A. Matthews, R. Kozdon, T. Ushikubo, and J. W. Valley. 2012. Seasonal resolution of eastern Mediterranean climate since 34 ka from Soreq Cave speleothem. *Geochimica et Cosmochimica Acta* 89: 240–255.

Orland, I. J., Y. Burstyn, M. Bar-Matthews, R. Kozdon, A. Ayalon, A. Matthews, and J. W. Valley. 2014. Seasonal climate signals (1998–2008) in a modern Soreq Cave stalagmite as revealed by high-resolution geochemical analysis. *Chemical Geology* 363: 322–333.

Özdoğan, M., N. Başgelen, and P. Kuniholm, eds. 2011a. *The Neolithic in Turkey: New Excavations and New Research. The Euphrates Basin*. Istanbul: Archaeology and Art Publications.

Özdoğan, M., N. Başgelen and P. Kuniholm, eds. 2011b. *The Neolithic in Turkey: New Excavations and New Research. The Tigris Basin*. Istanbul: Archaeology and Art Publications.

Rasmussen, S. O., K. K. Andersen, A. M. Svensson, J. P. Steffensen, B. M. Vinther, H. B. Clausen, M.-L. Siggaard-Andersen, S. J. Johnson, et al. 2006. A new Greenland ice core chronology for the last glacial termination. *Journal of Geophysical Research* 3. doi: 10. 1029/2005JD006079

Reimer, P. J., M. G. L. Baille, E. Bard, A. Bayliss, J. W. Beck, P. G. Blackwell, C. Bronk Ramsey, and C. E. Buck, et al. 2009. INTCAL09 and MARINE09 radiocarbon age calibration curves, 0–50,000 years cal BP. *Radiocarbon* 51: 1111–1150.

Robinson, S. A., S. Black, B. W. Sellwood, and P. J. Valdes. 2006. A review of palaeo-climates and paleoenvironments in the Levant and eastern Mediterranean from 25,000 and 5000 years BP: setting the environmental background for the evolution of human civilisation. *Quaternary Science Reviews* 25 (13–14): 1517–1542.

Rohling, E. J., J. Casford, R. Abu-Zied, S. Cooke, D. Mercone, J. Thomson, I. Croudace, F. J. Jorissen, H. Brinkhuis, J. Kallmeyer, and G. Wefer. 2002. Rapid Holocene climate changes in the eastern Mediterranean. In *Droughts, Food and Culture: Ecological Change and Food Security in Africa's Later Prehistory*, ed. F. A. Hassan. New York: Kluwer Academic/ Plenum Publishers, 35–46.

Rosen, A. 2007. *Civilizing Climate: Social Responses to Climate Change in the Ancient Near East*. Lanham, MD: Altamira Press.

Rosen, A. M., and I. Rivera-Collazo. 2012. Climate change, adaptive cycles, and the persistence of foraging economies during the late Pleistocene/Holocene transition in the Levant. *Proceedings of the National Academy of Sciences* 109: 3640–3645.

Rosenberg, M., and Redding, R. W. 2000. Hallan Çemi and early village organization in eastern Anatolia. In *Life in Neolithic Farming Communities: Social Organization, Identity, and Differentiation*, ed. I. Kujit. New York: Kluwer Academic/ Plenum, 39–62.

Rosenberg, M. N. 2011. Hallan Çemi. In *The Neolithic in Turkey: The Tigris Basin*, ed. M. Özdoğan, N. Başgelen, and P. Kuniholm. Istanbul: Archaeology and Art Publications, 61–78.

Rossignol-Strick, M. 1995. Sea-land correlation of pollen records in the eastern Mediterranean for the glacial-interglacial transition: biostratigraphy versus radio-metric time-scale. *Quaternary Science Reviews* 14: 893–915.

Simmons, A. H. 2007. *The Neolithic Revolution in the Near East: Transforming the Human Landscape*. Tucson: University of Arizona Press.

Smith, B. D. 2001. Low-level food production. *Journal of Archaeological Research* 9: 1–43.

Snir, A., D. Nadel, I. Groman-Yaroslavski, Y. Melamed, M. Sternberg, O. Bar-Yosef, and E. Weiss. 2015. The Origin of Cultivation and Proto-Weeds, Long Before Neolithic Farming. *PLoS ONE* 10(7) doi: 10.1371/journal.pone.0131422

Starkovich, B. M., and M. C. Stiner. 2009. Hallan Çemi Tepesi: High-ranked game exploitation alongside intensive seed processing at the Epipaleolithic-Neolithic transition in southeastern Turkey. *Anthropozoologica* 44: 41–61.

Stein, M., A. Torfstein, I. Gavrieli, and Y. Yechieli. 2010. Abrupt aridities and salt deposition in the post-glacial Dead Sea and their North Atlantic connection. *Quaternary Science Reviews* 29: 567–575.

Stein, M. 2014. The evolution of Neogene-Quaternary water-bodies in the Dead Sea Rift Valley. In *Dead Sea Transform Fault System: Reviews*, ed. Z. Garfunkel, Z.

Ben-Avraham, and E. Kagan. Modern Approaches in Solid Earth Sciences, vol. 6. Dordrecht, Netherlands: Springer.

Tanno, K.-I., G. Willcox, S. Muhesen, Y. Nishiaki, Y. Kanjo, and T. Akazawa. 2013. Preliminary results from analyses of charred plant remains from a burnt Natufian building at Dederiyeh Cave in Northwest Syria. In *Natufian Foragers in the Levant: Terminal Pleistocene Social Changes in Western Asia,* ed. O. Bar-Yosef and F. R. Valla. Ann Arbor, Michigan: International Monographs in Prehistory, 83–87.

Tchernov, E., and F. R. Valla. 1997. Two new dogs, and other Natufian dogs, from the southern Levant. *Journal of Archaeological Science* 24 (1): 65–95.

Torfstein, A., S. L. Goldstein, M. Stein, and Y. Enzel. 2013a. Impacts of abrupt climate changes in the Levant from Last Glacial Dead Sea levels. *Quaternary Science Reviews* 69: 1–7.

Torfstein, A., S. L. Goldstein, E. J. Kagan, and M. Stein. 2013b. Integrated multi-site U–Th chronology of the last glacial Lake Lisan. *Geochimica et Cosmochimica Acta* 104: 210–231.

Valla, F. R. 1996. L'animal "bon à penser": la domestication et la place de l'homme dans la nature. In M. Otte (ed.), *Nature et Culture II.* Liège: ERAUL.

Valla, F. R. 1999. The Natufian: a coherent thought? In *Dorothy Garrod and the Progress of the Palaeolithic,* ed. W. Davies and R. Charles. Oxford: Oxbow.

Valla, F. R. 1998. Natufian seasonality: a guess. In *Seasonality and Sedentism: Archaeological Perspectives from Old and New World Sites,* ed. O. Bar-Yosef and T. Rocek. Peabody Museum Bulletin, vol. 6. Cambridge, MA: Peabody Museum of Archaeology and Ethnology, Harvard University, 93-108.

Valla, F. R., H. Khalaily, H. Valladas, E. Kaltnecker, F. Bocquetin, T. Cabellos, D. Bar-Yosef Mayer, G. Le Dosseur, et al. 2007. Les fouilles de Ain Mallaha (Eynan) de 2003 à 2005: Quatrième rapport préliminaire. *Journal of the Israel Prehistoric Society* 37: 132–383.

Valla, F. R., H. Khalaily, N. Samuelian, and F. Bocquentin. 2010. What happened in the Final Natufian? *Journal of the Israel Prehistoric Society—Mitekufat Haeven* 40: 131–148.

Valla, F. R. 2009. Une énigme Natoufienne: les "mortiers" enterrés. In *De Méditerranée et d'Ailleurs . . . Mélanges offerts à Jean Guilaine,* ed. collectif. Toulouse: Archives d'Écologie Préhistorique, 751–760.

van Zeist, W., U. Baruch, and S. Bottema. 2009. Holocene palaeoecology of the Hula area, northeastern Israel. In *A Timeless Vale. Archaeological and Related Essays on the Jordan Valley in Honour of Gerrit van der Kooij on the Occasion of his Sixty-Fifth Birthday,* ed. E. Kaptijn and L. P. Petit. Leiden: University of Leiden, 29–64.

Verheyden S., F. H. Nader, H. Cheng, L. R. Edwards, and R. Swennen. 2008. Paleoclimate reconstruction in the Levant region from the geochemistry of a Holocene stalagmite from the Jeita cave, Lebanon. *Quaternary Research* 70: 368–381.

Vigne, J.-D., I. Carrère, F. Briois, and J. Guilaine. 2011. The early process of mammal domestication in the Near East: new evidence from the Pre-Neolithic and Pre-Pottery Neolithic in Cyprus. *Current Anthropology* 52 (Supplement 4): S255–S271.

Vigne J.-D. 2013. Domestication process and domestic ungulates: new observations from Cyprus. In *The Origins and Spread of Domestic Animals in Southwest Asia and Europe,* ed. S. Colledge, J. Conolly, K. Dobney, K. Manning, and S. Shennan. Walnut Creek, CA: Left Coast Press, 115–128.

Weinstein-Evron, M. 2009. *Archaeology in the Archives: Unveiling the Natufian Culture of Mount Carmel.* Boston: Brill.

Weinstein-Evron, M., R. Yeshurun, D. Kaufman, E. Eckmeie, and E. Boaretto. 2012. New ^{14}C dates for the early Natufian of el-Wad Terrace, Mount Carmel, Israel. *Radiocarbon* 54: 813–822.

Weiss, E., M. E. Kislev, and A. Hartmann. 2006. Autonomous cultivation before domestication. *Science* 312: 1608–1610.

Weiss, E., M. E. Kislev, O. Simchoni, D. Nadel, and H. Tschauner. 2008. Plant-food preparation area on an Upper Paleolithic brush hut floor at Ohalo II, Israel. *Journal of Archaeological Science* 35: 2400–2414.

Weiss, E., W. Wetterstrom, D. Nadel, and O. Bar-Yosef. 2004. The broad spectrum revisited: evidence from plant remains. *PNAS* 101: 9551–9555.

Weninger, B., L. Clare, E. J. Rohling, O. Bar-Yosef, U. Böhner, M. Budja, M. Bundschuh, A. Feurdean, et al. 2009. The impact of rapid climate change on prehistoric societies during the Holocene in the Eastern Mediterranean. *Documenta Praehistorica* 36: 7–59.

Weninger, B., and O. Jöris. 2008. A ^{14}C age calibration curve for the last 60ka: the Greenland-Hulu u/Th timescale and its impact on understanding the Middle to Upper Paleolithic transition in Western Eurasia. *Journal of Human Evolution* 55: 772–781.

Willcox, G. 2012. Searching for the origins of arable weeds in the Near East. *Vegetation History and Archaeobotany* 21: 163–167.

Willcox, G., S. Fornite, and L. Herveux. 2008. Early Holocene cultivation before domestication in northern Syria. *Vegetation History and Archaeobotany* 17: 313–325.

Willcox, G., R. Buxo, and L. Herveux. 2009. Late Pleistocene and early Holocene climate and the beginnings of cultivation in northern Syria. *The Holocene* 19: 151–158.

Willcox, G., and M. Savard. 2011. Botanical evidence for the adoption of cultivation in southeast Turkey. In *The Neolithic in Turkey,* ed. M. Özdoğan, N. Başgelen, and P. Kuniholm. Istanbul: Archaeology and Art Publications, 267–280.

Willcox, G., and D. Stordeur. 2012. Large-scale cereal processing before domestication during the tenth millennium BC cal. in northern Syria. *Antiquity* 86: 99–114.

Wolff, E. W., J. Chappellaz, T. Blunier, S. O. Rasmussen, and A. Svensson. 2010. Millennial-scale variability during the last glacial: the ice core record. *Quaternary Science Reviews* 29 (2010): 2828–2838.

Zohary, M. 1973. *Geobotanical Foundations of the Middle East.* Stuttgart: Springer Verlag.

Yoneda M., H. Nakata, M. Aoki, O. Kondo, Y. Nishiaki, and T. Akazawa. 2006. Age determinations at the Dederiyeh Cave, Syrian Arab Republic. *Anthropological Science* 114 (3): 251.

| 6600–6000 cal BC Abrupt Climate Change and Neolithic Dispersal from West Asia

BERNHARD WENINGER AND LEE CLARE

Recent advances in palaeoclimatological and meteorological research, combined with new radiocarbon data from western Anatolia and southeast Europe, lead us to formulate a new hypothesis for the temporal and spatial dispersal of Neolithic lifeways from their core areas of genesis. The new hypothesis, which we term the Abrupt Climate Change (ACC) Neolithization Model, incorporates a number of insights from modern vulnerability theory. We focus here on the Late Neolithic (Anatolian terminology), which is followed in the Balkans by the Early Neolithic (European terminology). From high-resolution ¹⁴C case studies, we infer an initial (very rapid) west-directed movement of early farming communities out of the Central Anatolian Plateau towards the Turkish Aegean littoral. This move is exactly in phase (decadal scale) with the onset of ACC conditions (~6600 cal BC). Upon reaching the Aegean coastline, Neolithic dispersal comes to a halt. It is not until some 500 years later—that is, at the close of cumulative ACC and 8.2 ka cal BP Hudson Bay cold conditions—that there occurs a second abrupt movement of farming communities into Southeast Europe, as far as the Pannonian Basin. The spread of early farming from Anatolia into eastern Central Europe is best explained as Neolithic communities' mitigation of biophysical and social vulnerability to natural (climate-induced) hazards.

Introduction

In recent years it has become increasingly apparent that the "8.2 ka cal BP climate event" triggered by the outflow of palaeo-lakes Agassiz and Ojibway through the Hudson Straight, is in fact embedded within a temporally much broader interval of climate fluctuations that occurred between 8.6 and 8.0 ka cal BP (Rohling & Pälicke 2005). Intriguingly, these latter climate fluctuations are far from unique; they belong to a well-defined set of recurring "Abrupt Climate Change" (ACC) Glacial and Holocene cold anomalies, of which the most recent is commonly referred to as the *Little Ice Age*. Here, we address the impacts of an ACC interval (6600–6000 cal BC) upon prehistoric communities in West Asia. Following an introduction of underlying climatic conditions during this period, as well as an overview of coeval Neolithic development in the eastern Mediterranean (eMed), our focus turns to the dispersal of early farming from central Anatolia to the Turkish Aegean Coast, and ultimately to Southeast Europe. A chronological guide is provided by a ¹⁴C-based site-by-site overview at the end of this chapter (fig. 2.5).

Abrupt Climate Change

Some ten years ago the term *Rapid Climate Change* (RCC) was introduced by Mayewski et al. (2004) to describe a set of recurring Holocene cold periods in the Northern Hemisphere, each with a typical duration of several centuries. We substitute *"Abrupt"* for the original *"Rapid"* to emphasize the challenges posed to contemporaneous Neolithic socioeconomic systems by frequently recurring and (for these communities) unprecedented climatic anomalies associated with *Abrupt Climate Change*.

ACC events are currently best documented (and best dated) by the Greenland GISP2 nss (non-sea salt) [K⁺] (potassium concentration) glaciochemical record. The GISP2 nss [K⁺] record shows a series of severe coolings that run systematically (with quasi-cyclicity of ca. 1450 yrs) through the entire Glacial, into the Holocene, and up to modern times (Mayewski et al. 1997, 2004). Each of these cold periods corresponds to a more pronounced Siberian High over Asia, leading to an increased occurrence and severity of winter outbreaks over Europe and the Eastern Mediterranean. The existence of cold anomalies is confirmed in a large number of terrestrial and marine records in the Eastern Mediterranean (Rohling et al. 2002, 2009; Meeker & Mayewski 2002; Marino et al. 2009). Based on the GISP2 nss (non-sea salt) [K⁺] record, it is inferred that strongest RCC conditions should be expected at about 10.2 ka cal BP, 8.6–8.0 ka cal BP, 6.0–5.2 ka cal BP, and about 3.0 ka cal BP. The most recent RCC is commonly referred to as the Little Ice Age (LIA; ca. AD 1450–1929).

Climate Records

As shown in the assemblage of Northern Hemisphere palaeoclimate records (fig. 2.1), these intervals of Abrupt Climate Change (ACC) are characterized by the deposition of unusually large amounts of nss K⁺ in the GISP2 ice core. This potassium stems from airborne dust from sources in Asia. Because the flow of airborne potassium over Greenland is related to an expansion and intensification of the Siberian High (SH), the GISP2 nss [K⁺] record provides an excellent proxy for changes in the strength and geographic expansion of SH. In general terms, the SH is a dominant atmospheric (anticyclone) circulation system in the lower troposphere that not only controls most of continental Asia, but also exerts

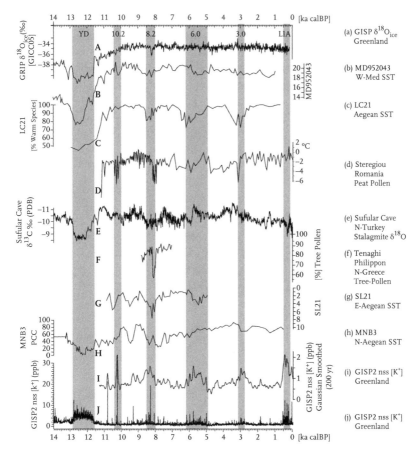

FIGURE 2.1 Northern Hemisphere palaeoclimate records of Holocene Abrupt Climate Change (ACC). *A.* Greenland GRIP ice-core $\delta^{18}O$ (Rasmussen et al. 2006); *B.* Western Mediterranean (Iberian Margin) core MD952043, sea-surface temperature (SST) C37 alkenones (Cacho et al. 2001; Fletcher & Goñi 2008); *C.* Eastern Mediterranean core LC21 (SST) fauna (Rohling et al. 2002); *D.* Steregiou Romania (peat core pollen), mean annual temperature of the coldest month (MTC, ºC; Feurdean et al. 2008); *E.* Sufular Cave Stalagmite $\delta^{13}C$, N. Turkey (Fleitmann et al. 2009); *F.* Tenaghi Philippon tree pollen, northern Greece (Pross et al. 2009); *G.* Eastern Aegean LS21 (SST) fauna (Marino et al. 2009); *H.* Northern Aegean Core MNB3 (Geraga et al. 2010); *I.* Greenland GISP2 potassium Gaussian smoothed (200 yr) (non-sea salt [K+]; ppb) ion proxy for the Siberian High (Mayewski et al. 1997; Meeker & Mayewski 2002); *J.* Greenland GISP2 potassium (non-sea salt [K+]; ppb) ion proxy for the Siberian High (Mayewski et al. 1997; Meeker & Mayewski 2002).

a significant influence on winter climate over mid- to high-latitude Eurasia. The genesis of the SH is related to a combination of mid- and high-latitude radiative cooling over Asia, teleconnections to the North Atlantic, and possibly to local variations in autumnal snow cover in its central area of origin (Cohen, Saito, & Entekhabi 2001). Modern meteorological analogies demonstrate that during periods with SH expansion, wide swathes of eastern Asia (especially Mongolia and China), the Ukrainian steppe zone, the Balkans, the Aegean (including Cyprus), and Anatolia, as well as SE Turkey, northern Mesopotamia, Syria, Israel, and Jordan all lie in the path of intensely cold and fast-flowing polar air masses. In Mongolia these anomalous cold air outbreaks are referred to as *dzuds* and are notoriously damaging to agriculture and livestock alike (Lau & Lau 1984; Begzsuren 2004; referenced in Tubi & Dayan 2013).

RCC Corridors

There are two main geographic corridors for the outflow of cold masses from the polar regions at times of pronounced Siberian High. While the first corridor extends westwards from Central Asia, running north of the Himalayas, crossing the North Pontic steppe, eventually entering Southeast Europe, the second corridor takes an easterly path across China and into the Pacific (Tubi & Dayan, 2013, with further references). The westward extension of the SH, on which we focus here, can lead to continental polar outbreaks over the Aegean, the Adriatic, and the Gulf of Lion (Rohling et al. 2002). These outbreaks are linked to orographic channelling of air masses at the northern Mediterranean margin, more commonly known as the Vardar, Bora, and Mistral winds (Casford et al. 2003).

Marine Core LC21 (situated east of Crete)

A key proxy for our present understanding of Holocene ACC conditions in the eastern Mediterranean is provided by the detailed, seasonal investigation and ^{14}C dating of marine microfossil assemblages in core LC21 (Rohling et al. 2009). The LC21 core (situated east of Crete) shows a distinct series of abrupt drops in Sea Surface Temperature (SST) that correlate well (fig. 2.1) with the above-mentioned periods of enhanced atmospheric dust flux. These drops in SST are attributed to the occurrence of northeasterly (ACC) winds, which before reaching the environs of the LC21 core would have blown across the surface of the Aegean Sea for some several hundred kilometers, thus contributing to Mediterranean deep water formation (Rohling et al. 2009). ACC winds are a winter/early spring phenomenon, usually lasting only a few days at a time. In light of these observations, the LC21 SST record provides strong indications that ACC periods saw the regular and intensive influx of polar air masses into the northern Aegean. During such periods the eastern Mediterranean would have been bathed in abnormal quantities of extremely cold air.

Hudson Bay Outflow

Notably, the (atmospheric) SH mechanism is not the only source of extreme climate perturbations in the Northern Hemisphere during the early Holocene.

In particular, the final centuries of the 9th millennium cal BP also feature the (marine) 8.2 ka cal BP (Hudson Bay) Event (see, for example, Alley & Ágústsdóttir 2005). As stressed by Rohling & Pälicke (2005), even though the two mechanisms (ACC and Hudson Bay) are unrelated, their accumulated impact may well have triggered some of the most extreme climate anomalies of the Holocene. The outflow of meltwater from the Laurentide ice sheet into the Labrador Sea from lakes Agassiz and Ojibway resulted in the interruption of heat transfer from the ocean to the atmosphere and deep-water formation in the North Atlantic, thus leading to a significant cooling of the Northern Hemisphere for a period of approximately two centuries (ca. 8.2–8.0 ka cal BP).

To conclude, the GISP2 nss [K+] record shows a series of severe cooling episodes that run systematically (with quasi-cyclicity of about 1450 years) through the entire Glacial, into the Holocene, and up to modern times (Mayewski et al. 1997, 2004). Each of these episodes is associated with a more pronounced Siberian High over Asia, which in turn would have led to the increased occurrence and severity of winter outbreaks over Europe and in the eastern Mediterranean. The existence of cold anomalies is further confirmed by a large number of terrestrial and marine records from the eastern Mediterranean (Rohling et al. 2002, 2009; Meeker & Mayewski 2002; Marino et al. 2009). Based on the GISP2 nss (non-sea salt) [K+] record, strongest ACC conditions are inferred for the time-intervals ca. 10.2 ka cal BP, 8.6–8.0 ka cal BP, 6.0–5.2 ka cal BP, and ca. 3.0 ka cal BP. The most recent ACC-event is commonly referred to as the Little Ice Age (LIA: ca. AD 1450–1929). Additionally, cold conditions observed during the 8.6–8.0 ka cal BP interval were amplified during its later stages (8.2.–8.0 ka cal BP) by the Hudson Bay event.

Bipartite Vulnerability Model

On the basis of aforementioned insights from palaeoclimatological research, we posit that archaeological climate-culture analysis for the age range 6600–6000 cal BC is best studied in terms of a two-component (bipartite) chronological model. This model comprises an earlier Phase A (ACC only) and a later Phase B (ACC amplified by Hudson Bay impact). The two climate mechanisms would have resulted in similar environmental impacts with partial temporal overlap (Rohling & Pälicke 2005).

The Younger Model Component (Phase B)

The most illustrative archaeological case study for considering the cumulative impacts of ACC and the 8.2 ka cal BP (Hudson Bay) Event is the abandonment of the Neolithic settlement at Çatalhöyük East. It is widely acknowledged that the abandonment of this site and the onset of the Hudson Bay outflow are exactly contemporaneous and that a return to climatically milder conditions saw the foundation of a new settlement at adjacent Çatalhöyük West (fig. 2.2). Certainly, we are not inferring that this site desertion and/or relocation were directly *caused* by Hudson Bay. Rather, we make reference to Modern Vulnerability Theory, which stresses that the prevailing *properties* of afflicted communities are equally responsible for what might be termed *catastrophes* or *disasters* (Blaikie et al. 1994). Indeed, it is the underlying socioeconomic system that can turn an otherwise only weakly acting natural event into a crisis (Dikau 2008). As to Çatalhöyük East, although site abandonment in the younger model phase B is perhaps the most clearly visible societal reaction in the face of adverse climate conditions, there is probably more to learn—in the sense of "beyond collapse" (cf. Weiss 2000)—from studies of preceding settlement phases.

The Older Model Component (Phase A)

The older ACC model component (Phase A) begins with the onset of (atmospheric) cold conditions at around ca. 8.6 ka cal BP. To stay with the Çatalhöyük example, and according to our bipartite modeling, the relevant stratigraphic unit for ACC impact studies at Çatalhöyük East should not to be sought on the eroded site surface, as is sometimes assumed, but instead from Level VI (coinciding with the onset of ACC conditions) and in the subsequent Levels V–II (for the duration of ACC). For this reason, we suggest that Çatalhöyük East is *not* a prime candidate for the study of impacts associated with the 8.2 ka cal BP event. Rather, because of its long settlement history (with Levels XI–VII dating prior to ACC), the site is far better suited to the study of running societal impact of ACC (Phase A) conditions.

In addition to Çatalhöyük East in central Anatolia, we have studied societal impacts of ACC conditions at archaeological sites in Pisidia (Turkish Lakes District), SW Anatolia. According to available ¹⁴C dates (fig. 2.5), many of these sites were only occupied at some later stage of the time interval 8.6 ka to 8.3 ka cal BP (ACC Phase A). In contrast, newly available ¹⁴C data from key sites Ulucak and Çukuriçi Höyük show that the arrival of farming communities on the Turkish West Coast is coincident (decadal scale) with the beginning of the ACC interval (fig. 2.6). As is shown in the chronological overview (fig. 2.5), the farming communities on the Turkish West Coast actually predate the earliest known sites in Pisidia—an observation that is subject to change when results from new excavations become available. For now, we interpret the later foundation of Pisidian sites, as well as the earlier appearance of Neolithic communities on the Turkish Aegean, as representative of the established "Go West" model (Özdoğan 2011), albeit with a clear climatic component.

Overview

Phase A: Eastern Mediterranean (6600–6300 cal BC)

Major developments in the Eastern Mediterranean during the first model phase (ACC: 6600–6300 cal BC) are summarized in Figure 2.3. In this phase there

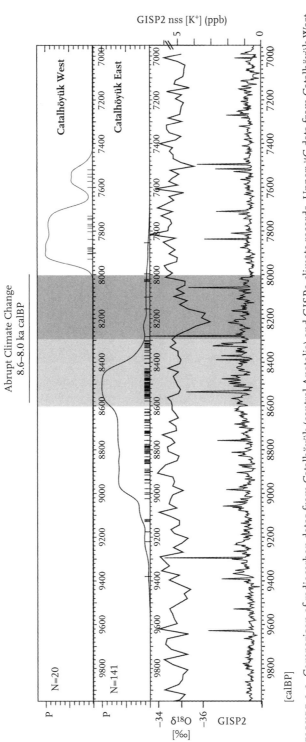

FIGURE 2.2 Comparison of radiocarbon dates from Çatalhöyük (central Anatolia) and GISP2 climate records. Upper: ¹⁴C data from Çatalhöyük West and Çatalhöyük East (Böhner & Schyle 2008; Higham et al. 2007). Lower: Greenland GISP2 ice-core δ¹⁸O (Grootes et al. 1993); GISP2 potassium (non-sea salt [K⁺]; ppb) ion proxy for the Siberian High (Mayewski et al. 1997; Meeker & Mayewski 2002). Adapted from Weninger et al. 2009 (fig.18).

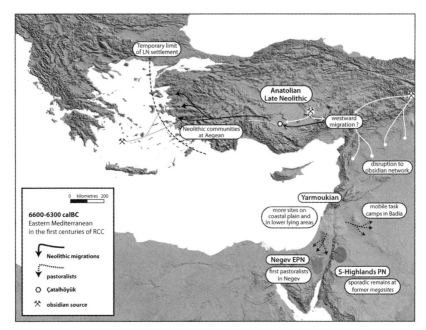

FIGURE 2.3 Anatolia and the Near East in the Pottery Neolithic. Summary of Events in the 1st Model Phase A (6600–6300 cal BC) of the ACC Interval (Negev EPN: Negev Early Pottery Neolithic; S-Highlands PN: Southern Highlands Pottery Neolithic)

is manifold evidence for population movements not only within Anatolia, but also in coastal and low-lying locations in the northern and the southern Levant. The onset of ACC (about 8.6 ka cal BP) in the southern Levant coincides with the first appearance of pottery-bearing communities, commonly referred to as Yarmoukian culture, and intensified settlement activities on the coastal plain. These trends coincide with the gradual decline of LPPN megasites in the Jordanian Highlands (Gebel 2004). As coastal and low-lying areas would have been less affected by typical ACC impacts (for example, summer drought in combination with severe winters), we posit that the widely observed habitat tracking to milder regions, such as the coastal Levant and the Turkish west coast, may well be attributed to the very same climate mechanism.

Rubble Layers in the Southern Levant

Previously, the collapse of LPPNB (Late Pre-Pottery Neolithic B) *megasites* in the highlands of the Southern Levant has been erroneously correlated with the 8.2 ka cal BP Hudson Bay Event (that is, ACC Phase B). Meanwhile, closer consideration of ^{14}C ages shows that depopulation of these regions commenced at the onset of ACC Phase A. Significantly, in these same areas we observe a large number of sites that feature *Yarmoukian Rubble Layers,* so-called by Rollefson & Kafafi (1994: 11) because of the substantial presence in these strata of very dense concentrations of fist-sized angular limestone rubbles, which also often contain Neolithic (Yarmoukian) pottery sherds. The number of Levantine Neolithic sites that are known to feature such layers is remarkable and includes 'Ain Ghazal, 'Ain Jammam, 'Ain Rahub, Abu Suwwan, Ba'ja, Basta, es-Sifiya, Jebel Abu Thawwab, Umm Meshrat I and II, and Wadi Shu'eib (Rollefson 2009; Zielhofer et al. 2012).

Recent geomorphological studies at 'Ain Ghazal have shown that rubble levels accumulated not only during the Yarmoukian but also at different times in the middle Holocene. Further, multiple rubble accumulation events are confirmed from observations at Basta, where FPPNB (Final PPNB) structures have been attested within the Lower Rubble Layer, while only the Upper Rubble Layer contains Pottery Neolithic material (Gebel 2009). Meanwhile, sediment analyses at 'Ain Ghazal have revealed that water was not determinant in rubble accumulation (Zielhofer et al. 2012), which refutes our earlier hypothesis that Rubble Layers were the result of landslides triggered by flashfloods (Weninger 2009). Rather, new insights suggest that Rubble Layers typically derive from the slow decay of houses following site abandonment (Gebel 2009; Zielhofer et al. 2012). One common *climatic* feature of sites with rubble layers remains, however: nearly all sites were abandoned in the Pottery Neolithic (no later than ca. 8.0 ka cal BP) and are located predominantly in landscapes that experienced depopulation during the ACC interval (8.6–8.0 ka cal BP). Although the direct impact of ACC climate perturbations (such as flashfloods) on rubble-layer genesis now appears unlikely, the postulated role of ACC in the abandonment of these settlements remains valid.

Phase B: Eastern Meditteranean (6300–6000 cal BC)

Major developments in the second model phase (that is, combined ACC and Hudson Bay: 6300–6000 cal BC) are summarized in Figure 2.4. These centuries are synonymous with unprecedented societal perturbations in the entire eastern Mediterranean. In the Southern Levant this phase is referred to as the *Late Yarmoukian Crisis* (Clare 2013). It sees the widespread abandonment of settlements in the Transjordanian Highlands and the Lower Jordan Valley (south of Lake Galilee), including the major Yarmoukian site of Sha'ar Hagolan. Notably, subsequent Jericho IX culture sites (7900–7600 cal BP/ 5900–5600 cal BC) are limited to the southern coastal plain, the Jezreel Valley and the Hula Basin. Once again, this trend appears linked to strategies aimed at the mitigation of ACC impacts.

Remarkably, eastern Anatolian and Syrian data testify to a similar *instability period*. A number of sites in these regions provide substantial evidence for temporal or complete abandonment of settlements, including among them Shir (Bartl 2010), Tell Sabi Abyad (Akkermans et al. 2006; van der Plicht et al. 2011), Çayönü (Özdoğan 1999), Akarçay Tepe (Özbaşaran & Duru 2011), Aşıklı Höyük (Özbaşaran 2011), and Mersin-Yumuktepe (Caneva & Köroğlu 2010). Furthermore, in the Amuq plain we observe a potential hiatus in the ceramic sequence between phases Amuq A and Amuq B (Balossi 2004), while in the Rouj Basin (Tell el-Kerkh) we see a shift in burial practices with the first appearance of a demarcated burial ground (cemetery) and evidence of the use of central

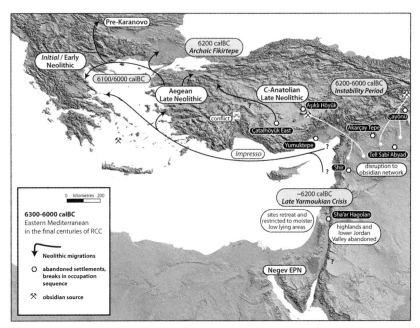

FIGURE 2.4 Anatolia and the Near East in the Pottery Neolithic. Summary of Events in the 2nd Model Phase B (6300–6000 cal BC) of the ACC Interval (Negev EPN: Negev Early Pottery Neolithic)

areas for *public* purposes (Tsuneki 2010). Cumulatively, the evidence suggests potentially significant upheavals and changes to prevailing social systems at this time. For example, as indicated in Figure 2.4, the observed introduction of impresso pottery elements in the Aegean could be related to the arrival of groups from the northern Levant via a marine route (Çilingiroğlu 2010; Brami & Heyd 2011). In the northern Levant, this development appears to go hand in hand with the widespread appearance of Proto-Halaf culture sites (see, for example, Cruells 2008). Finally, in northwest Anatolia (Marmara region), Phase B (6300–6000 cal BC) of the ACC interval sees the genesis of the Fikirtepe culture. Generally speaking, one of the most astounding aspects of the archaeological evidence reviewed in figures 3.3 and 3.4 is the increase in cultural (and presumably also demic) mobility in many regions of the Near East and Anatolia.

Nevertheless, there are many regional and temporal differences. In the southern Levant, increases in mobility from the early ninth millennium cal BP onwards are associated with the development of nomadic pastoralism as an important subsistence strategy (Köhler-Rollefson 1992). In comparison, higher levels of mobility in Anatolia (from ca. 8600 cal BP) are clearly targeted to the settlement of presumably uninhabited regions of the Turkish Lakes District and the Aegean coast. Irrespective of other factors that may have led to higher levels of mobility, the trend appears to be associated with the inherent *advantages* and *disadvantages* of various landscapes in the southern Levant and, in particular, in central Anatolia.

Climatic Geography of Anatolia

The central plain of Turkey receives relatively low amounts of precipitation, close to the threshold of rainfed agriculture (approx. 250 mm/yr), which renders it considerably more vulnerable to drought than areas further to the west. In the Aegean, for example, values of 600–800 mm/year are not uncommon, with even higher amounts observed in northwest Turkey. Higher average winter temperatures and, specifically, the less significant impact of snow would have further enhanced coastal locations for traditional (pre-industrial) Neolithic communities. While in the central plain average January temperatures frequently drop below freezing, coastal areas are considerably milder. Copious amounts of snow can restrict mobility for periods of days, weeks, or even months. It is this restricted mobility that is especially hazardous to any system that is reliant on material resources dispersed over large territories, as we might expect for the *megasite* Çatalhöyük at the center of the Konya plain.

The biophysical and social vulnerability of pre-industrial settlement systems in central Anatolia is well illustrated in a historical report by E. Neumann, an engineer who travelled the Konya region in 1890 on behalf of the Istanbul-Baghdad railway project. He describes one of the most severe documented human catastrophes in the region:

> *One of the worst famines known in this region from modern history occurred from winter 1873 to spring 1875. It most strongly affected the Vilayets (provinces) Kastamuni [Kastamonu], Angara [Ankara] and Kaiseri [Kayseri]. The great drought of 1873 produced a crop failure, and from November and December there occurred torrential rainfall, followed in January and February 1874 by quite extraordinary snowfall. The small food reserves of the snowed-in villages were soon exhausted, and—since the extreme winter had disrupted all paths of communication—widespread death soon ensued. It is reported that altogether some 150.000 people and 100.000 head of cattle died in a very brief time. The loss of sheep and goats is estimated at 40%.*
>
> —Neumann 1893, cited in Christiansen & Weniger 1964;
> translation from German by the authors.

Neumann's account highlights that it is the combination of natural hazards with a highly vulnerable socioeconomic system—not only an isolated climate event—that eventually leads to disaster. Moreover, the Little Ice Age scenario suggests what could as well have occurred in the Neolithic ACC period.

Çatalhöyük (Konya Plain)

One of the best-known prehistoric sites in the highly sensitive Konya Plain is Çatalhöyük. As we have already seen, impacts from ACC can be expected at this site beginning in Level VI (fig. 2.2), when the settlement reached it maximum size of some 13 hectares, with an estimated several thousand inhabitants (Düring 2011: 118). The growth coincides with increased levels of competition between groups (Asouti 2005) and a rise of the autonomous household as the dominant social unit (Düring & Marciniak 2006), indicated by public feasting, as well as by increased evidence for craft specialization and the differentiation of male and female work and identity (Hodder 2006). These changes in LN society are further reflected in the general plan of the settlement, which sees an increase in the amount of non-built open space through time, especially between level VIa and V, when open rubbish areas (middens) were replaced by genuine courtyards. At the same time, rooftop movement became less important and the number of communal ovens decreased (cf. Cutting 2005: 159).

The supraregional importance of Çatalhöyük has been related to its function as an important thoroughfare for obsidian in Anatolia, it even being suggested that some of its inhabitants acted as middlemen or itinerant traders in this commodity (Carter & Shackley 2007). Likewise, this central, and indeed singular, settlement in the Konya plain in the LN would also conceivably have served as a *gateway* for neolithization of regions to its west. The potentially swelling groups on their trek westwards (fig. 2.4) would, however, have placed unprecedented stress on local resources in the Konya plain, especially in light of the increased likelihood of climate/environmental hazards associated with the onset of the ACC interval.

Pisidia (Turkish Lakes District)

Moving to the settlement region west of the Konya plain, in Pisidia (Turkish Lakes District) we find convincing signs of contemporaneous social turbulence in the large-scale destruction of a *religious complex* at Höyücek Höyük, with a subsequent chronological hiatus (Duru 1995: 486–487; Duru & Umurtak 2005: 230). The conflagration responsible for the destruction of this complex at the end of the so-called *Shrine Phase* ensured the preservation of large numbers of sling missiles on the floor of one of the structures (building 3). Further burnt structures have been excavated at Bademağacı (ENII/3), where one of the buildings contained the remains of eight individuals (Duru 2005). One of the victims shows evidence for perimortem cranial trauma, thought to be indicative of a blow to the head (Erdal & Erdal 2012). Also at Bademağacı (ENII/4–3), there is evidence for the existence of a fortification structure (Duru 2004; 2005; Umurtak 2007), though it is not yet securely dated to the ACC interval (6600–6000 cal BC). Further conflagrations have been observed at Hacilar VI (Mellaart 1970; Brami & Heyd 2011).

It is conceivable that the conflicts in Pisidia might have had an immediate climatic background. Yet from ethnographic studies we are aware that social conflicts are seldom overtly linked to (environmentally induced) resource shortfalls, though periods with documented climate stress are indeed known to correlate (weakly) with pronounced outbreaks of intergroup violence. For now, at least two types of conflict may be distinguished in the dispersal of Neolithic lifeways from their core areas of genesis: (1) clashes arising from politically motivated territorial expansion, which may have been favoured by climate-induced stress, and (2) power imbalances created by group displacements,

SITE	PERIOD [calBC]		N	MAP NR
		RCC — RCC+Hudson		
		Late Mesolithic — Early Neolithic		
		8000 7500 7000 6500 6000 5500 5000		
Schela Cladovei	Iron Gorges		N=46	44
Vlasac	[Meso/Neo]		N=46	43
Ecsegfalva 2			N=38	42
Szolnok-Szanda	North East Hungary		N=11	41
Nagyköru-Tsz			N=9	40
Dudeşti Vechii			N=4	39
Foeni-Gaz			N=1	38
Foeni-Sălaş	Roumania		N=2	37
Gura Baciului			N=2	36
Miercurea			N=7	35
Stara Zagora-Okr.Boln			N=13	34
Elesnica			N=9	33
Karanovo			N=43	32
Slatina			N=12	31
Cavdar	Bulgaria		N=28	30
Galabnik			N=12	29
Džhulyunica			N=31	28
Kovačevo			N=5	27
Donja Branjevina			N=7	26
Blagotin	Serbia		N=3	25
Divostin			N=10	24
Anzabegovo	FYROM		N=29	23
Nea Nikimedeia	North Greece		N=15	22
Dikili Tas			N=19	21
Aşağı Pınar			N=23	20
Hoca Çeşme	North West Anatolia		N=14	19
Ilıpınar			N=66	18
Menteşe			N=11	17
Ege Gübre			N=10	16
Ulucak Höyük	West Coast		N=36	15
Çukuriçi Höyük			N=15	14
Kuruçay Höyük			N=5	13
Hacılar	Lakes District		N=11	12
Höyücek			N=7	11
Erbaba			N=7	10
Bademağacı			N=7	9
Çatalhöyük West			N=25	8
Çatalhöyük East			N=117	7
Can Hasan I			N=13	6
Tepecik	Central Anatolia		N=6	5
Can Hassan III			N=16	4
Suberde			N=11	3
Musular			N=9	2
Aşıklı Höyük			N=46	1

[calBC] 8500 8000 7500 7000 6500 6000 5500 5000
PERIOD Epi   PPN   EPN   LN   ECh   MCh

FIGURE 2.5 Overview of ¹⁴C ages (total N = 857) for 44 archaeological sites. Site numbers refer to the map, fig. 3.6. Using the Barcode Method of ¹⁴C-age calibration each vertical line represents the calibrated median value of one ¹⁴C age (CalPal software; Weninger & Jöris 2008) and INTCAL09 calibration data (Reimer et al. 2009). Shaded areas show Rapid Climate Change (RCC) interval 8.6–8.9 ka BP (per Rohling et al. 2002 and Meeker & Mayewski 2002), with Hudson Bay outflow interval set schematically to ~6200–6000 cal BC. The position of the 9.3 ka BP ACC interval (per Fleitmann et al. 2009) is included for explorative purposes. Abbreviations: Epi = Epipalaeolithic, PPN = Pre-Pottery-Neolithic, EPN = Early-Pottery-Neolithic, LN = Late Neolithic, ECh = Early Chalcolithic, MCh = Middle Chalcolithic, FYROM = Former Yugoslav Republic of Macedonia.

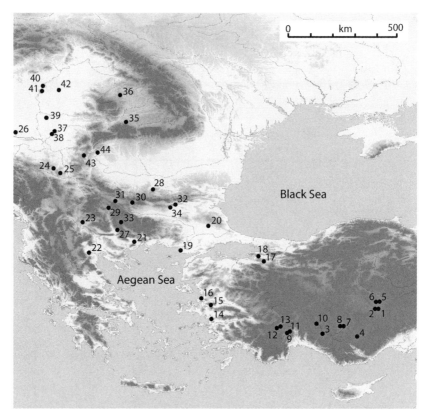

FIGURE 2.6 Geographical distribution of archaeological study sites in Anatolia, southeast Europe, and eastern Central Europe during and after the ACC interval (6600–5700 cal BC). Corresponding site data is given in Fig. 2.5. Site data sources are 1–20: Clare & Weninger 2014; 21: Lespez et al. 2013; 22: Weninger et al. 2006; Linick, 1977; 23: Tasić 1988; 25: Bogdanović 2008; 26: Biagi & Spataro 2005, Luca, Diaconescu, & Suciu 2008; 27: Lichardus-Itten et al. 2002; 28: Krauß et al. in press; 29–34: Görsdorf & Bojadžiev 1966; 35–39: Biagi & Spataro 2005, Luca, Diaconescu, & Suciu 2008; 40–42: Oross & Siklósi 2012; 43–44: Borić 2011.

which could have been caused by widespread crop failure and epidemics intensified under ACC conditions (Clare & Weninger 2016).

Turkish West Coast/Southeast Europe

The establishment of an exact date for the arrival of the first farming communities on the Turkish west coast is essential for validating (or falsifying) our proposed abrupt climate change neolithization model. If it were the case that farming had already been introduced to coastal regions of western Turkey prior to the onset of ACC, the model would be difficult to endorse. In this respect, recent excavation results from the sites of Ulucak (Çilingiroğlu 2011; Çilingiroğlu,

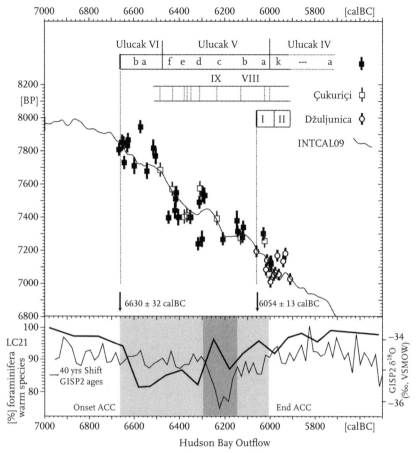

FIGURE 2.7 Comparison of ¹⁴C age models for archaeological sites Ulucak (W. Turkey, site nr.15), Çukuriçi (W. Turkey, site nr.14), and Džuljunica (Bulgaria, site nr. 28), each based on (assume linear) stratigraphic wiggle matching by sample depth. Data bars: ± 1 σ. Thin line: INTCAL09 calibration curve (Reimer et al. 2009). The site chronologies are compared with (bottom) Greenland GISP2 ice-core δ¹⁸O record (Grootes et al. 1993) as proxy for North Atlantic air temperature, and LC21 foraminifera data (percent warm species) as proxy for Aegean Sea Surface Temperature (Rohling et al. 2002). Arrows indicate foundation of Ulucak at 6630 ± 32 cal BC and Džuljunica at 6054 ± 13 cal BC.

Cevik, & Çilingiroğlu 2012) and Çukuriçi Höyük (Horejs et al. 2015) provide a most welcome *experimentum crucis* for evaluating the coastal refugium model of Weninger et al. 2014 (wherein "Go to the Coast" = "Go West").

Figure 2.7 shows the chronological positions of Ulucak and Çukuriçi Höyük, set in relation to available ACC proxies and to Džuljunica, a site in Northeast Bulgaria assigned to the initial Neolithic in this part of southeast Europe (Krauss et al. 2014). As shown, the initial Neolithic occupation at Ulucak (first arrow at Phase VI) dates to 6630 ± 32 cal BC—that is, to the interval of ACC onset at 6700–6600 cal BC indicated in marine core LC21.

As shown by the second arrow, the earliest Neolithic at Džuljunica is dated to 6054 ± 13 cal BC—that is, to the immediate aftermath of the 8.2 cal BP Hudson Bay event, which is near the end of ACC conditions. From these two key dates we infer that, at present, the spread of early farming from Anatolia to southeast Europe is best explained by a bipartite migration model that combines an initial *refugium colonization* of the Turkish West Coast (well-dated at Ulucak: 6630 ± 32 cal BC) with a subsequent, delayed *secondary expansion* into southeast Europe (well-dated at Džuljunica to 6054 ± 13 cal BC). The extent of the delay is reflected by the difference (6630 ± 32– 6054 ± 13 cal BC = 576 ± 35 years) in the earliest arrival times of the Neolithic at the two sites.

The younger age is independently confirmed by ceramics from the oldest settlement layer at Džuljunica (Dž-I), which show clear similarities with material from the west Anatolian Late Neolithic, especially from the Izmir region (Krauss et al., in press) and specifically Ulucak V^a and early IV Çukuriçi Höyük and Yeşilova (Çilingiroğlu 2009, 2011; Galik/Horejs 2011; Derin 2011). Pottery from Dž-I represents the very beginning of the southeast European Neolithic cultural sequence (Krauss 2011). Nevertheless, neolithization of the central Balkans a few decades before the earliest occupation at Džuljunica still cannot be ruled out. As is well known, available ^{14}C data from Thessaly suggest that neolithization commenced slightly earlier in Greece than in the Balkans, where the river valleys of the Vardar/Axios and Morava would have provided natural routes for the northwards dispersal of the new form of subsistence. Brami and Heyd (2011) have discerned additional routes recently based on pottery-style distributions.

In summary, the initial dispersal of Neolithic lifeways from the Aegean to central Eastern Europe was completed within ca. 200 years (fig. 3.5). This conclusion will not be unexpected among archaeologists working in southeast Europe (for example, Biagi, Shennan, & Spataro 2005; Forenbaher & Miracle 2005; Borić 2011; Brami & Heyd 2011; Krauss 2011). However, the implied rapidity with which the dispersal occurred contradicts long-established and still frequently advocated wave-of-advance models based on physical transport equations (for example, Ammerman & Cavalli-Sforza 1984; Gkiasta et al. 2003; Pinhasi, Fort, & Ammerman 2005; Lemmen, Gronenborn, & Wirtz 2011; Bocquet-Appel et al. 2012; Fort 2012). Wave-of-advance models imply that the neolithization process was inherently slow, most often gradual and continuous, and characterized by an average rate of spread in the range of 1–3 km/year. Yet, as Budja (2009: 128) points out, the assumed gradual diffusion model within the Balkans from south to north and, in particular, the millennium timespan calculated for the process find no confirmation in the available ^{14}C dates (fig. 2.5). We may also extend this conclusion to cover the development of the Körös culture, much further north, in the northeastern corner of the Pannonian Basin (fig. 2.6).

Conclusions

The sharing of knowledge over considerable distances has recently been referred to as the *esteemed value* of the Pre-Pottery Neolithic (PPN) without which primary neolithization could not have been accomplished (Özdoğan 2010: 54). Simultaneously, this particular value is considered nonconducive with social stress, rivalry, and conflict between communities. While Pre-Pottery-Neolithic (PPN) values were based on peaceful coexistence, thus promoting the initial spark of neolithization, at the transition to the Pottery Neolithic (PN) in the early 7th millennium cal BC this paradigm appears to lose its validity. Özdoğan (2010: 55) concludes that indications of some sort of social turbulence were apparent in most core areas of the Neolithic at this time. This period of turbulence has been the focus of our contribution.

As indicated by the unfiltered radiocarbon data (fig. 2.5), beginning with the Pre-Pottery-Neolithic (ca. 7500 ca BC), the dispersal of Neolithic lifeways from core areas to central Eastern Europe appears to have been established in a stepwise manner. In the Late Neolithic (the ACC interval 6600–6000 cal BC), we identify two major steps. The first was the dispersal of the Neolithic from the Near East to the Aegean at ca. 6600 cal BC (or the onset of ACC conditions). This step appears to have been accomplished within a few decades. The second step, which takes the Neolithic lifestyle from the Aegean littoral all the way to northeastern Hungary, begins at ca. 6100 cal BC (that is, the end of ACC conditions) and is completed within 200 years.

Acknowledgments

We would particularly like to thank Barbara Horejs (Austrian Academy of Sciences, Vienna/Austria) and Raiko Krauss (University of Tübingen/ Germany) for their interest in our studies and frequent advice. Special thanks are due to Amit Tubi and Uri Dayan (University of Jerusalem/Israel) for invaluable information relating to the meteorology of Abrupt Climate Change.

References

Akkermans, P. M. M. G., R. Cappers, C. Cavallo, and O. Nieuwenhuyse. 2006. Investigating Early Pottery Neolithic of Northern Syria: New Evidence from Tell Sabi Abyad. *American Journal of Archaeology* 110: 123–156.

Alley, R. B., and A. M. Ágústsdóttir. 2005. The 8k Event. Cause and Consequences of a Major Holocene Abrupt Climate Change. *Quaternary Science Reviews* 24: 1123–1149.

Ammerman, A. J., and L. L. Cavalli-Sforza. 1984. *The Neolithic Transition and the Genetics of Populations in Europe*. Princeton University Press. Princeton.

Asouti, E. 2005. Group identity and politics of dwelling at Neolithic Çatalhöyük. In *Çatalhöyük Perspectives: Reports from the 1995–99 Seasons*, ed. I. Hodder. Cambridge: McDonald Institute for Archaeological Research and British Institute of Archaeology at Ankara, 75–91.

Balossi, F. 2004. New Data for the Definition of the DFBW Horizon and Its Internal Developments. *Anatolica* XXX: 109–149.

Bartl, K. 2010. Shir, West Syria. *Neo-Lithics* 1/10: 92–93.

Begzsuren, S., J. E. Ellis, D. S. Ojima, M. B. Coughenour, and T. Chuluun. 2004. Livestock responses to droughts and severe winter weather in the Gobi Three Beauty National Park, Mongolia. *Journal of Arid Environments* 59, 4: 785–796.

Biagi, P., S. Shennan, and M. Spataro. 2005. Rapid rivers and slow seas? New data for the radiocarbon chronology of the Balkan Peninsula. In *Prehistoric Archaeology and Anthropological Theory and Education*, ed. L. Nikolova and J. Higgins. RPRP 6–7: 43–51.

Biagi, P., and M. Spataro. 2005. New observations on the radiocarbon chronology of the Starčevo-Criş and Körös Cultures. In *Prehistoric Archaeology and Archaeological Theory and Education*, ed. L. L. Nikolova and Jude Higgins. RPRP 6–7: 35–52.

Blaikie, P., T. Cannon, I. Davis, and B. Wisner. 1994. *At Risk. Natural Hazards, People's Vulnerability, and Disasters*. Routledge. London and New York.

Bocquet-Appel, J. P., S. Naji, M. Vander Linden, and J. Kozlowski. 2012. Understanding the rates of expansion of the farming system in Europe. *J. Arch.Sci.* 39(2): 531–546.

Bogdanović, M. 2008. *Grivac: Settlements of Proto-Starčevo and Vinča Culture*. Center for Scientific Research of Serbian Academy of Sciences and Arts.

Böhner, U., and D. Schyle. 2006. *Radiocarbon CONTEXT Database 2002–2006*. doi:10.1594GFZ.CONTEXT.Ed1

Borić, D. 2011. Adaptations and transformations of the Danube Gorges foragers (c.13000–5500 BC): an overview. In *Beginnings–New Research in the Appearance of the Neolithic between Northwest Anatolia and the Carpathian Basin; Papers of the International Workshop 8th–9th April 2009, Istanbul*, ed. R. Krauss. Rahden/Westf.: Verlag Marie Leidorf, 157–203.

Brami, M., and V. Heyd. 2011. The origins of Europe's first farmers: the role of Hacilar and western Anatolia, fifty years on. *Prähistorische Zeitschrift* 86: 165–206.

Budja, M. 2009. Early Neolithic pottery dispersals and demic diffusion in Southeastern Europe. *Documenta Praehistorica* 36: 117–137.

Cacho, I., J. O. Grimalt, M. Canals, L. Sbaffi, N. J. Shackleton, J. Schoenfeld, and R. Zahn. 2001. Variability of the western Mediterranean Sea surface temperature during the last 25000 years and its connection with the Northern Hemisphere climatic changes. *Paleoceanography* 16: 40.

Caneva, I., and G. Köroğlu. 2010. *Yumuktepe: A Journey Through Nine Thousand Years*. Istanbul: Ege Yayınları.

Carter, T., and M. S. Shackley. 2007. Sourcing obsidian from neolithic Çatalhöyük (Turkey) using energy dispersive x-ray fluorescence. *Archaeometry* 49 (3): 437–454.

Casford, J. S. L., E. J. Rohling, R. Abu-Zied, C. Fontanier, F. J. Jorissen, M. J. Leng, G. Schmiedl, and J. Thomson. 2003. A dynamic concept for eastern Mediterranean circulation and oxygenation during Sapropel formation. *Palaeogeography, Palaeoclimatology, Palaeoecology* 190: 103–119.

Christiansen-Weniger, F. 1964. Gefährdung Anatoliens durch Trockenjahre und Dürrekatastrophen. *Zeitschrift für Ausländische Landwirtschaft* 3: 133–147.

Çilingiroğlu, A., O. Cevik, and Ç. Çilingiroğlu. 2012. Towards understanding the early farming communities of middle west Anatolia. In *The Neolithic in Turkey. New Excavations and New Research. Western Turkey*, ed. M. Özdoğan, N. Başgelen, and P. Kuniholm. Istanbul: Archaeology & Art Publications. Istanbul, 139–175.

Çilingiroğlu, Ç. 2009. "Central-West Anatolia at the End of the 7th and Beginning of 6th Millennium BCE in the Light of Pottery from Ulucak (Izmir)." Unpublished doctoral thesis. Eberhard-Karls-Universität Tübingen.

Çilingiroğlu, Ç. 2010. The appearance of impressed pottery in the neolithic Aegean and its implications for maritime networks in the eastern Mediterranean. *Tüba-Ar* 13: 9–22.

Çilingiroğlu, Ç. 2011. The Current State of Neolithic Research at Ulucak, Izmir. In *Beginnings–New Research in the Appearance of the Neolithic between Northwest Anatolia and the Carpathian Basin; Papers of the International Workshop 8th–9th April 2009, Istanbul*, ed. R, Krauss. Rahden/Westf.: Verlag Marie Leidorf, 67–76.

Clare, L. 2016. *Culture Change and Continuity in the Eastern Mediterranean and During Rapid Climate Change. Assessing the Vulnerability of Neolithic Communities to a "Little Ice Age" in the Seventh Millennium cal* BC. Kölner Studien zur Prähistorischen Archäologie Vol 7, 269 pp. Verlag Marie Leidorf GmbH.

Clare, L., and B. Weninger, 2014. The dispersal of Neolithic lifeways: absolute chronology and rapid climate change. In *The Neolithic in Turkey*, vol. 6, ed. M. Özdoğan, N. Başgelen, and P. Kuniholm. Istanbul: Archaeology and Art Publications, 1–65.

Clare, L., and B. Weninger, 2016. Early Warfare and Its Contribution to Neolithisation and Dispersal of First Farming Communities in Anatolia. In *Palaeoenvironment and the Development of Early Settlements*. Studien aus den Forschungsclustern des Deutschen Archäologischen Instituts. Cluster 1, Band 14. Ed. M. Reindel, K. Bartl, F. Lüth, and N. Benecke. Rahden: Verlag Marie Leidorf, 29–49.

Cohen, J., K. Saito, and D. Entekhabi. 2001. The role of the Siberian High in Northern Hemisphere climate variability. *Geophysical Research Letters* 28: 299–302.

Cruells, W. 2008. The Proto-Halaf: Origins, definition, regional framework and chronology. In *Proceedings of the 5th International Congress on the Archaeology of the Ancient Near East*, vol. 3, ed. J. Mª Córdoba, M. Molist, Mª C. Pérez, I. Rubio, and S. Martínez. Madrid: Centro Superior de Estudios sobre el Oriente Próximo y Egipto, 671–689.

Cutting, M. 2005. The architecture of Çatalhöyük: continuity, household and settlement. In *Çatalhöyük Perspectives: Reports from the 1995–99 Seasons*, ed. I. Hodder. Cambridge/London: McDonald Institute for Archaeological Research and British Institute of Archaeology at Ankara, 151–169.

Derin, Z. 2011. Yeşilova Höyük. In *Beginnings–New Research in the Appearance of the Neolithic between Northwest Anatolia and the Carpathian Basin; Papers of the International Workshop 8th–9th April 2009, Istanbul*, ed. R. Krauss. Rahden/ Westf.: Verlag Marie Leidorf, 95–106.

Dikau, R. 2008. Katastrophen–Risiken–Gefahren: Herausforderungen für das 21. Jahrhundert. In *Umgang Mit Risiken: Katastrophen–Destabilisierung–Sicherheit*, ed. E. Kulke and H. Popp. Bayreuth/Berlin: Deutsche Gesellschaft für Geographie, 47–68.

Düring, B. S. 2011. *The Prehistory of Asia Minor: From Complex Hunter-Gatherers to Early Urban Societies.* Cambridge: Cambridge University Press.

Düring, B. S., and A. Marciniak. 2006. Households and communities in the central Anatolian Neolithic. *Archaeological Dialogues* 12(2): 165–187.

Duru, R. 1995. Höyücek Kazıları [Excavations}, 1991–1992. *Belleten* 59: 447–490.

Duru, R. 2004. Bademağacı kazıları [excavations at Bademağacı] in 2003. *ANMED*: 15–20.

Duru, R. 2005. Bademağacı kazıları 2002 & 2003 [excavations at Bademağacı, preliminary report]. *Belleten* 68: 519–560.

Duru, R., and G. Umurtak. 2005. *Höyücek: Results of the Excavations 1989–1992.* Ankara: Türk Tarih Kurumu.

Erdal, Y. S., and Ö. D. Erdal. 2012. Organized violence in Anatolia: a retrospective research on the injuries from the Neolithic to Early Bronze Age. *International Journal of Paleopathology* 2(2–3): 78–92.

Feurdean, A., S. Klotz, V. Mosbrugger, and B. Wohlfarth. 2008. Pollen-based quantitative reconstructions of Holocene climate variability in NW Romania. *Palaeogeography, Palaeoclimatology, Palaeoecology* 260 (3–4): 494–504.

Fleitmann, D., H. Cheng, S. Badertscher, R. L. Edwards, M. Mudelsee, O. M. Göktürk, A. Fankhauser, R. Pickering, C. C. Raible, A. Matter, J. Kramers, and O. Tüysüz. 2009. Timing and climatic impact of Greenland interstadials recorded in stalagmites from northern Turkey. *Geophysical Research Letters* 36 (19): L19707.

Fletcher, W. J., and M. F. S. Goñi. 2008. Orbital- and sub-orbital climate impacts on vegetation of the western Mediterranean basin over the last 48,000 years. *Quaternary Research* 70(3): 451–464.

Forenbaher, S., and P. T. Miracle. 2005. The spread of farming in the eastern Adriatic. *Antiquity* 79: 514–528.

Fort, J. 2012. Synthesis between demic and cultural diffusion in the Neolithic transition in Europe. *Proc. Natl. Sci. USA* 109(46): 18669–18673.

Galik, A., and B. Horejs. 2011. Çukuriçi Höyük—Various aspects of its earliest settlement phase. In *Beginnings–New Research in the Appearance of the Neolithic between Northwest Anatolia and the Carpathian Basin; Papers of the International Workshop 8th–9th April 2009, Istanbul,* ed. R. Krauss. Rahden/Westf.: Verlag Marie Leidorf, 83–94.

Gebel, H. G. K. 2004. Central to what? The centrality issue of the LPPNB mega-site phenomenon in Jordan. In *Central Settlements in Neolithic Jordan: Proceedings of a Symposium Held in Wadi Musa, Jordan, 21st–25th of July 1997,* ed. H. D. Bienert, H. G. K. Gebel, and R. Neef. Studies in Early Near Eastern Production, Subsistence, and Environment 5. Berlin: ex Oriente, 1–19.

Gebel, H. G. K. 2009. The Intricacy of Neolithic Rubble Layers. The Ba'ja, Basta, and 'Ain Rahub Evidence. *Neo-Lithics* 1/09: 33–48.

Geraga, M., C. Ioakim, V. Lykousis, S. Tsaila-Monopolis, and G. Mylona. 2010. The high-resolution palaeoclimatic and palaeoceanographic history of the last 24000 years in the central Aegean Sea, Greece. *Palaeogeography, Palaeoclimatology, Palaeoecology* 287: 101–115.

Gkiasta, M., T. Russell, S. Shennan, and J. Steele. 2003. Neolithic transition in Europe: the radiocarbon record revisited. *Antiquity* 77: 45–62.

Görsdorf, J., and J. Bojadžiev. 1966. Zur absoluten Chronologie der bulgarischen Urgeschichte. *Eurasia Antiqua* 2: 105–173.

Grootes, P. M., M. Stuiver, J. W. C. White, S. Johnson, and J. Jouzel. 1993. Comparison of oxygen isotope records from the GISP2 and GRIP Greenland ice cores. *Nature* 366: 552–554.

Higham, T. F. G., C. Bronk Ramsey, F. Brock, D. Baker, and P. Ditchfield. 2007. Radiocarbon dates from the Oxford AMS system: archaeometry datelist 32. *Radiocarbon* 49: S1–S60.

Hodder, I. 2006. *The Leopard's Tale: Revealing the Mysteries of Çatalhöyük*. First edition. London: Thames & Hudson.

Horejs, B., Milić, B., Ostmann, F., Thanheiser, U., Weninger, B., Galik, A.. 2015. The Aegean in the Early 7th Millennium BC: Maritime Networks and Colonization, Journal of World Prehistory 28, 2015, 289–330.

Köhler-Rollefson, I. 1992. A model for the development of nomadic pastoralism on the Transjordanian Plateau. In *Pastoralism in the Levant: Archaeological Materials in Anthropological Perspective*, ed. O. Bar-Yosef and A. Khazanov. Monographs in World Archaeology 10. Madison, WI: Prehistory Press, 11–18.

Krauss, R. 2011. On the "monochrome" Neolithic in southeast Europe. In *Beginnings–New Research in the Appearance of the Neolithic between Northwest Anatolia and the Carpathian Basin; Papers of the International Workshop 8th–9th April 2009, Istanbul,* ed. R. Krauss. Rahden/Westf.: Verlag Marie Leidorf, 109–125.

Krauss, R., N. Elenski, B. Weninger, L. Clare, C. Çakırlar, and P. Zidarov. 2014. Beginnings of the Neolithic in Southeast Europe: the early Neolithic sequence and absolute dates from Džuljunica-Smărdeš (Bulgaria). *Documenta Praehistorica* 41: 51–77.

Lau, N. C., and K.-M. Lau. 1984. Structure and energetics of midlatitude disturbances accompanying cold air outbreaks over East Asia. *Monthly Weather Review* 112(7): 1309–1327.

Lemmen, C., D. Gronenborn, and K. W. Wirtz. 2011. A simulation of the Neolithic transition in western Eurasia. *Journal of Archaeological Science* 38: 1–14.

Lespez, L., Z. Tsirtsoni, P. Darcque, H. Koukouli-Chryssanthraki, D. Malamidou, R. Treuil, and C. Oberlin. 2013. The lowest levels at Dikili Tash, northern Greece: a missing link in the early Neolithic of Europe. *Antiquity* 87: 30–45.

Lichardus-Itten, M., J.-P. Demoule, L. Perničeva, M. Grebska-Kulova, I. and Kulov. 2002. The site of Kovačevo and the beginnings of the Neolithic period in south-western Bulgaria. The French-Bulgarian excavations 1986–2000. In *Beiträge zu jungsteinzeitlichen Forschungen in Bulgarien,* ed. M. Lichardus-Itten, J. Lichardus, and V. Nikolov. Bonn: Habelt-Verlag: 99–158.

Linick, T. W. 1977. La Jolla Natural Radiocarbon Measurements VII. *Radiocarbon* 19: 19–48.

Luca, S. A., D. Diaconescu, and C. I. Suciu. 2008. Archaeological research in Miercurea Sibiului-Petriş (Sibiu Country, Romania): the Starčevo-Criş level during 1997–2005 (a preliminary report). *Documenta Praehistorica* 35: 325–343.

Marino, G., E. J. Rohling, F. Sangiorgi, A. Hayes, J. L. Casford, A. F. Lotter, M. Kucera, and H. Brinkhuis. 2009. Early and middle Holocene in the Aegean Sea: interplay between high and low latitude climate variability. *Quaternary Science Reviews* 28: 3246–3262.

Mayewski, P., L. D. Meeker, M. S. Twickler, S. Whitlow, Q. Yang, W. B. Lyons, and M. Prentice. 1997. Major features and forcing of high latitude Northern Hemisphere circulation using a 110,000-year-long glaciochemical series. *Journal of Geophysical Research* 102: 26345–26366.

Mayewski, P. A., E. J. Rohling, J. C. Stager, W. Karlén, K. A. Maasch, L. D. Meeker, E. A. Meyerson, F. Gasse, S. van Kreveld, K. Holmgren, J. Lee-Thorp, G. Rosqvist, F. Rack, M. Staubwasser, R. R. Schneider, and E. J. Steig. 2004. Holocene climate variability. *Quaternary Research* 62: 243–255.

Meeker, L. D., and P. A. Mayewski. 2002. A 1400-year high-resolution record of atmospheric circulation over the north Atlantic and Asia. *The Holocene* 12(3): 257–266.

Mellaart, J. 1970. *Excavations at Hacılar*. Occasional Publications of the British Institute at Ankara 10. Edinburgh: Edinburgh University Press.

Naumann, E. 1893. *Vom Goldenen Horn zu den Quellen des Euphrat. Reisebriefe, Tagebuchblätter und Studien über die asiatische Türkei und die anatolische Bahn*. Munich/Leipzig: R. Oldenbourg.

Oross, K., and Zs. Siklósi. 2012. Relative and absolute chronology of the Early Neolithic in the Great Hungarian Plain. In *The First Neolithic Sites in Central/South-East European Transect, Vol. III, The Körös Culture in Eastern Hungary*, ed. A. Anders and Zs. Siklósi. B.A.R. Int. Ser. 2334: 129–159.

Özbaşaran, M. 2011. Re-Starting at Aşıklı. *Anatolia Antiqua* 19: 27–37.

Özbaşaran, M., and G. Duru. 2011. Akarçay Tepe. A PPNB and PN settlement in Middle Euphrates-Urfa. In *The Neolithic in Turkey, Vol. 2: The Euphrates Basin*, ed. M. Özdoğan, N. Başgelen, and P. Kuniholm. Istanbul: Archaeology & Art Publications, 165–202.

Özdoğan, A. 1999. Çayönü. In *Neolithic in Turkey: The Cradle of Civilization. New Discoveries, ed*. M. Özdoğan and N. Başgelen. Istanbul: Arkeoloji ve sanat yayınları, 35–63.

Özdoğan, M. 2010. The Neolithic medium: warfare due to social stress or state of security through social welfare. *Neo-Lithics* 1(10): 54–55.

Özdoğan, M. 2011. An Anatolian perspective on the neolithization process in the Balkans: new questions, new perspectives. In *Beginnings–New Research in the Appearance of the Neolithic between Northwest Anatolia and the Carpathian Basin; Papers of the International Workshop 8th–9th April 2009, Istanbul*, ed. R. Krauss. Rahden/Westf.: Verlag Marie Leidorf, 23–33.

Pinhasi, R., J. Fort, and A. J. Ammerman. 2005.Tracing the Origin and Spread of Agriculture in Europe. *PLoS Biol* 3(12): e410. doi:10.1371/journal.pbio.0030410

Pross, J., U. Kotthoff, U. C. Müller, O. Peyron, I. Dormoy, G. Schmiedl, S. Kalaitzidis, and A. M. Smith. 2009. Massive perturbation in terrestrial ecosystems of the eastern Mediterranean region associated with the 8.2 Kyr BP climatic event. *Geology* 37: 887–890.

Rasmussen, S. O., K. K. Andersen, A. M. Svensson, J. P. Steffensen, B. M. Vinther, H. B. Clausen, M.-L. Siggaard-Andersen, S. J. Johnsen, L. B. Larsen, D. Dahl-Jensen, M. Bigler, R. Röthlisberger, H. Fischer, K. Goto-Azuma, M. E. Hansson, and U. Ruth. 2006. A new Greenland ice core chronology for the last Glacial Termination. *Journal of Geophysical Research* 111: D06102. doi:10.1029/2005JD006079

Reimer, P. J., M. G. L. Baillie, E. Bard, A. Bayliss, J. W. Beck, P. J. Blackwell, C. Bronk Ramsey, C. E. Buck, G. S. Burr, R. L. Edwards, M. Friedrich, P. M. Grootes, et al. 2009. IntCal09 and Marine09 radiocarbon age calibration curves, 0–50,000 years cal BP. *Radiocarbon* 51: 1111–1150.

E. J. Rohling, P. Mayewski, R. Abu-Zied, J. Casford, and A. Hayes. 2002. Holocene atmosphere-ocean interactions: records from Greenland and the Aegean Sea. *Climate Dynamics* 18(7): 587–593.

Rohling, E. J., and H. Pälike. 2005. Centennial-scale climate cooling with a sudden cold event around 8,200 years ago. *Nature* 434: 975–979.

Rohling, E. J., R. Abu-Zied, J. L. Casford, A. Hayes, and B. A. A. Hoogakker. 2009. The marine environment: present and past. In *The Physical Geography of the Mediterranean*, ed. J. Woodward. Oxford: Oxford University Press, 33–68.

Rollefson, G., and Z. Kafafi. 1994. The 1993 season at 'Ain Ghazal: preliminary report. *Annual of the Department of Antiquities of Jordan* 38: 11–32.

Rollefson, G. O. 2009. Slippery slope: the late Neolithic rubble layer in the southern Levant. *Neo-Lithics* 1/09: 12–18.

Tasić, N. 1988. Comparative C-14 dates for the Neolithic settlements in Serbia. In *The Neolithic of Serbia: Archaeological Research 1948–1988*, ed. D. Srejović. Belgrade: University of Belgrade.

Tsuneki, A. 2010. A newly discovered Neolithic cemetery at Tell el-Kerkh, northwest Syria. In P. Matthiae, F. Pinnock, L. Nigro, and N. Marchetti (eds), *Proceedings of the 6th International Congress on the Archaeology of the Ancient Near East, Vol. 2*. Wiesbaden: Harrassowitz Verlag, 697–713.

Tubi, A., and U. Dayan, U. 2012. The Siberian high: teleconnections, extremes and association with the Icelandic Low. *International Journal of Climatology* 33: 1357–1366.

Umurtak, G. 2007. Die jungsteinzeitlichen Siedlungen im südwestanatolischen Seengebiet. In C. Lichter (ed.), *Vor 12.000 Jahren in Anatolien: Die Ältesten Monumente der Menschheit. Begleitbuch zur Großen Landesausstellung, im Badischen Landesmuseum Schloss, Karlsruhe*. Stuttgart: Konrad Theiss, 139–149.

van der Plicht, J., P. Akkermans, O. Nieuwenhuyse, A. Kaneda, and A. Russell. 2011. Tell Sabi Abyad, Syria: Radiocarbon chronology, cultural change, and the 8.2 Ka event. *Radiocarbon* 53(2): 229.

Weiss, H. 2000. Beyond the Younger Dryas: collapse as adaptation to abrupt climate change in ancient West Asia and the eastern Mediterranean. In G. Bawden and M. Reycraft (eds), *Environmental Disaster and the Archaeology of Human Response*. Anthropological Papers 7. Albuquerque, NM: Maxwell Museum of Anthropology, 75–98.

Weninger, B., E. Alram-Stern, E. Bauer, L. Clare, U. Danzeglocke, O. Jöris, C. Kubatzki, G. O. Rollefson, H. Todorova, and T. van Andel. 2006. Climate forcing due to the 8200 cal BP event observed at early Neolithic sites in the eastern Mediterranean. *Quaternary Research* 66: 401–420.

Weninger, B., L. Clare, E. J. Rohling, O. Bar-Yosef, U. Böhner, M. Budja, M. Bundschuh, A. Feurdean, H. G. K. Gebel, O. Jöris, J. Linstädter, P. Mayewski, T. Mühlenbruch, A. Reingruber, G. O. Rollefson, D. Schyle, L. Thissen, H. Todorova, and C. Zielhofer. 2009. The impact of rapid climate change on prehistoric societies during the Holocene in the eastern Mediterranean. *Documenta Praehistorica* 36: 7–59.

Weninger, B. Clare, L., Gerritsen, F., Horejs, B., Krauß, R., Linstädter, J., Özbal, R., and Rohling, E.J., 2014. Neolithisation of the Aegean and Southeast Europe during the 6600-6000 cal BC period of Rapid Climate Change. Documenta Praehistorica XLI, 1–31.

Weninger, B., and O. Jöris. 2008. A ^{14}C age calibration curve for the last 60 ka: the Greenland-Hulu U/Th timescale and its impact on understanding the Middle to Upper Paleolithic transition in Western Eurasia. *Journal of Human Evolution* 55: 772–781.

Zielhofer, C., L. Clare, G. O. Rollefson, S. Wächter, D. Hoffmeister, G. Bareth, C. Roettig, H. Bullmann, B. Schneider, H. Berke, and B. Weninger. 2012. The decline of the early Neolithic population centre of 'Ain Ghazal and corresponding earth surface processes, Jordan Rift Valley. *Quaternary Research* 78: 427–441.

| 4.2 ka BP Megadrought and
the Akkadian Collapse

HARVEY WEISS

*The Akkadians, of southern Mesopotamia, created the first empire ca. 2300
BC with the conquest and imperialization of southern irrigation agriculture
and northern Mesopotamian dry-farming landscapes. The Akkadian Empire
conquered and controlled a territory of roughly 30,000 square kilometers and,
importantly, its wealth in labor and cereal crop-yields. The Empire maintained
a standing army, weaponry, and a hierarchy of administrators, scribes,
surveyors, craft specialists, and transport personnel, sustainable and profitable
for about one hundred years. Archaeological excavations indicate the empire
was still in the process of expansion when the 2200–1900 BC/4.2–3.9 ka
BP global abrupt climate change deflected or weakened the Mediterranean
westerlies and the Indian Monsoon and generated synchronous megadrought
across the Mediterranean, West Asia, the Indus, and northeast Africa. Dry-
farming agriculture domains and their productivity across West Asia were
reduced severely, forcing adaptive societal collapses, regional abandonments,
habitat-tracking, nomadization, and the collapse of the Akkadian Empire.
These adaptive processes extended across the hydrographically varied landscapes
of west Asia and thereby provided demographic and societal resilience in the
face of the megadrought's abruptness, magnitude, and duration.*

The Physical Setting

Mesopotamia is the lowland drainage, alluvial plain of the Tigris and
Euphrates Rivers and extends from the Anatolian plateau, source of the
Tigris and Euphrates Rivers, to the Persian Gulf (fig. 3.1). The Tigris and the
Euphrates are fed by the winter precipitation of the cyclonic Mediterranean
westerlies, which also provide 200 to 500 millimeters of precipitation along
the southern edge of the plateau where the alluvial plain begins, with an
interannual variability of about 20 percent. Hence, high-yield cereal dry
farming is practiced within the valleys of Anatolia and extensively across
the lowland plains that extend from the base of the plateau to the limits of

dry-farming precipitation at the 200 to 300 millimeter isohyet. The region extends some 150 to 200 kilometers southward from the base of the plateau and encompasses northern Mesopotamia, including northeastern and north-western Syria (Wirth 1971).

As the Tigris and Euphrates drop from the Anatolian plateau onto the northern Mesopotamian plains below, they incise their riverbeds to tens of meters and become inaccessible for plain-level irrigation. However, as the rivers course through their near conjunction at ancient Sippar and modern Baghdad, the elevation of the plains gradually drops to just 37 meters above sea level and slows their flow. Still farther south, where precipitation is well below 300 millimeters per annum, the rivers begin to approach plain level and become available for irrigation agriculture (Buringh 1960). Southern irrigation agriculture cereal yields are one-and-a-half to two times greater per unit cultivated than northern dry-farming yields, but they are limited in extent to thin ribbons of canal-watered fields (Weiss 1986). Mesopotamia's agricultural setting, then, encompasses three regions: extensive lowland dry farming to the north, intensive irrigation agriculture to the south, and the semi-arid steppe between that serves the seasonal transhumance of the region's pastoral nomadic populations and their flocks.

Early Irrigation Agriculture

Abrupt century-scale megadroughts occurred across West Asia at nearly millennial intervals of the Holocene and, by their abruptness, magnitude, and duration, severely affected the productivity, socioeconomic sustainability, and social and economic interactions of the three Mesopotamian regions' populations. The earliest abrupt climate change occurred at 8.2 ka BP and was a two-century long global cold and dry period, notable culturally for Anatolian and southeastern European neolithization (Weninger & Clare, ch. 2, this volume). The impact of the 8.2 ka BP event in Mesopotamia, apart from adaptive responses at such sites as Sabi Abyad in the Balikh River drainage (van der Pflicht et al. 2011), can only be suggested at this time, as little fieldwork in Mesopotamia has recently been devoted to this period. Nevertheless, it was during this period that the enigmatic early settlement of southern Mesopotamia occurred, to judge from the radiocarbon dates at Tell Oueilli (Valladas, Evin, & Arnold 1996), suggesting that the two-century megadrought may have pushed central Mesopotamian dry farmers to the refugium of southern irrigation-agriculture domains, the riverine area extending south of Baghdad to the head of the Persian Gulf (Staubwasser & Weiss 2006).

The subsequent Ubaid period (ca. 6500–3800 BC) saw the growth of villages and small towns where the irrigation agriculture was controlled by small, temple-centered, chiefdom-level societies with relatively little centralization of agricultural surplus. Early developments in the Uruk period (ca. 4000–3000 BC) appear to mark the transition from chiefdom to state, with social stratification and an urbanized landscape. By 3500 BC urban Late Uruk society flourished in Sumer, southernmost Mesopotamia, and adjacent irrigation realms

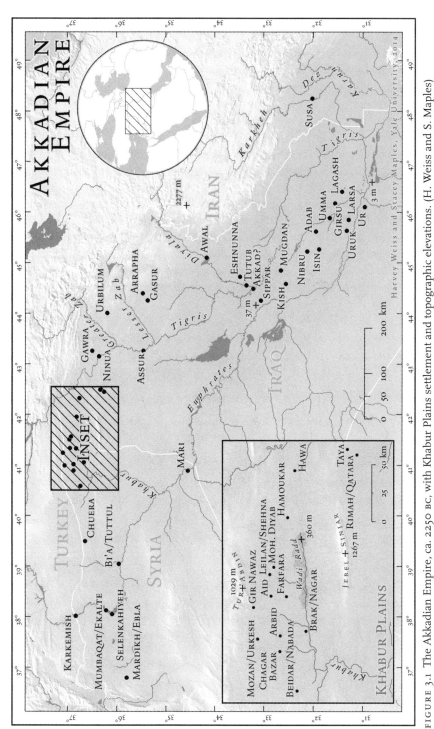

FIGURE 3.1 The Akkadian Empire, ca. 2250 BC, with Khabur Plains settlement and topographic elevations. (H. Weiss and S. Maples)

that extended south from the area of modern Baghdad. This early "urban revolution" comprised Sumerian cities as large as 250 hectares, or approximately 50,000 persons (Finkbeiner 1991), sustained by high-yield cereal irrigation agriculture and low-cost harvest transport via canals (Weiss 1986; Hruška 2007; Algaze 2008).

In the Late Uruk period, increasingly complex exchanges in Sumer between urban institutional managers and their agricultural workers were recorded in accounts rendered with numerical and pictographic notations (Nissen, Damerow, & Englund 1993). At the same time, Late Uruk "colonists" (perhaps long-distance traders pursuing exotic materials) continued a Middle Uruk tradition of settling dispersed communities across the dry-farming plains and plateau valleys of adjacent Iran, Anatolia, and Syria (Petrie 2014; Rothman, ed. 2001). A few northern Mesopotamian towns under dry-farming regimes seem to have grown during this period into ca. 100 hectare, arguably urban, settlements (Brustolon & Rova 2007; Oates et al. 2007), with one even growing to 300 diffuse hectares (al-Quntar et al. 2011).

These Uruk-period colonies and the few large northern settlements, however, were suddenly depopulated or abandoned at ca. 3200–3000 BC. A similar quick retraction or consolidation occurred in southern cities and towns (Postgate 1986). No explanation of this occurrence has been forthcoming but for its coincidence with a severe megadrought at ca. 3200–3000 BC/5.2–5.0 ka BP (Weiss 2003; Charles, Pessin, & Hald 2010). Intriguing as well, but essentially unexplored, is the coincidence of the megadrought and Late Uruk collapse with major social and political innovations in southern Mesopotamia at this time. Reduced Euphrates flow may have generated considerable social and institutional stress and reinforced population agglomeration at some urban sites (Adams 1981; Staubwasser & Weiss 2006). Among the most significant developments was the collapse of the temple authority that regulated Ubaid and Uruk urban society for three thousand years and its innovative replacement by secular, palace-based authorities that now owned and controlled all city-state land and agricultural production (Visicato 2000).

Precipitation rose again by 2800 BC, and in southernmost Sumerian Mesopotamia some cities grew to more than 300 hectares during the Early Dynastic period (ca. 2900–2350 BC). Judging from their temples, palaces, and cemeteries and their highly urbanized riverine landscapes (Adams 1981), these cities reached an apogee of regional extractive wealth accumulation by ca. 2600 BC. For uncertain reasons, Northern Mesopotamian dry-farming landscapes did not undergo this early urban growth and remained small, dispersed villages and towns during the early third millennium—that is, the early Ninevite 5 period, ca. 2900–2600 BC (Weiss 2003).

The Second Urban Revolution

Also yet to be explained is the sudden development of large urban centers at around 2600–2500 BC across the dry-farming landscapes of northern

Mesopotamia and western Syria (Weiss 1990; Akkermans & Schwartz 2008). Emulating the administrative technologies and iconographies of contemporary southern Sumerian cities, many of these northern and western cities grew to the 90 to 120 hectare range, with surrounding towns and villages, and were arrayed evenly across the dry-farming plains. Their pattern of settlement, with dependent village distributions, suggests that the cities were located to maximize the high-yield cereal agriculture potential of the extensive dry-farming plains available to them at such sites as Taya, Nineveh, and Erbil on the Assyrian plains and Leilan, Mozan, Brak, and Hamoukar on the Khabur Plains. To the Sumerians in the south, this region as a whole was likely known as Subir. The region's inhabitants, as we know from the Ebla and Tell Beidar cuneiform archives, spoke Semitic languages and used the southern Mesopotamian cuneiform writing system initially developed to record Sumerian (Sallaberger & Pruss 2015). The agricultural wealth and potential of these dry-farming cities, with populations already highly organized for agricultural production, were soon to be the target of nascent southern Mesopotamian imperialism.

Akkadian Imperialization and Collapse

In the early 24th century BC, a period of warring among southern city-states terminated with the ascent of one "lord of the land." Lugalzaggisi, the king of the city-state of Umma, emerged from decades-long battles to control many, if not all, of Sumer's other city-states, including Nibru, Adab, and Uruk—the first, dozens of kilometers distant from Umma (Almamori 2014). The few known surrounding events include Lugalzaggisi's conquest of Mari on the central Euphrates and royal travel as far as the Mediterranean Sea.

Lugalzaggisi's rule gave way in the immediately succeeding decades to a quantitative and qualitative leap in Sumer's supremacy. The next ruler, according to the Sumerian King List, was Sargon, the founder of a five-generation Akkadian dynasty that created a capital city at Akkad (or Akkade), likely near Sippar (but still unlocated) and spoke and wrote the early Semitic language Akkadian. Within two generations the Akkadians embarked upon an imperial venture that was exponentially more extensive and extractive than envisioned by Lugalzaggisi's "lord of the land." At ca. 2200 BC, only about one hundred years after its launch and full-blown development, this first imperial effort, the Akkadian Empire, was truncated by natural forces at the 4.2 ka BP megadrought. Yet in spite of, or even perhaps because of, this abrupt rupture, Akkadian imperial successes and ideology were emulated and venerated by succeeding empires for the next thousand years. Epigraphic and archaeological data document three stages in the establishment of Akkadian imperial power, marked by the rules of Sargon, his sons Rimush and Manishtushu, and his grandson Naram-Sin, dated in Table 3.1 using the "Middle Chronology," with a range of ca. 60 years (Sallaberger & Schrakamp 2015).

TABLE 3.1 Stages of Akkadian Imperial Power

Stage 1	Sargon (ca. 2324–2285 BC) extended the united realm from Akkad in the north to Ur in the south—ca. 5000 square kilometers—and embarked upon a series of long-distance conquests up the Euphrates, to Mari, Tuttul, and Ebla, respectively ca. 500, 700, and 900 kms distant). This was a departure from regional southern Mesopotamian city conquest, and a qualitative leap from both Mari's and Ebla's more limited efforts towards regional hegemony. Little apparent state building followed Sargon's conquests: with no controlling fortresses left in the conquered cities, they provided only the "primitive", immediate, acquisition of plunder.
Stage 2	Rimush and Manishtushu (ca. 2284–2262 BC), in short order, extended Akkadian imperial power into northern Mesopotamia. This stage marks the beginning of Akkadian designs on the dry-farming cities of the north, at Tell Brak, Nineveh, and Tell Leilan—where the earliest Akkadian texts of the Khabur Plains were retrieved within an enigmatic Akkadian scribal room (deLillis Forrest et al. 2007).
Stage 3	Naram-Sin (ca. 2261–2206 BC), Sargon's grandson, took another qualitative leap, conquering the urbanized landscapes of adjacent dry-farming regions in southwestern Iran, northeastern Iraq, and northeastern Syria. The most detailed examples of this conquest and imperialization process have been retrieved recently on the Khabur Plains, where the Akkadians installed themselves within palace fortresses at Leilan, Mozan, and Brak and ruled the dry-farming lands extending north across Mesopotamia to Taya, Nineveh, Awal, and Gasur/Kirkuk and east to Susa. Henceforth, the Akkadians extracted and deployed revenues from both the rain-fed and the irrigation-agriculture regions of Mesopotamia. This imperialization across Mesopotamia, recorded in Akkadian provincial archives (see, for example, Visicato 1999; Foster 1982; Brumfield 2013), also marked the deployment of a new imperial metrology in the collection of both raw commodities and finished products (Glassner 1986; Powell 1990) as well as land surveying and agrimensorial innovations (Foster 2011; Høyrup 2011).

In southern Mesopotamia the Akkadians created enormous new estates from domains seized or purchased, and then extracted revenues for the empire's administrators and elite officialdom and for imperial projects, such as new temple construction and maintenance of the empire's armies (Westenholz 1987, 1999). The large imperial revenues were dispatched to the capital by water transport: as one record attests, tow barges from southern Adab were loaded with 885,000 liters, or about 539 metric tons, of barley (Maiocchi 2009: 78).

The revenues of the northern dry-farming areas were gathered and retrieved from imperialized regional cities, a deployment made possible in no small part by the state's ability to exploit a large, dependent labor force. At Gasur, near Kirkuk, for instance, agricultural workers were provided with barley rations (*še-ba*) for their agricultural labor, while estate harvests were directed, as in the south, to the local Akkadian administration (Foster 1982), where skilled artisans, as well as dependent and levied labor, were compensated with rations measured to imperial standards (Westenholz 1987). On the Khabur Plains, an Akkadian bulla retrieved from a small, 5-hectare town at Chagar Bazar, alongside the now dry wadi Khanzir, likely records the water-borne transport of as much as 160 gur, or 379 metric tons, of emmer (Chagar Bazar A. 391, Brumfield 2012). One imperial Akkadian account records more than 45,000 liters, or over 30 metric tons, of barley and emmer, probably transported by river from 50-hectare Nagar/Tell Brak, on the Khabur Plains, to Sippar near Baghdad (Sommerfeld, Archi, & Weiss 2004; cf. Englund 2015) In the dry-farming areas, imperial targets were the urban-dominated agricultural landscapes, which comprised a pre-adaptation for Akkadian imperialism. Northern imperialization targets, other than cereal production, are difficult to identify, since the northern plains lack other extractable resources and the Akkadian imperial fortresses did not extend to the adjacent Anatolian and Iranian plateaus' potential precious metal sources—even though silver and cedar were famously retrieved from the Amanus mountains far to the northwest. In fact, the Akkadians ignored the Hakkari sources on the plateau and obtained their copper from Oman and Kerman (Potts 2007).

The 4.2 ka BP Megadrought

In the midst of imperial success and expansionary activity, at ca. 2230 BC in the Leilan radiocarbon chronology, probably in the reign of Shar-kali-sharri, Naram-Sin's son and successor, the abrupt onset of the 4.2 ka BP global megadrought desiccated the dry-farming agricultural landscape of the Mediterranean, West Asia, and northern Mesopotamia with 30–50 percent precipitation reductions and colder temperatures. The chronology and global extent of this abrupt climate change are now well documented and frame the collapse of the Akkadian Empire.

Much has been gleaned about the 4.2 ka BP event from climate science, which traces the direction and intensity of the cyclonic North Atlantic westerlies that are controlled by the North Atlantic Oscillation (Kushnir & Stein 2010; Cullen et al. 2002) and delivered through the Mediterranean trough to West Asia (Lionello, Malanotte-Rizzoli, & Boscolo 2013). The North Atlantic Oscillation's boundaries are reflected in abundant and synchronous Mediterranean westerlies proxy records and the yet inexplicably linked Indian Monsoon paleoclimate proxy records that are plotted in figure 3.2 within the winter season moisture transport. Two high-resolution data sets define the 4.2 ka BP event's chronology and magnitude: (1) Icelandic lake sediment records (Geirsdóttir et al. 2013; Blair, Geirsdóttir, & Miller 2015), and (2) a Greenland lake-sediment record linked to the North Atlantic Oscillation-index and derived from tree-ring and speleothem records (Olsen, Anderson, & Knudsen 2012).

Most of these paleoclimate proxies document the 4.2–3.9 ka BP abrupt climate change, but some anomalous records exist. These anomalies include Jeita Cave, Lebanon (54), surrounded by prominent 4.2 ka BP event proxies, Sofular Cave (110) at the Black Sea, the only Anatolian region that receives precipitation throughout the year, Qunf Cave (77) in the Intertropical Convergence Zone, and Lake Bosumtwi (86), situated between the well-documented 4.2 ka BP event proxies at Lake Yoa (76) and the Gulf of Guinea (96). A few older, low-resolution proxies also do not display the 4.2 ka BP event, among them Bouara, Syria (53), where there may be an analysis error, and the older Greek proxy records surrounded by new high-resolution speleothem and marine core records, such as the Alepotrypa cave speleothem on the Peloponnese (114). A few Anatolian lake proxies, such as Söğütlü Marsh (31) and the poorly dated Eski Açigöl core (33) also miss the event, though they are surrounded by recent high-resolution Anatolian proxies with prominent 4.2 ka BP event proxy excursions, such as Nar Lake (112) and Gulf of Gemlik (113) and the high-resolution speleothem core at Gol-e Zard, Iran (111). The 4.2 ka BP event Mediterranean westerlies proxies, usually ^{14}C and U-Th isotope dated, extend across the seven sub-regions in Table 3.2 and are listed by number in the Appendix.

The 4.2–3.9 ka BP event is now well documented in East African and Indus paleoclimate proxies (fig. 3.2), which detail synchronous abrupt alterations for both Nile flow and Indus precipitation and river flow, functions of the Indian Monsoon as it passes across the Arabian Sea between the sub-continent and the Horn of Africa. Marine, lake, and speleothem cores indicate that the 4.2 ka BP event disruption of the Indian Monsoon (Berkelhammer et al. 2012; Dixit, Hoddell, & Petrie 2014; Prasad et al. 2014) was approximately coincident with disrupted Harappan urbanization along the Indus River, though temporal details await refinement (Ponton et al. 2012), while mitigating cropping strategies may have prevailed in dry-farming regions (Petrie et al. 2016). The weakening of the Indian Monsoon also diminished northeast African precipitation (Marshall et al. 2011; Revel et al. 2014; Davis & Thompson 2006) and consequent Nile flow (Blanchet et al. 2013; Hassan & Tassie 2006; Bernhardt, Horton, & Stanley 2012; Welc & Marks 2014; Revel et al. 2014), coincident with the collapse of the Old Kingdom in Egypt and introduction of the First Intermediate Period (Ramsey et al. 2010).

In Africa, West African and Saharan precipitation were also disrupted (Marchant & Hooghiemstra 2004), as at Lake Yoa (76) and Jikariya Lake (81)—an aridification and dust event that terminated the African Humid Period (Lézine 2009) and likely created the sources of synchronous African dust in Tuscany (17). The same megadrought event is observed in central West African lake cores (90) and Gulf of Guinea marine cores (96). North to south, the event is recorded from coastal Algeria (95) to the 30-degree latitude in southern Africa (Chase et al. 2010; Schefuss et al. 2011).

The abundant West Asian proxy records are also linked to central Asia, the Himalayas (Nakamura et al. 2016), and Mongolia (Yang et al. 2015). In southeastern

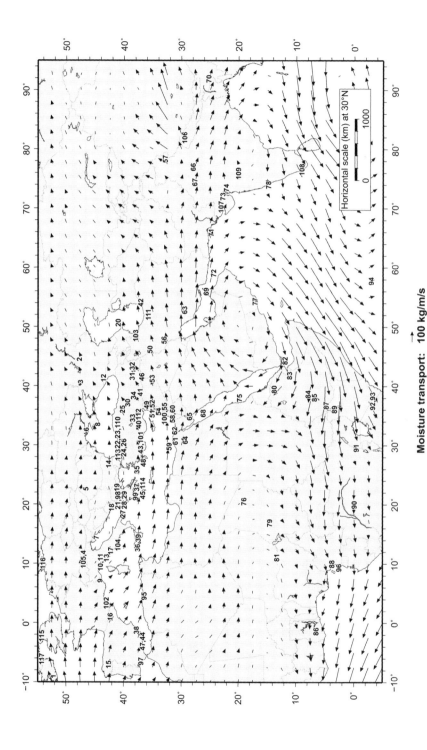

Moisture transport: 100 kg/m/s

FIGURE 3.2 Paleoclimate proxy sites for 4.2 ka BP event with NOAA Moisture Vectors: Europe, Mediterranean, North Africa, West Asia, and Indus. The vertically integrated moisture transport, as estimated for December 1949 through February 2014, from the National Centers for Environmental Prediction—National Center for Atmospheric Research Reanalysis. The units are in kg/m/s, with the reference vector shown at bottom. (Moisture vectors map by R. Seager)

TABLE 3.2 Mediterranean Westerlies 4.2 ka BP event paleoclimate proxy records. (Numbers refer to figure 3.2 and the Appendix.)

Coastal Spain and France
Doñana National Park (97): Sierra de Gador (47); Cova da Arcoia (15); Borreguiles de la Virgen (44); Lake Montcortès (16); Puerto de Mazzarón (38); Lac Petit (9).

Central Mediterranean Italian lakes
Lago di Pergusa (39); Bucca della Renella (10); Lago Alimini Piccolo (27); Maar lakes (17); Lake Accessa (13); Lago Preolo (36); Corchia Cave (11).

Greece and the Balkans
Lake Lerna (37); Osmananga Lagoon (Pylos) (45); Lake Vrana (7); Lake Prespa (98): Leng et al. 2010; Lake Shkodra (18); Lake Ohrid (21); Lake Dojran (19); Aegean Sea (35); Kos basin, south Aegean Sea (48); Kotychi Lagoon (99): Haenssler et al. 2014; Rezina Marsh (28); Gramousti Lake (29).

Levant and Red Sea
Acre (100): Kaniewski et al. 2014; Tweini (52); Dead Sea (60); Zeelim, Dead Sea; Sedom, Dead Sea; Soreq Cave (58); Shaban Deep (68); northern Red Sea and Gulf of Aqaba (65); central Red Sea (75); Lake Hula (55); Jeita Cava (54); Ghab Valley (51); Tell Mardikh (49).

Anatolian plateau and northern Mesopotamia
Konya lakes (40); Göl Hissar Gölü (43); Eski Açigol (33); Koçain Cave (101): Göktürk 2011; Abant Gölü (23); Yeniçaga Gölü (22); Sogutlu Marsh (31); Nar Lake (112); Lake Iznik (24); Kaz Gölü (25); Gulf of Gemlik (113); Yenişehir (26); Arslantepe (34); Göbekli Tepe (41); Lake Van (32); Lake Tecer (30); Tell Leilan (46).

Persian Gulf
Gulf of Oman (72); Awafi (69); Qunf Cave (77).

Black Sea, Caspian Sea, Iranian Plateau
Lake Zeribar (50); Lake Mirabad (56); Lake Maharlu (63); Black Sea (8); Sofular Cave (110); Caspian Sea (20); southeastern Caspian Sea (42); Gol-e Zard Cave (111).

Tibet, millet crop agriculture persisted to 4.2 ka BP, when drier and colder conditions forced regional settlement abandonment until the adoption of wheat-barley agriculture ca. 3.5 ka BP, while the arrival of wheat and barley ca. 4 ka BP on the northeastern Tibetan plateau allowed for uninterrupted occupation (d'Alpoim Guedes et al. 2016; Wang et al. 2015). In eastern China, numerous 4.2–3.9 ka BP records document East Asian Monsoon instabilities that disturbed late Neolithic settlement systems (Cai et al. 2010; Donges et al. 2015; Dykoski et al. 2005; Liu and Feng 2012; Lu et al 2015) and extended to interruption of the Indonesian-Australian Summer Monsoon (Rosenthal, Linskey & Oppo 2013; Deniston et al. 2013).

In North America, seven glacial, speleothem, and lake-core proxy records of the event cross the continent from New Jersey to the Yukon (Dean 1997; Zhang & Hebda 2005; Booth et al. 2005; Li, Yu, & Kodama 2007; Fisher 2011; Hardt et al. 2010; Menounos et al. 2008). Additionally, now available is the annual-resolution Great Basin tree-ring record (Salzer et al. 2014), which documents the 4.2 ka BP event at the introduction of maize agriculture to the US southwest (Merrill et al. 2009) and the Yucatan (Torrescano-Valle & Islebe 2015). In South America, the well-known glacial record (Davis & Thompson 2006) is now supplemented with other Andean proxy records (Baker et al. 2009; Licciardi et al. 2006, 2009; Schittek et al. 2015) to suggest a causal linkage between 4.2 ka BP and the poorly understood rise and fall of contemporary Peruvian Late Pre-ceramic cities (Sandweiss et al. 2009). 4.2 ka BP proxy records extend southwards to 44° S in Chilean Patagonia (dePorras et al. 2014) and Antarctic glacial cores (Peck et al. 2015).

The Multi-Proxy Stack

The multi-proxy stack (fig. 3.3) portrays several currently available high- to low-chronological resolution paleoclimate proxies at 4.2–3.9 ka BP across the Mediterranean and West Asia (Weiss et al. 2012; Walker et al. 2012). Linear interpolation across uranium-thorium (U/Th) or radiocarbon (^{14}C) dated points is provided for measured quantities of precipitation and temperature proxies, such as stable isotopes, arboreal and other pollen, diatoms, carbonates, lake levels, and magnetic susceptibility. The ranges of chronological resolution in each record are reproduced here within two standard deviations around 4.2–3.9 ka BP. The standard deviation bars illustrate two important qualities of the 4.2 ka BP paleoclimate record. First, the dating is quite variable, extending from very low to very high resolution. Second, chronological resolution is dependent not only upon sampling and dating intervals, but on the standard deviation of the radiometric datings. For comparative purposes, the coincident high-resolution proxies at Mawmluh Cave (Berkelhammer et al. 2012) and Mount Logan (Fisher 2011) represent global records of the 4.2 ka BP event (Walker et al. 2012).

At Lac Petit, France, in the southern Alps, an abrupt detrital pulse triggered by more intense or more frequent rainfall marks a major shift in diatom assemblages at 4300–4100 BC, according to ^{14}C dating (Brisset et al. 2013). The Koçain Cave, Turkey, speleothem (Göktürk 2011) provides high-resolution U/Th dates that constrain abrupt decreases and increases of δ^{18}O (oxygen isotope ratio). The Eski Açigöl, Turkey, lake core (Roberts et al. 2001) has no radiocarbon dates during a rise in lake-core charcoal misinterpreted as anthropogenic deforestation (Turner, Roberts, & Jones 2008), while the Göl Hissar, Turkey, lake-core carbonate spike and rise in δ^{18}O are framed by radiocarbon dates 2000 years apart (Eastwood et al. 2007). The Lake Van, Turkey, core (Lemcke & Sturm 1997) displays a quartz spike understood as a dust proxy and is dated by varve counts with slight errors (see Kuzucuoğlu et al. 2011). The dense sampling intervals for the Soreq Cave, Israel, speleothem (Bar-Matthews & Ayalon 2011) δ^{18}O and δ^{13}C values are linked to U/Th dates, but with large standard deviations and, therefore, a labile chronology. The Dead Sea lake levels (Kagan et al 2015; Litt et al. 2012; Migowski et al. 2006) are estimated to have dropped abruptly by 45 meters at ca. 4.2 ka BP (see Frumkin 2009). At the Red Sea Shaban Deep core (Arz, Lamy, & Pätzold 2006), 15-year diatom sampling

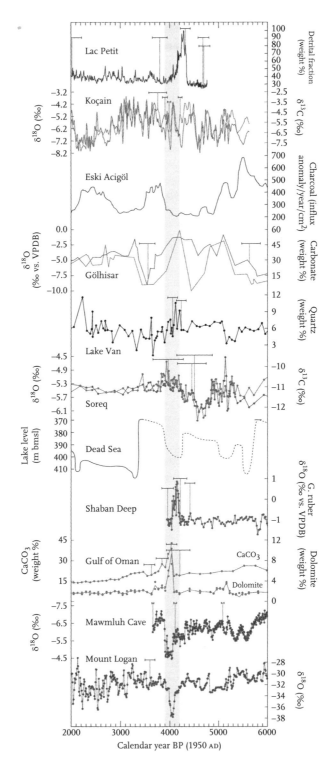

FIGURE 3.3 Multi-proxy stack of Mediterranean westerlies and related paleoclimate proxies displaying the 4.2 ka BP abrupt climate change event within marine, lake, speleothem, and glacial records of varying chronological resolution, with 2-standard deviations dating indicated (Weiss et al. 2012; Walker et al. 2012). (H. Weiss & M. Besonen)

intervals are constrained by high-resolution radiocarbon dates with a marine reservoir correction (see Edelman-Furstenberg, Almogi-Labin, & Hemleben 2009). The Gulf of Oman marine core (Cullen et al. 2000) has dolomite and calcium carbonate (dust) spikes framed by ^{14}C dates and is tephra-linked to Tell Leilan chronostratigraphy (Weiss et al. 1993).

The Mawmluh Cave, India, speleothem (Berkelhammer et al. 2012) provides 6-year $\delta^{18}O$ sampling intervals constrained with U/Th dates and links the Nile flow reductions (see Hassan & Tassie 2006; Bernhardt, Horton, & Stanley 2012), east African lake-level reductions (see Gasse 2000), and the Indian Summer Monsoon deflection (see Dixit, Hodell, & Petrie 2014). The Mount Logan, Yukon, glacial core (Fisher 2011) is cross-dated with the NorthGRIP-core and dated tephra records and exemplifies the 4.2 ka BP event's North American expression, second in magnitude to the 8.2 ka BP event. Illustrative of the abrupt climate-change synchronicity in the Mediterranean are the Lac Petit, France, core (9), and other recent and adjacent cores in Spain, Lake Shkodra, Albania (18), Lake Accessa, Italy (13), and Acre, Israel (100), dated ca. 4.2–3.9 ka BP, or, in some cases, at lower resolution, ca. 4.3–3.8 ka BP (see Figure 3.2, Table 3.2, and Appendix 3.1).

Effects of the 4.2 ka BP Megadrought

The effects of the global 4.2 ka BP abrupt climate change varied regionally and climatically, of course, and the integration of these extensive data comprises a major new research task. In West Asia and northern Mesopotamia, the dry-farming areas contracted when precipitation dropped 30–50 percent at ca. 4.2 ka BP (Bar-Matthews & Ayalon 2011; Frumkin 2009; Staubwasser & Weiss 2006: 380–382, figs. 4–6.) Regional aggregate cereal yields plummeted, and most of the Khabur Plains of northeastern Syria fell below the minimal 200–300 millimeter isohyet necessary for dry farming (fig. 3.4). Similar drought conditions prevailed across the Mediterranean, western Syria, and northern Iraq, regions in which precipitation was a function of the same Mediterranean westerlies, and paleoclimate records indicate that the megadrought extended across the Mediterranean to Anatolia, the proximate source of northern Mesopotamian winter precipitation.

The Akkadian Empire's investment in the conquest, control, and manipulation of northern Mesopotamia had included a standing army, weaponry, and a hierarchy of administrators, scribes, surveyors, craft specialists, and transport personnel across a territory of roughly 30,000 square kilometers. This imperial system had proven both sustainable and profitable for about one hundred years (Ristvet 2012). The megadrought, however, eliminated dry-farming cereal cultivation across the Khabur Plains and the north Mesopotamian and Syrian plains to the east as well as the west. The flow of northern imperial agricultural and finished product levies to provincial centers and to the Akkadian capital terminated.

The effects of a 30–50 percent reduction in Tigris-Euphrates flow upon southern Akkadian agriculture can only be estimated. Although the flow always seems to have exceeded the demands of irrigation agriculture, such a reduction would have substantially diminished canal extent and irrigated field areas. At the onset of the megadrought, with reduced Euphrates flow, aggregate Akkadian yields likely fell precipitously. In the course of the megadrought flow reduction, the successor Ur III state redesigned canal systems into linear paths in an attempt to counter channel meandering (Adams 1981: 164). However, in spite of considerable epigraphic documentation for southern Akkadian and Ur III agriculture, it has not yet proven possible to convincingly compare Akkadian and Ur III period agricultural production (van de Mieroop 1999: 125).

Although its economy remains to be quantified, the Empire's collapse was swift. That the capital city, Akkad, has neither been located nor excavated remains a central challenge to our understanding of events in the Akkadian heartland, but the extant cuneiform record is itself graphic in this instance. "Who was king, who was not king?" records the Sumerian King List, perhaps written at the time or shortly thereafter, when "Akkade was defeated and kingship was taken to Uruk" (Black et al. 2004). The "Curse of Akkade" poetically describes the populace's drought-stricken wails when food was scarce and the "canal bank tow-paths' grass grew long," trailed by the invading Gutian mountain neighbors, "with the brains of dogs and the faces of apes," who were brought by divine force to conquer Akkad (Cooper 1983; Black et al. 2004). A series of petty kings followed. Ur III dynasty successors would rule irrigation-agriculture southern Mesopotamia for a hundred years, but they never reclaimed the abandoned, drought-stricken northern realms. Three hundred and fifty years after the collapse, the fall of Akkad was still an exalted event (Grayson 1987: 53). The collapse of the Empire is unquestioned; nevertheless, its demise was ecologically more complicated than "The Fall of Akkad" epigrammatically suggests.

Ecological Variability of Collapse

Adaptive responses to megadrought varied within the ecological zones across West Asia, the dry-farming zone, the riverine irrigation-agriculture zones, and the semi-arid steppe. In the dry-farming zone, the 30–50 percent reduction in precipitation made dry-farming impossible, and region-wide site abandonments quickly followed. These abandonments are most visible now in the excavated sites and regional surveys on the Khabur Plains of northeastern Syria and in the plains of southwestern Turkey, western Syria, and the Levant (fig. 3.4).

Khabur Plains

The collapsed Akkadian Empire's abandonment of the Khabur Plains was swift and sudden, and most of the indigenous regional population departed with the Akkadians. Three major urban settlements—Brak, Leilan, and Hamoukar— and their surrounding towns and villages were abandoned synchronously and completely, while a fourth major settlement at Mozan was 80 percent deserted (Buccellatti & Buccellatti 2000; Pfälzner 2012). These Khabur Plains

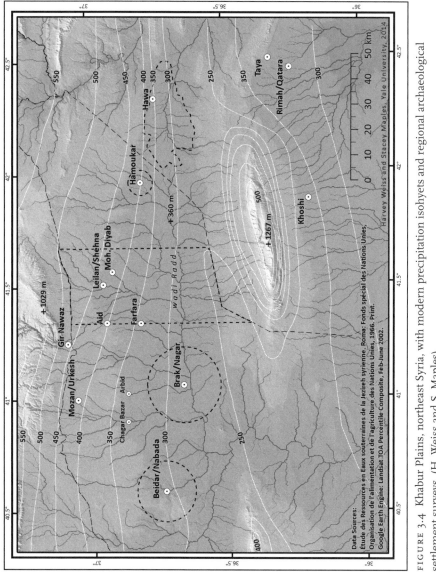

Data Sources:
Étude des Ressources en Eaux souterraines de la Jezireh syrienne. Roma: Fonds spécial des Nations Unies,
Organisation de l'alimentation et de l'agriculture des Nations Unies, 1966. Print.
Google Earth Engine: Landsat TOA Percentile Composite, Feb-June 2002.

Harvey Weiss and Stacey Maples, Yale University, 2014

FIGURE 3.4 Khabur Plains, northeast Syria, with modern precipitation isohyets and regional archaeological settlement surveys. (H. Weiss and S. Maples)

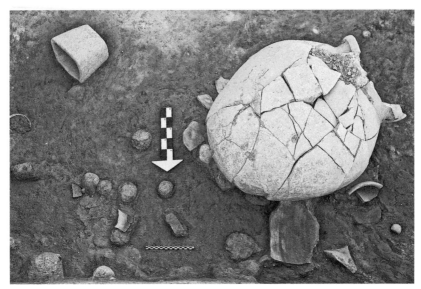

FIGURE 3.5 Clay balls for tablet manufacture; uninscribed clay tablets, cereal storage jar, and basalt 2-liter ration measure, abandoned on Room 12 floor, terminal occupation, Akkadian Administrative Building, Tell Leilan, end period IIb1; AMS radiocarbon dated 2254–2220 BC (68.2 percent) (H. Weiss)

abandonments included the cities' lower town areas, which were each probably populated by upwards of 20,000 indigenous agricultural workers.

The Unfinished Buildings on the Khabur Plains

In three instances, major building projects were abandoned in mid-construction. "The Unfinished Buildings" included the Akkadians' Naram-Sin fortress at Tell Brak, a gateway city at the southern edge of the Khabur Plains. This massive structure was probably intended to serve as a regional grain store, but it was abandoned with unfinished floors and walls, stamped "Naram-Sin" on their lower course bricks (Mallowan 1947).

At Tell Leilan, in the heart of the eastern Khabur Plains, about a hundred years of large-scale grain storage, processing, and redistribution took place in the Akkadian Administrative Building on the Leilan Acropolis. The Akkadians suddenly departed at ca. 2230 BC, however, leaving clay balls for tablet manufacture, uninscribed clay tablets, a large storage vessel, and a 2-liter ground basalt measure on the terminal room 12 floor (fig. 3.5). Across the stone-paved street, The Unfinished Building at Tell Leilan had rough-dressed basalt block walls yet without brick, and some walls still only three or four courses high upon a mud-set sherd layer (fig. 3.6). A semi-circle of partially dressed blocks awaited finishing and wall placement and a line of basalt blocks extended west to the edge of the Leilan Acropolis. At its desertion, the string-impressed clay sealing of the imperial Akkadian minister, "Haya-abum, šabra" (L93–66; fig. 3.7), was left on The Unfinished Building construction surface (Weiss et al. 2012; McCarthy 2012).

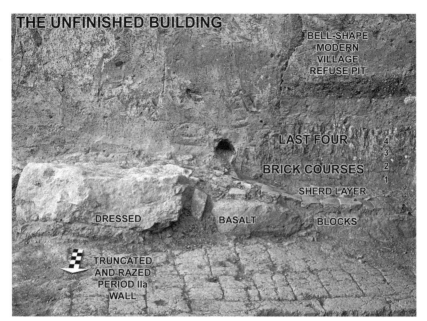

FIGURE 3.6 Tell Leilan 1999, The Unfinished Building, 44W16 south stratigraphic section: Period IIa razed brick wall, Period IIb1 incomplete line of dressed basalt blocks, mud pack, sherd layer, and four courses of calcic horizon mudbrick. Modern village pit halted at calcic horizon mudbrick. (H. Weiss)

FIGURE 3.7 Tell Leilan 1993, object 66, 44W15, southern Mesopotamian imperial Akkadian seal-impression fragment with inscription "Hayabum, shabra," retrieved from working floor of The Unfinished Building at corner of north and west basalt block walls. Reverse: string impressions. (H. Weiss)

FIGURE 3.8 Post-Akkadian four-room house built around a courtyard, Tell Leilan period IIc, AMS radiocarbon dated 2233–2196 BC (68.2 percent). The house was occupied briefly after the Leilan period IIb Akkadian site abandonment and is the only post-Akkadian occupation located at Tell Leilan to date. (H. Weiss)

Two similar but fragmentary buildings were abandoned mid-construction at Tell Mohammed Diyab, eight kilometers east of Tell Leilan (Nicolle 2006: 64, 133). Taken together, The Unfinished Buildings at Tell Brak, Leilan, and Mohammed Diyab document patterns of both imperial success and expansionary designs up to the very moment of the administrators' decision to abandon the Khabur Plains.

Post-Akkadian Settlement on the Khabur Plains

Following the abandonments, residential and very short-term post-Akkadian occupations are known from excavations at four sites—Brak, Leilan, Arbid, and Chagar Bazar—each of which terminated at ca. 2200 BC, as determined by high-resolution AMS radiocarbon dating (see fig. 3.8; Weiss et al. 2012: 175). Similarly, the Leilan Region Survey, a 1650 square kilometer transect through the center of the eastern Khabur Plains, documents an 87 percent reduction in settled area at the termination of Akkadian imperialization in the post-Akkadian Leilan IIc period (ca. 2230–2200 BC). This brief remnant occupation was followed by an approximately 250-year abandonment of the region, until the return of pre-megadrought precipitation (fig. 3.9; Arrivabeni 2012; Colantoni 2012).

Dry-farming Western Syria, Turkey, the Levant, and the Aegean

Distant from direct Akkadian imperialization, the dry-farming plains of the upper Euphrates drainage near Urfa and Harran in Turkey, including urban sites with cities such as Tilbeşar, Titriş and Kazane, were similarly abandoned

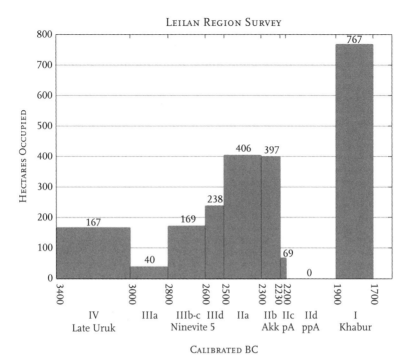

FIGURE 3.9 Leilan Region Survey (1650 sq km) histogram of total settlement hectares/period spans. The end period IIb Akkadian collapse reduced post-Akkadian period IIc settlement by 87 percent at ca. 2230 BC, and by 100 percent at ca. 2200 BC. Major resettlement by formerly pastoralist Amorite populations (Khabur period occupations) is evident by ca. 1950 BC. (M. Arrivabeni, L. Ristvet, E. Rova, & H. Weiss).

at 2200 BC. In western Syria, south of Aleppo, the rebuilt Early Bronze IVB city at Ebla was reduced in size, while its Archaic Palace, singularly fitted with water cisterns, remained unfinished like The Unfinished Buildings at Leilan and Brak and Mohammed Diyab (Matthiae 2013). On the Jabbul Plain, the 20-hectare town at Umm el-Marra was abandoned suddenly (Schwartz et al. 2012), along with Rawda and its environs further south in the semi-arid marginal steppe (Brochier in press; Barge, Castel & Brochier 2014). Synchronous and similar scale abandonments occurred across the dry-farming southern Levant (Haiman 1996; D'Andrea 2012; Harrison 2012; Finkelstein & Langgut 2014), Anatolia (Boyer, Roberts, & Baird 2006), the eastern Mediterranean (Weiberg & Finné 2013; Davis 2013; Weiss 2000: 89–90), and as far east as Turkmenistan (P'yankova 1994). In addition to regional abandonments, Anatolian excavations document conflagration and inter-settlement conflict among the social forces unleashed within the megadrought (Massa 2014; Massa & Şahoğlu 2015).

Riverine Refugia: The Euphrates and Orontes Rivers

In western Syria (fig. 3.10), the karst-fed Orontes River system encompassing the Ghab valley swamp and the Amuq Plain (Voûte 1961) was the habitat-tracking target that attracted and sustained large agricultural populations at such new urban sites as Qatna, Nasriyah, and Acharné (al-Maqdissi 2010; Yener 2005; Morandi Bonacossi 2009). To the west of this karst plateau, along the fertile littoral, springs provided for the town at Tell Arqa and its villages and Tell Sukas. Meanwhile, Ugarit and Byblos, lacking karstic springs, were subject to population reductions and site abandonments (Weiss 2014). In the southern Levant, the period IIId settlement at Tell es-Sultan/Jericho provides an illuminating example of a karstic spring refugium for sedentarizing pastoralists (Nigro 2013), and walled Khirbet Iskander continued to be occupied because of its location along a major perennial wadi (Cordova & Long 2010).

Euphrates River flow during this period, though diminished, still provided for irrigation agriculture in central and southern Mesopotamia. Hence, habitat tracking from desiccated dry-farming areas to irrigation agriculture Euphrates and Orontes River refugia was the adaptive response of dry-farming agriculturalists and Hanaean/Amorite pastoralists (Coope 1979; Eldredge 1985). In southern Mesopotamia, this population movement and its subsequent population doubling within a century generated the hypertrophic Ur III dynasty cities aligned along the Euphrates River (Adams 1981). Urban settlement also flourished and expanded during this post-Akkadian shakkanaku-period at such central Euphrates cities as Mari, Terqa, Tuttul, Emar, Carchemish, and Samsat (Butterlin 2007).

As noted, we lack the data with which to understand the effects of mega-drought onset upon the Akkadian imperial agricultural economy in southern Mesopotamia. Euphrates flow alone is estimated to sustain irrigation for an area of 8000 square kilometers (Adams 1981) and a total Tigris-Euphrates flow sufficient to irrigate 30,000 square kilometers (Wilkinson 2003: 76). The extent and size distribution of Akkadian and Ur III period settlement, however, remains uncertain, since diagnostic ceramic indicators have been revised, along with epigraphic reconstructions of settlement areas. Estimates of southern Akkadian harvest are therefore not available. Estimates of aggregate Ur III cultivated area do seem nearly attainable (Nissen, Damerow, & Englund 1994: 142) given the extensive data available for Ur III shipments of, for instance, 30 tons of barley to Nippur (Sharlach 2004: 329). However, we do not know if, or to what extent, these aggregate data resulted from early Ur III attempts to straighten the Euphrates and Tigris meanders, likely generated by 4.2 ka BP/terminal Akkadian reduced river flow (Adams 1981: 164). We may perhaps assume that cereal yield per unit cultivated (Postgate 1984; Maekawa 1984) would not have been affected and that the straightening of Ur III canals could have corrected for potential aggregate yield reductions. Irresolvable, given present data constraints, is the significance of northern dry-farming exports for the wealth and resilience of the southern imperial economy. We cannot quantify the truncation at 2200 BC from either the north or the south from the epigraphic documentation for southern grain imperialization. Its effect was, nevertheless, real, and the constrained successor southern states did not imperialize the desiccated northern domains.

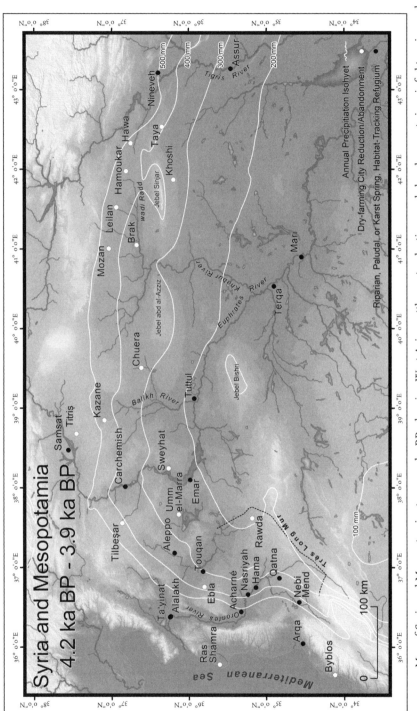

FIGURE 3.10 Map of Syria and Mesopotamia at 4.2–3.9 ka BP, showing West Asian settlement reductions and abandonments in rainfed terrains and riparian, paludal, and karstic-spring refugia. The "Très Long Mur" may have protected the new Orontes River urban refugia from Amorite nomad incursions, much as did its contemporary analog, "The Repeller-of-the-Amorites Wall," in southern Mesopotamia (H. Weiss and S. Maples).

The Steppe and Jebel Bishri

The steppe, with its 100–300 millimeters precipitation per annum, separates the dry-farming northern plains from the irrigation-agriculture south and provided for the base and transit camps of seasonal flock-foraging Hanaean/ Amorite sheep-goat pastoralists. In the epigraphic and archaeological records, it is the mountainous Jebel Bishri region, in the steppe south of the Euphrates River, that features as a prominent pastoral nomad landmark, a nexus from which regional nomadization seems to have occurred in three phases.

The first phase of nomadization, at the onset of megadrought, was the interruption of seasonal pastoralist transhumance between the Euphrates River and the Khabur Plains. The drought would have rendered the Khabur Plains inhospitable to seasonal flock forage, forcing pastoralists into the steppe and the adjacent banks of the Euphrates, and then into southern Mesopotamia for Euphrates-fed foraging. Epigraphic documentation for this period of nomadization remains scant, of course, but the impressive tomb cemeteries at the Jebel Bishri point to Amorite pastoral populations (Ohnuma 2010).

The second phase was marked by the construction in southern Mesopotamia of city walls, such as "The Repeller of the Amorites Wall" in the Ur III period, intended to thwart the steppe nomads (Gasche 1990; Sallaberger 2009). The walls proved essentially porous and futile, as the pastoralist presence would only increase.

In the third phase of nomadization, the sedentarized pastoralists emerge as the controlling dynasts of southern Mesopotamian cities (Finkelstein 1966) and, with the opportunistic resettlement ca. 1950 BC that accompanied a return of pre-4.2 ka BP precipitation, undergo sedentarization in the north as well (Heimpel 2003; Ristvet 2008). It was this process that led to the ascent of Shamshi-Adad I and his Amorite kingdom in Upper Mesopotamia, with new capital cities at Ekallatum and Shubat Enlil (Weiss et al. 2012; Ristvet & Weiss 2013; fig. 3.11).

Steppic nomadization remains difficult to quantify, but is nevertheless evident in the Jebel Bishri cemeteries and in the detailed epigraphic record. Significantly, pastoral nomad excursions beyond Jebel Bishri-based camps forced Aramaean-Assyrian conflicts in the twelfth to ninth centuries BC, during the 3.2 ka BP aridification event (Postgate 1992; Kirleis & Herles 2007; Pappi 2006). In the cross-cultural archaeological record, nomadization is evident as a drought response to the conclusion of the African Humid Period in central and southwestern Sahara (Manning & Timpson 2014) and to the Tiwanaku collapse in Bolivia (Dillehay & Kolata 2004).

Conclusions

The causal weight of the global 4.2–3.9 ka BP megadrought in the regional abandonments, collapses, habitat tracking, and nomadization of West Asia

FIGURE 3.11 Tell Leilan, Acropolis Northeast, Period I, Building Level II temple, north façade, erected during the reign of Shamshi-Adad and his successors, ca. 1850–1725 BC. After the pastoralists' resettlement of the Khabur Plains, beginning ca. 1950 BC, the region would be transformed into Shamshi-Adad's north Mesopotamian kingdom, with a capital city, Shubat Enlil, at Tell Leilan (H. Weiss).

is clear, placing the Akkadian imperial collapse within a regional and global frame of abrupt climate change that crossed both ecological zones and continents. Counterfactually, these synchronous West Asian and adjacent collapse and abandonment events and processes would not have occurred without the 4.2 ka BP megadrought.

Five objections to these data and analyses may be noted here. Karl Butzer (2012) dismissed the 4.2–3.9 ka BP event paleoclimate proxies, regional and global, as uncertain and uncertainly dated. In contradiction of the archaeological and epigraphic records, and the high-resolution radiocarbon dating of the Khabur region Akkadian imperialization and abandonments, Butzer believed the apparent widespread decline is rooted in the fact that Akkadian imperialization under Sargon and Naram-Sin destroyed the cities of western Syria and thereby destroyed an interlinked world economy of urban societies and trade networks from the Aegean to the Indus. Unfortunately, Butzer's version of late third millennium history misdates the collapse by about 150 years in southern Mesopotamia, on the Khabur Plains, across western Syria, and across the Mediterranean and ignores the famous Akkadian long-distance trade that even brought boats filled with exotic goods from as far as the Indus Valley to the Akkadian capital.

Neil Roberts et al. (2011) argue that successful urban adaptations during this period disprove the "environmental determinism" of those who quantify the regional abandonments caused by the megadrought's effects upon rainfed and irrigation-agriculture production. Yet the sites purported by Roberts et al. to be successful urban adaptations, such as Brak and Rawda, either famously did collapse or were located in riparian and karstic refugia, such as Mari on the Euphrates River and Qatna along the Orontes River basin. Similarly, Glenn Schwartz (2007) imagines an isotropic, uniform landscape across Syria, thereby missing the region's hydrologic variability, and also mistakes 4.2–3.9 ka BP habitat tracking to riparian Orontes and Euphrates refugia for evidence of stable population centers.

Tony Wilkinson et al. (2007) have hypothesized that the agricultural economies of the Mesopotamian dry-farming zone were "brittle" and susceptible to collapse during brief arid "spells" like those of the instrumental record. The independent urban economies of northern Mesopotamia thrived, however, for almost 300 years without collapse before the Akkadians targeted their success, extracted still greater agricultural surplus for an additional 100 years, and planned further imperial expansion, until the well-documented megadrought made regional dry-farming impossible.

According to another recent disclaimer, the major Khabur Plains cities of Brak, Hamoukar, and Leilan collapsed at different times during the late third millennium BC and for different reasons (Ur 2015). The argument goes that the climate of northern Mesopotamia "[m]ore likely . . . experienced a gradual aridification" (p. 85), and that the farmers of Brak and Hamoukar were able to forestall disaster—unlike their singularly imperialized neighbors in Leilan—by fertilizing their fields with organic domestic refuse that included broken household pottery. Thus, while Leilan was abandoned ca. 2200 BC, Brak and Hamoukar apparently remained occupied. Ur's argument ignores the wealth of global, high-resolution paleoclimate records for the 4.2 ka BP megadrought's abruptness, magnitude, and duration, which distinguish the global event from the modern instrumental drought record for West Asia (Weiss 2012; Weiss et al 2012). Instead, Ur deems relevant an explicitly discounted idea (Weiss et al. 1993) that the megadrought was a product of volcanism and an improbable link between Leilan and Brak pre-Akkadian destruction debris and an extra-terrestrial event (Courty 2001).

The evidence for Ur's hypothesis that Leilan farmers did not fertilize their fields while Brak and Hamoukar farmers did is the relatively limited density of off-site sherd scatters at Leilan (Ristvet 2005) compared to Ur's off-site collections at Brak and Hamoukar. The off-site sherd scatters are hypothesized to be late third millennium BC residues from household manures mixed with potsherds. The sherd-scatter-as-remnant-manure hypothesis has been disconfirmed repeatedly, however (Alcock, Cherry & Davis 1994; Wilkinson 1990: 76–78; Styring et al 2017). Settlement and artifact distributions at Tell Leilan were constrained within its third and second millennium city wall, a feature absent at Brak and at Hamoukar's diffuse settlement, which likely explains the differences in the densities of off-site sherd scatters.

Missing from Ur's discussion is the Akkadian imperialization of northern Mesopotamia. The extensive evidence reviewed above indicates successful agro-production imperialization that was suddenly truncated, not only at Leilan but also across the Khabur and Assyrian Plains, from Brak to Nineveh. That evidence includes The Unfinished Buildings, probably granaries, at Brak, Leilan, and Mohammed Diyab that were suddenly abandoned at ca. 2200 BC along with the already occupied monumental buildings at Brak, Leilan and Mozan (Weiss 2012).

The archaeological record tells us that the abandonment at ca. 2200 BC was region-wide, not limited to Leilan, as Ur posits. The high-resolution radiocarbon dating of terminal occupations at Brak (Emberling 2012), Arbid (Kolinski 2012), and Leilan (Weiss et al. 2012) and the Leilan Region Survey (Ristvet 2012; Arrivabeni 2012) defines the synchronous two-stage abandonments of these cities and other settlement across the Khabur Plains at ca. 2200 BC—that is, at the end of the approximately thirty-year post-Akkadian period—at which time such ceramics also seal the occupation at Hamoukar (Gibson 2012). High-resolution radiocarbon dating places the collapse and abandonments of towns in western Syria, at Umm el-Marra (Schwartz et al. 2012), and Rawda (Brochier, in press) at the same time, coincident with abandonments across the Mediterranean and West Asia, as detailed above.

Finally, Ur's argument that the good citizens of Leilan inexplicably failed to fertilize their fields and thereby suffered collapse disregards the region's millennia-old integrated farming system of cereal cultivation and ruminant stubble grazing, documented as early as the eighteenth century BC (Matthews 1978: 90). Ur discounts sheep manure as oxidizing too rapidly in "semi-arid climates" to be an effective fertilizer, an assumption derived from an informal comment about the "Middle East" (Keen 1946: 48) that was subsequently applied to "arid and hot regions" (Buringh 1960: 253) and then to semi-arid northern Mesopotamia (Wilkinson 1982). However, modern observations of field nitrification of urine and feces by sheep grazing in north Syria, although difficult to quantify, suggest perduring fertilizing effects (Thomas et al. 2006; White et al. 1997), especially as plowing may rapidly follow grazing (Hirata, Fujita, Miyazaki 1998).

The collapse and abandonment of Akkadian imperialized dry-farming settlement and the synchronous collapse of the Akkadian Empire occurred alongside abandonment of adjacent dry-farming domains in West Asia and the Aegean, and the collapse of the Old Kingdom in Egypt due to Nile flow failure. The collapses and abandonments were a direct effect of 4.2 ka–3.9 ka BP megadrought's abruptness (less than 5 years), magnitude (30–50 percent precipitation reduction) and duration (200–300 years), which altogether reduced dry-farming agriculture to unsustainable societal limits and reduced aggregate irrigation-agriculture production. In the absence of technological innovation, or region-wide subsistence relief, the dry-farming region adaptations across West Asia were collapse, abandonment, habitat tracking to agricultural refugia, and nomadization, each a form of demographic and societal resilience. In the steppic zone bordering the Euphrates, the adaptive pastoral response was forage-driven, step-wise movement into southern Mesopotamia, which ultimately generated brief, hypertrophic Ur III-period urbanism. In Euphrates and Orontes riverine environments, both habitat-tracking and urban growth ensued, while imperial reorganization was restricted to the realigned irrigation zone of Ur III dynastic control, where the descendants of Amorite steppe pastoralists eventually achieved state power. The 4.2 ka BP abrupt megadrought provides, therefore, an explanatory causal force behind the dramatic archaeological and epigraphic record of the Akkadian Empire and its collapse.

Acknowledgments

The Directorate-General of Antiquities and Museums, Syrian Arab Republic, graciously provided administrative support for the Yale University Tell Leilan Project. Research funding was provided by the National Science Foundation, National Endowment for the Humanities, Yale University, Malcolm Wiener, Roger and Barbara Brown, Raymond and Beverly Sackler Foundation, and the late Leon Levy. Richard Seager (Lamont-Doherty Earth Observatory), Mark Besonen (Texas A&M University, Corpus Christi), and Stace Maples (Stanford University) offered their essential technical assistance.

References

Adams, R. McC. 1981. *Heartland of Cities*. Chicago: University of Chicago Press.

Akkermans, P. M. M. G., and G. M. Schwartz. 2008. *The Archaeology of Syria: From Complex Hunter-Gatherers to Early Urban Societies (c. 16,000–300 BC)*. Cambridge:Cambridge University Press.

Alcock, S. E., J. F. Cherry, and J. L. Davis. 1994. Intensive survey, agricultural practice and the classical landscape of Greece. In *Classical Greece: Ancient Histories and Modern Archaeologies*, ed. I. Morris. Cambridge: Cambridge University Press.

Algaze, G. 2008. *Ancient Mesopotamia at the Dawn of Civilization: The Evolution of an Urban Landscape*. Chicago: University of Chicago Press.

Almamori, H. O. 2014. Gissa (Umm al-Aqarib), Umma (Jokha) and Lagash in the Early Dynastic III Period. *al-Rāfidān* 35: 1–38.

Al-Maqdissi, M. 2010. Matériel pour l'étude de la ville en Syrie (deuxième partie): Urban planning in Syria during the SUR (Second Urban Revolution) (mid-third millennium BC). In *Formation of Tribal Communities: Integrated Research in the Middle Euphrates, Syria*, ed. K. Ohnuma. Special Issue of *al-Rāfidān*. Tokyo: Kokushikan University, 131–146.

Al-Quntar, S., L. Khalidi, and J. Ur. 2011. Proto-urbanism in the late 5th millennium BC: survey and excavations at Khirbat al-Fakhar/Hamoukar, northeast Syria. *Paléorient* 37: 151–75.

Arrivabeni, M. 2012. Post-Akkadian settlement distribution in the Leilan Region Survey. In *Seven Generations since the Fall of Akkad*, ed. H. Weiss. Studia Chaburensia, vol. 3. Wiesbaden: Harrassowitz, 261–278.

Arz, H. W., F. Lamy, and J. Pätzold. 2006. A pronounced dry event recorded around 4.2 ka in brine sediments from the northern Red Sea. *Quaternary Research* 66: 432–441.

Baker, P. A., S. C. Fritz, S. J. Burns, E. Ekdahl, and C. A. Rigsby. 2009. The nature and origin of decadal to millennial scale climate variability in the southern tropics of South America: the Holocene record of Lago Umayo, Peru. In *Past Climate Variability in South America and Surrounding Regions: From the Last Glacial Maximum to the Holocene*, ed. F. Vimeux, F. Sylvestre, and M. Khodri. Dordrecht: Springer, 301–22.

Bar-Matthews, M., and A. Ayalon. 2011. Mid-Holocene climate variations revealed by high-resolution speleothem records from Soreq Cave, Israel and their correlation with cultural changes. *The Holocene* 21: 163–171.

Barge, O., C. Castel, and J. É. Brochier. 2014. Human impact on the landscape around Al-Rawda (Syria) during the Early Bronze IV: evidence for exploitation, occupation and appropriation of the land. In *Settlement Dynamics and Human-Landscape Interaction in the Dry Steppes of Syria*, ed. D. Morandi-Bonacosi. Studia Chaburensia, vol. 4. Wiesbaden: Harrassowitz, 173–185.

Baruch, U., and S. Bottema. 1999. A new pollen diagram from Lake Hula: vegetational, climatic, and anthropogenic implications. In *Ancient Lakes: Their Cultural and Biological Diversity, ed.* H. Kawanabe, G. W. Coulter, and A. C. Roosevelt. Ghent: Kenobi Productions, 75–86.

Berkelhammer, M., A. Sinha, L. Stott, H. Cheng, F. S. R. Pausata, and K. Yoshimura. 2012. An abrupt shift in the Indian Monsoon 4000 years ago. In *Climates, Landscapes, and Civilizations*, ed. L. Giosan, D. Q. Fuller, K. Nicoll, R. K. Flad, and P. D. Clift. Geophysical Monograph Series, vol. 198. Washington, D. C.: American Geophysical Union, 75–88.

Bernhardt, C. E., B. P. Horton, J.-D. Stanley. 2012. Nile delta vegetation response to Holocene climate variability. *Geology* 40 (7): 615–618.

Black, J. A., G. Cunningham, E. Robson, and G. Zólyomi. 2004. *The Literature of Ancient Sumer*. Oxford: Oxford University Press.

Blanchet, C. L., R. Tjallingii, M. Frank, J. Lorenzen, A. Reitz, K. Brown, T. Feseker, and W. Brückmann. 2013. High- and low-latitude forcing of the Nile River regime during the Holocene inferred from laminated sediments of the Nile deep-sea fan. *Earth and Planetary Science Letters* 364: 98–110.

Blair, C. L., Á. Geirsdóttir, and G. H. Miller. 2015. A high-resolution multi-proxy lake record of Holocene environmental change in southern Iceland. *Journal of Quaternary Science* 30: 281–292.

Booth, R. K., S. T. Jackson, S. L. Forman, et al. (2005). A severe centennial-scale drought in mid-continental North America 4200 years ago and apparent global linkages. *The Holocene* 15: 321–328.

Bottema, S. 1997. Third Millennium climate in the Near East based upon pollen evidence. In *Third Millennium BC Climate Change and Old World Collapse*, ed. H. N. Dalfes, G. Kukla, and H. Weiss. NATO ASI Series I: Global Environmental Change, vol. 49. Berlin: Springer, 489–515.

Bottema, S., H. Woldring, and I. Kayan. 2001. The late Quaternary vegetation history of western Turkey. In *The Ilıpınar Excavations II*, ed. J. J. Roodenberg and L. C. Thissen. PIHANS, vol. 93. Leiden: Netherlands Institute for the Near East, 327–354.

Boyer, P., N. Roberts, and D. Baird. 2006. Holocene environment and settlement on the Çarşamba alluvial fan, south-central Turkey: integrating geoarchaeology and archaeological field survey. *Geoarchaeology* 21: 675–698.

Brisset, E., C. Miramont, F. Guiter, E. J. Anthony, K. Tachikawa, J. Poulenard, F. Arnaud, C. Delhon, J.-D. Meunier, E. Bard, and F. Suméra. 2013. Non-reversible geosystem destabilisation at 4200 cal. BP: sedimentological, geochemical and botanical markers of soil erosion recorded in a Mediterranean alpine lake. *The Holocene* 23: 1863–1874.

Brochier, J. É. In press. A cataclysm in the steppe? Environmental history of al-Rawda, an ephemeral town in the arid Syrian margins at the end of the third millennium. In *Origins, Structure, Development and Sociology of Circular Cities of Early Bronze Age Syria*, ed. J.-W. Meyer, P. Quenet and C. Castel. Brepols: Subartu.

Brumfield, S. 2012. CDLI no. P212515 Chagar Bazar A.391. *Cuneiform Digital Library*. UCLA.

Brumfield, S. 2013. Imperial Methods: Using Text Mining and Social Network Analysis to Detect Regional Strategies in the Akkadian Empire. Ph.D. dissertation, University of California, Los Angeles. escholarship.org/uc/item/0cr156f6

Brustolon, A., and E. Rova. 2007. The late chalcolithic period in the Tell Leilan region: a report on the ceramic material of the 1995 survey. *Kaskal* 4: 1–42.

Buccellati, G., and M. Kelly-Buccellati. 2000. The royal palace of Urkesh. Report on the 12th season at Tell Mozan/Urkesh: excavations in area AA, June-October 1999. *Mitteilungen der Deutschen Orient-Gesellschaft zu Berlin* 132: 133–183.

Butterlin, P. 2007. Mari, les Shakkanakkû et la crise de la fin du troisième millénaire. In *Sociétés Humaines et Changement Climatique à la Fin du Troisième Millénaire: Une Crise a-t'elle eu Lieu en Haute Mésopotamie? Actes du Colloque de Lyon, 5–8 Décembre 2005*, ed. C. Kuzucuoğlu and C. Marro. Istanbul: Institut Français d/Études Anatoliennes Georges Dumézil, 227–245.

Buringh, P. 1960. *Soils and Soil Conditions in Iraq*. Baghdad: Ministry of Agriculture.

Butzer, K. 2012. Collapse, environment, and society. *Proceedings of the National Academy of Sciences* 109: 3632–3639. doi: 10.1073/pnas.1114845109

Cai, Y., L. Tan, H. Cheng, Z. An, R. L. Edwards, M. J. Kelly, X. Kong, and X. Wang. 2010. The variation of summer monsoon precipitation in central China since the last deglaciation. *Earth and Planetary Science Letters* 291: 21–31.

Charles, M., H. Pessin, and M. M. Hald. 2010. Tolerating change at Late Chalcolithic Tell Brak: responses of an early urban society to an uncertain climate. *Environmental Archaeology* 15: 183–198.

Chase, B. M., M. E. Meadows, A. S. Carr, and P. J. Reimer. 2010. Evidence for progressive Holocene aridification in southern Africa recorded in Namibian hyrax middens: implications for African Monsoon dynamics and the "African Humid Period." *Quaternary Research* 74: 36–45.

Colantoni, C. 2012. Touching the void: the post-Akkadian period viewed from Tell Brak. In *Seven Generations Since the Fall of Akkad, ed.* H. Weiss. Studia Chaburensia, vol. 3. Wiesbaden: Harrassowitz, 45–64.

Coope, G. R. 1979. Late Cenozoic fossil Coleoptera: evolution, biogeography, and ecology. *Annual Review of Ecology and Systematics* 10: 247–267.

Cooper, J. 1983. *The Curse of Agade*. Baltimore: Johns Hopkins University Press.

Cordova, C. E., and P. H. Lehman. 2005. Holocene environmental change in southwestern Crimea (Ukraine) in pollen and soil records. *The Holocene* 15: 263–277.

Cordova, C. E., and J. C. J. Long. 2010. Khirbat Iskandar and its modern and ancient environment. In *Khirbat Iskandar: Final Report on the Early Bronze IV Area C Gateway and Cemeteries*, ed. S. Richard, J. C. J. Long, P. Holdorf, and G. Peterman. ASOR Archaeological Reports 14. Boston, MA: American Schools of Oriental Research, 21–35.

Courty, M.-A. 2001. Evidence at Tell Brak for the Late EDIII/Early Akkadian Air Blast Event (4 kyr BP). In *Excavations at Tell Brak. Vol. 2: Nagar in the Third Millennium BC*, ed. D. Oates, J. Oates, and H. McDonald. London: McDonald Institute for Archaeology/British School of Archaeology in Iraq, 367–372.

Cullen, H. M., P. B. deMenocal, S. Hemming, G. Hemming, F. H. Brown, T. Guilderson, and F. Sirocko. 2000. Climate change and the collapse of the Akkadian Empire: evidence from the deep sea. *Geology* 28: 379–382.

Cullen, H. M., A. Kaplan, P. A. Arkin, and P. B. deMenocal. 2002. Impact of the North Atlantic Oscillation on Middle Eastern climate and streamflow. *Climatic Change* 55: 315–338.

d'Alpoim Guedes, Jade, Sturt W. Manning, and R. Kyle Bocinsky. 2016. A 5,500-Year Model of Changing Crop Niches on the Tibetan Plateau. *Current Anthropology* 57:517-522. doi: 10.1086/687255

D'Andrea, M. 2012. The early Bronze IV period in South-Central Transjordan: reconsidering chronology through ceramic technology. *Levant* 44: 17–50.

Davis, J. L. 2013. "Minding the gap": a problem in eastern Mediterranean chronology, then and now. *American Journal of Archaeology* 117: 527–533.

Davis, M. and L. Thompson. 2006. An Andean ice-core record of a Middle Holocene An Andean ice-core record of a Middle Holocene mega-drought in North Africa and Asia. *Annals of Glaciology* 43:34-41. doi.org/10.3189/172756406781812456

Dean, W. E. 1997. Rates, timing, and cyclicity of Holocene eolian activity in north-central United States: evidence from varved lake sediments. *Geology* 25: 331–334.

deLillis Forest, F., L. Milano, and L. Mori. 2007. The Akkadian occupation in the northwest area of the Acropolis. *KASKAL* 4: 43–64.

Deniston, R. F., K.-H. Wyrwoll, V. J. Polyak, J. R. Brown, Y. Asmerom, A. D. Wanamaker Jr., Z. LaPointe, R. Ellerbroek, M. Barthelmes, D. Cleary, J. Cugley, D. Woods, and W. F. Humphreys. 2013. A stalagmite record of Holocene Indonesian-Australian summer monsoon variability from the Australian tropics. *Quaternary Science Reviews* 78: 155–168.

de Porras, M. E., A. Maldonado, F. A. Quintana, A. J. Martel-Cea, O. Reyes, and C. Méndez. 2014. Environmental and climatic changes in central Chilean Patagonia since the Late Glacial (Mallín El Embudo, 44° S). *Climate of the Past* 10: 1063–1078. doi:10.5194/cp-10-1063-2014

Dillehay, T. D., and A. L. Kolata. 2004. Long-term human response to uncertain environmental conditions in the Andes. *Proceedings of the National Academy of Sciences* 101: 4325–4330.

Di Rita, F., and D. Magri. 2009. Holocene drought, deforestation and evergreen vegetation development in the central Mediterranean: a 5500-year record from Lago Alimini Piccolo, Apulia, southeast Italy. *The Holocene* 19: 295–306.

Dixit, Y., D. A. Hodell, and C. A. Petrie. 2014. Abrupt weakening of the summer monsoon in northwest India ~4100 yr ago. *Geology* 42: 339–342.

Djamali, M., J.-L. De Beaulieu, N. F. Miller, V. Andrieu-Ponel, P. Ponel, R. Lak, N. Sadeddin, H. Akhani, and H. Fazeli. 2009. Vegetation history of the SE section of the Zagros Mountains during the last five millennia; a pollen record from the Maharlou Lake, Fars Province, Iran. *Vegetation History and Archaeobotany* 18 (2): 123–136.

Donges, J. F., R. V. Donner, N. Marwan, S. F. M. Breitenbach, K. Rehfeld, and J. Kurths. 2015. Non-linear regime shifts in Holocene Asian monsoon variability: potential impacts on cultural change and migratory patterns. *Climates of the Past* 11: 709–741.

Drysdale, R., G. Zanchetta, J. Hellstrom, R. Maas, A. Fallick, M. Pickett, I. Cartwright, and L. Piccini. 2006. Late Holocene drought responsible for the collapse of Old World civilizations is recorded in an Italian cave flowstone. *Geology* 34: 101–104.

Dykoski, C., R. L. Edwards, H. Cheng, D. Yuan, Y. Cai, M. Zhang, Y. Lin, Z. An, and J. Revenaugh. 2005. A high-resolution, absolute-dated Holocene and deglacial Asian monsoon record from Dongge Cave, China. *Earth and Planetary Science Letters* 233: 71–86.

Eastwood, W. J., M. J. Leng, N. Roberts, and B. Davis. 2007. Holocene climate change in the eastern Mediterranean region: a comparison of stable isotope and pollen data from Lake Gölhisar, southwest Turkey. *Journal of Quaternary Science* 22: 327–341.

Edelman-Furstenberg, Y., A. Almogi-Labin, and C. Hemleben. 2009. Palaeoceanographic evolution of the central Red Sea during the late Holocene. *The Holocene* 19: 117–127.

Ehrmann, W., G. Schmiedl, Y. Hamann, T. Kuhnt, C. Hemleben, and W. Siebel. 2007. Clay minerals in late glacial and Holocene sediments of the northern and southern Aegean Sea. *Palaeogeography, Palaeoclimatology, Palaeoecology* 249: 36–57.

Eldredge, N. 1985. *Time Frames: The Evolution of Punctuated Equilibria.* Princeton, NJ: Princeton University Press.

Emberling, G., H. McDonald, J. Weber, and H. Wright. 2012. After collapse: the post-Akkadian occupation in the Pisé building, Tell Brak. In *Seven Generations Since the Fall of Akkad,* ed. H. Weiss. Studia Chaburensia, vol. 3. Wiesbaden: Harrassowitz, 65–88.

Englund, R. 1994. *Archaic Administrative Texts from Uruk: The Early Campaigns.* ATU, vol. 5. Berlin: Mann.

Englund, R. 2015. CDLI no. P212952 Sippar BM 080452. *Cuneiform Digital Library.* UCLA.

Finkbeiner, U. 1991. *Uruk Kampagne 35–37, 1982–1984. Die archäologische Oberflächenuntersuchung (Survey).* Deutsches Archäologisches Institut, Ausgrabungen in Uruk-Warka: Endberichte, vol. 4. Darmstadt: Philipp von Zabern.

Finkelstein, I., and D. Langgut. 2014. Dry climate in the Middle Bronze I and its impact on settlement patterns in the Levant and beyond: new pollen evidence. *Journal of Near Eastern Studies* 73: 219–234.

Finkelstein, J. J. 1966. The genealogy of the Hammurapi dynasty. *Journal of Cuneiform Studies* 20: 95–118.

Fisher, D. A. 2011. Connecting the Atlantic-sector and the north Pacific (Mt Logan) ice core stable isotope records during the Holocene: the role of El Niño. *The Holocene* 21: 1117–1124.

Fleitmann, D., S. J. Burns, M. Mudelsee, U. Neff, J. Kramers, A. Mangini, and A. Matter. 2003. Holocene forcing of the Indian monsoon recorded in a stalagmite from southern Oman. *Science* 300: 1737–1739.

Foster, B. R. 1982. Archives and record-keeping in Sargonic Mesopotamia. *Zeitschrift für Assyriologie* 72: 1–27.

Foster, B. R. 2011. The Sargonic period: two historiographical problems. In *Akkade is King*, ed. G. Barjamovic, J. L. Dahl, U. S. Koch, W. Sommerfeld, and J. G. Westenholz. Leiden: Nederlands Instituut voor het Nabije Oosten, 127–137.

Francke, A., B. Wagner, M. J. Leng, and J. Rethemeyer. 2013. A Late Glacial to Holocene record of environmental change from Lake Dojran (Macedonia, Greece). *Climate of the Past* 9: 481–498.

Frumkin, A. 2009. Stable isotopes of a subfossil tamarix tree from the Dead Sea region, Israel, and their implications for the Intermediate Bronze Age environmental crisis. *Quaternary Research* 71: 319–328.

Gasche, H. 1990. Mauer (mur) B. Archäologisch. In *Reallexikon der Assyriologie und Vorderasiatischen Archäologie* 7: 591–595.

Gasse, F. 2000. Hydrological changes in the African tropics since the Last Glacial Maximum. *Quaternary Science Reviews* 19: 189–211.

Geirsdóttir, Á., G. H. Miller, D. J. Larsen, and S. Ólafsdóttir. 2013. Abrupt Holocene climate transitions in the northern North Atlantic region recorded by synchronized lacustrine records in Iceland. *Quaternary Science Reviews* 70: 48–62.

Gibson, M. 2001. Hamoukar. *Oriental Institute 2000–2001 Annual Report*. Chicago, IL: University of Chicago, 77–83.

Glassner, J.-J. 1986. *La Chute d'Akkadé: l'Événement et Sa Mémoire*. Berliner Beiträge zum Vorderen Orient. Vol. 5. Berlin: Dietrich Reimer.

Göktürk, O. M. 2011. Climate in the Eastern Mediterranean through the Holocene Inferred from Turkish Stalagmites. Ph.D. dissertation, Universität Bern.

Grayson, A. K., G. Frame, D. Frayne, and M. P. Maidman. 1987. *Assyrian Rulers of the Third and Second Millennia BC (to 1115 BC)*. Vol. 1. Toronto: University of Toronto Press.

Haenssler, E., I. Unkel, W. Dörfler, and M.-J. Nadeau. 2014. Driving mechanisms of Holocene lagoon development and barrier accretion in Northern Elis, Peloponnese, inferred from the sedimentary record of the Kotychi Lagoon. *Quaternary Science Journal* 63: 60–77.

Haiman, M. 1996. Early Bronze Age IV settlement pattern of the Negev and Sinai Deserts: view from small marginal temporary sites. *Bulletin of the American Schools of Oriental Research* 303: 1–32.

Hardt, B., H. D. Rowe, G. S. Springer, H. Cheng, and R. L. Edwards. 2010. The seasonality of east central North American precipitation based on three coeval Holocene speleothems from southern West Virginia. *Earth and Planetary Science Letters* 295: 342–348.

Harrison, T. P. 2012. The Southern Levant. In *A Companion to the Archaeology of the Ancient Near East*, ed. D. T. Potts. Malden, MA: Wiley-Blackwell, 629–646.

Hassan, F., G. Tassie, R. Flower, M. Hughes, and M. Hamden. 2006. Modelling environmental and settlement change in the Fayum. *Egyptian Archaeology* 29: 37–40.

Heimpel, W. 2003. *Letters to the Kings of Mari: A New Translation, with Historical Introduction, Notes, and Commentary.* Winona Lake, IN: Eisenbraun's.

Hirata, M., H. Fujita, and A. Miyazaki. 1998. Changes in grazing areas and feed resources in a dry area of north-eastern Syria. Journal of Arid Environments 40: 319–329.

Høyrup, J. 2011. Written mathematical traditions in ancient Mesopotamia: knowledge, ignorance, and reasonable guesses. Traditions of Written Knowledge in Ancient Egypt and Mesopotamia Conference. Frankfurt am Main, December 3–4. Preprint.

Hruška, B. 2007. Agricultural techniques. In *The Babylonian World*, ed. G. Leick. London: Routledge, 54–65.

Jahns, S. 1993. On the Holocene vegetation history of the Argive Plain (Peloponnese, southern Greece). *Vegetation History and Archaeobotany* 2 (4): 187–203.

Jiménez-Moreno, G., and R. S. Anderson. 2012. Holocene vegetation and climate change recorded in alpine bog sediments from the Borreguiles de la Virgen, Sierra Nevada, southern Spain. *Quaternary Research* 77: 44–53.

Kagan, E. J., D. Langgut, E. Boaretto, F. H. Neumann, and M. Stein. 2015. Dead Sea levels during the Bronze and Iron Ages. *Radiocarbon* 57: 237–252.

Kaniewski, D., E. Paulissen, E. Van Campo, M. al-Maqdissi, J. Bretschneider, and K. Van Lerberghe. 2008. Middle East coastal ecosystem response to middle-to-late Holocene abrupt climate changes. *Proceedings of the National Academy of Sciences* 105: 13941–13946.

Kaniewski, D., E. Van Campo, J. Guiot, S. Le Burel, T. Otto, and C. Baeteman. 2013. Environmental roots of the Late Bronze Age crisis. *PLOS One* August 14. doi: 10.1371/journal.pone.0071004

Kaniewski, D., E. Van Campo, C. Morhange, J. Guiot, D. Zviely, I. Shaked, T. Otto, and M. Artzy. 2014. Early urban impact on Mediterranean coastal environments. *Scientific Reports* 3: 3540. doi:10.1038/srep03540

Keen, B. A. 1946. The Agricultural Development of the Middle East. London: His Majesty's Stationery Office.

Kirleis, W., and M. Herles. 2007. Climatic change as a reason for Assyro-Aramaean conflicts? Pollen evidence for drought at the end of the 2nd Millennium BC. *State Archives of Assyria Bulletin* 16: 7–37.

Koliński, R. Generation count at Tell Arbid, sector P. In *Seven Generations since the Fall of Akkad*, ed. H. Weiss. Wiesbaden: Harrassowitz.

Kröpelin, S., D. Verschuren, A.-M. Lézine, H. Eggermont, C. Cocquyt, P. Francus, J.-P. Cazet, M. Fagot, et al. 2008. Climate-driven ecosystem succession in the Sahara: the past 6000 years. *Science* 320: 765–768.

Kuhnt, T., G. Schmiedl, W. Ehrmann, Y. Hamann, and N. Andersen. 2008. Stable isotopic composition of Holocene benthic foraminifers from the Eastern Mediterranean Sea: past changes in productivity and deep water oxygenation. *Palaeogeography, Palaeoclimatology, Palaeoecology* 268: 106–115.

Kushnir, Y., and M. Stein. 2010. North Atlantic influence on 19th–20th century rainfall in the Dead Sea watershed, teleconnections with the Sahel, and implication for Holocene climate fluctuations. *Quaternary Science Reviews* 29: 3843–3860.

Kuzucuoğlu, C., W. Dörfler, S. Kunesch, and F. Goupille. 2011. Mid- to late-Holocene climate change in central Turkey: the Tecer Lake record. *The Holocene* 21: 173–188.

Lemcke, G., and M. Sturm. 1997. $\delta^{18}O$ and trace element measurements as proxy for the reconstruction of climate changes at Lake Van (Turkey): preliminary results. In *Third Millennium BC Climate Change and Old World Collapse*, ed. H. N. Dalfes, G. Kukla, and H. Weiss. NATO ASI Series I: Global Environmental Change 49. Berlin: Springer, 653–678.

Leng, M. J., I. Baneschi, G. Zanchetta, C. N. Jex, B. Wagner, and H. Vogel. 2010. Late Quaternary palaeoenvironmental reconstruction from Lakes Ohrid and Prespa (Macedonia/Albania Border) using stable isotopes. *Biogeosciences* 7(3): 3815–3853.

Leng, M. J., N. Roberts, J. M. Reed, and H. J. Sloane. 1999. Late Quaternary palaeohydrology of the Konya Basin, Turkey, based on isotope studies of modern hydrology and lacustrine carbonates. *Journal of Paleolimnology* 22: 187–204.

Leroy, S. A. G., L. López-Merino, A. Tudryn, F. Chalié, and F. Gasse. 2014. Late Pleistocene and Holocene palaeoenvironments in and around the middle Caspian Basin as reconstructed from a deep-sea core. *Quaternary Science Reviews* 101: 91–110.

Leroy, S. A. G, F. Marret, E. Gibert, F. Chalié, J. L. Reyss, and K. Arpe. 2007. River inflow and salinity changes in the Caspian Sea during the last 5500 years. *Quaternary Science Reviews* 26: 3359–3383.

Lézine, A.-M. 2009. Timing of vegetation changes at the end of the Holocene Humid Period in desert areas at the northern edge of the Atlantic and Indian monsoon systems. *Compte Rendus Geoscience* 341: 750–759.

Li, Y.-X., Z. C. Yu, and K. P. Kodama. 2007. Sensitive moisture response to Holocene millennial-scale climate variations in the mid-Atlantic region, USA. *The Holocene* 17: 3–8.

Licciardi, J. M., M. D. Kurz, and J. M. Curtice. 2006. Cosmogenic 3He production rates from Holocene lava flows in Iceland. *Earth and Planetary Science Letters* 246: 251–264. doi: 10.1016/j.epsl.2006.03.016

Licciardi, J. M., J. M. Schaefer, J. R. Taggart, and D. C. Lund. 2009. Holocene glacier fluctuations in the Peruvian Andes indicate northern climate linkages. *Science* 325: 1677–1679.

Lionello, P., P. Malanotte-Rizzoli, and R. Boscolo (eds.). 2013. *Mediterranean Climate Variability: 4 (Developments in Earth and Environmental Sciences)*. Amsterdam: Elsevier.

Litt, T., C. Ohlwein, F. H. Neumann, A. Hense, and M. Stein. 2012. Holocene climate variability in the Levant from the Dead Sea pollen record. *Quaternary Science Reviews* 49: 95–105.

Liu, F., and Z. Feng. 2012. A dramatic climatic transition at ~4000 cal. yr BP and its cultural responses in Chinese cultural domains. *The Holocene* 22: 1181–1197.

Lu, R., F. Jia, S. Gao, Y. Shang, J. Li, and C. Zhao. 2015. Holocene aeolian activity and climatic change in Qinghai Lake basin, northeastern Qinghai-Tibetan Plateau. *Palaeogeography, Palaeoclimatology, Palaeoecology* 430: 1–10.

Maekawa, K. 1984. Cereal cultivation in the Ur III period. *Bulletin on Sumerian Agriculture* 1: 73–96.

Magny, M., J.-L. de Beaulieu, R. Drescher-Schneider, B. Vannière, A.-V. Walter-Simonnet, Y. Miras, L. Millet, G. Bossuet, O. Peyron, E. Brugiapaglia, and A. Leroux. 2007. Holocene climate changes in the central Mediterranean as recorded by lake-level fluctuations at Lake Accesa (Tuscany, Italy). *Quaternary Science Reviews* 26: 1736–1758.

Magny, M., B. Vannière, C. Calo, L. Millet, A. Leroux, O. Peyron, G. Zanchetta, T. La Mantia, and W. Tinner. 2011. Holocene hydrological changes in south-western Mediterranean as recorded by lake-level fluctuations at Lago Preola, a coastal lake in southern Sicily, Italy. *Quaternary Science Reviews* 30: 2459–2475.

Magny, M., N. Combourieu-Nebout, J. L. de Beaulieu, V. Bout-Roumazeilles, D. Colombaroli, S. Desprat, A. Francke, S. Joannin, et al. 2013. North-south palaeohydrological contrasts in the central Mediterranean during the Holocene: tentative synthesis and working hypotheses. *Climate of the Past* 9: 2043–2071. doi: 10.5194/cp-9-2043-2013

Magri, D., and I. Parra. 2002. Late Quaternary Western Mediterranean pollen records and African winds. *Earth and Planetary Science Letters* 200: 401–408.

Magri, D., and L. Sadori. 1999. Late Pleistocene and Holocene pollen stratigraphy at Lago di Vico, central Italy. *Vegetation History and Archaeobotany* 8: 247–260.

Maiocchi, M. 2009. *Classical Sargonic Tablets Chiefly from Adab in the Cornell University Collection.* Cornell University Studies in Assyriology and Sumerology. Vol. 13. Bethesda, MD: CDL Press.

Maley, J., and R. Vernet. 2015. Populations and climatic evolution in north tropical Africa from the end of the Neolithic to the dawn of the Modern Era. *African Archaeological Review* 32: 179–232.

Mallowan, M. E. L. 1947. Excavations at Brak and Chagar Bazar. *Iraq* 9: 89–259.

Manning, K., and A. Timpson. 2014. The demographic response to Holocene climate change in the Sahara. *Quaternary Science Reviews* 101: 28–35.

Marchant, R., and H. Hooghiemstra. 2004. Rapid environmental change in African and South American tropics around 4000 years before present: a review. *Earth Science Reviews* 66: 217–260.

Marshall, M. H., H. F. Lamb, D. Huws, S. J. Davies, R. Bates, J. Bloemendal, J. Boyle, M. J. Leng, M. Umer, and C. Bryant. 2011. Late Pleistocene and Holocene drought events at Lake Tana, the source of the Blue Nile. *Global and Planetary Change* 78: 147–161.

Masi, A., L. Sadori, G. Zanchetta, I. Baneschi, and M. Giardini. 2013. Climatic interpretation of carbon isotope content of mid-Holocene archaeological charcoals from eastern Anatolia. *Quaternary International* 303: 64–72.

Massa, M. 2014. Destructions, abandonments, social reorganisation and climatic change in west and central Anatolia at the end of the third millennium BC. In *Regional Studies in Archaeology Symposium Proceedings,* ed. B. Erciyas and E. Sökmen. Series IV. Istanbul: Ege Yayınları, 100–123.

Massa, M., and V. Şahoğlu. 2015. The 4.2 ka BP climatic event in west and central Anatolia: combining palaeo-climatic proxies and archaeological data. In *2200 BC. A Climatic Breakdown as a Cause for the Collapse of the Old World?* ed. H. Meller, H. W. Arz, R. Jung, and R. Risch. Halle: Landesmuseum für Vorgeschichte, 61–78.

Matthews, V. H. 1978. Pastoral Nomadism in the Mari Kingdom: Ca. 1830–1760 BC. Baltimore: American Schools of Oriental Research.

Matthiae, P. 2013. The Third Millennium in northwestern Syria: stratigraphy and architecture. In *Archéologie et Histoire de la Syrie I: La Syrie de l'époque néolithique à l'âge du fer,* ed. W. Orthmann, M. al-Maqdissi, and P. Matthiae. Wiesbaden: Harrassowitz, 181–198.

McCarthy, A. 2012. The end of empire: Akkadian and post-Akkadian glyptic in the Jezirah, the evidence from Tell Leilan in context. In *Seven Generations Since the Fall of Akkad,* ed. H. Weiss. Studia Chaburensia, vol. 3. Wiesbaden: Harrassowitz, 217–224.

Menounos, B., J. J. Clague, G. Osborn, B. H. Luckman, T. R. Lakeman, and R. Minkus. 2008. Western Canadian glaciers advance in concert with climate change circa 4.2 ka. *Geophysical Research Letters* 35(7): L07501. doi: 10.1073/pnas.0906075106

Merrill, W. L., R. J. Hard, J. B. Mabry, G. J. Fritz, K. R. Adams, J. R. Roney, and A. C. MacWilliams. 2009. The diffusion of maize to the southwestern United States and its impact. *Proceedings of the National Academy of Sciences* 106: 21019–21026. doi: 10.1073/pnas.0906075106

Migowski, C., M. Stein, S. Prasad, J. F. W. Negendank, and A. Agnon. 2006. Holocene climate variability and cultural evolution in the Near East from the Dead Sea sedimentary record. *Quaternary Research* 66(3): 421–431.

Morandi Bonacossi, D. 2009. Tell Mishrifeh and its region during the EBA IV and the EBA–MBA transition: a first assessment. In *The Levant in Transition: Proceedings of a Conference Held at the British Museum on 20–21 April 2004,* ed. P. J. Parr. Leeds: Maney, 56–68.

Nakamura, A., Yokoyama, Y., H. Maemoku, H. Yagi, M. Okamura, H. Matsuoka, N. Miyake, T. Osada, D. Adhikari, V. Dangol, M. Ikehara, Y. Miyairi, and H. Matsuzaki. 2016. Weak monsoon event at 4.2 ka recorded in sediment from Lake Rara, Himalayas. *Quaternary International.* doi: 10.1016/j.quaint.2015.05.053

Nicolle, C. 2012. Pre-Khabur occupations at Tell Mohammed Diyab (Syrian Jezireh). In *Seven Generations Since the Fall of Akkad,* ed. H. Weiss. Studia Chaburensia, vol. 3. Wiesbaden: Harrassowitz, 129–144.

Nigro, L. 2013. Jericho. In D. M. Master (ed.), *The Oxford Encyclopedia of the Bible and Archaeology, vol. II.* Oxford: Oxford University Press, 1-8.

Nissen, H. J., P. Damerow, and R. K. Englund. 1993. *Archaic Bookkeeping: Early Writing and Techniques of Economic Administration in the Ancient Near East.* Trans. Paul Larsen. Chicago: University of Chicago.

Oates, J., A. McMahon, P. Karsgaard, S. Al Quntar, and J. Ur. 2007. Early Mesopotamian urbanism: a new view from the north. *Antiquity* 81: 585–600.

Ohnuma, K. (ed.). 2010. *Formation of Tribal Communities: Integrated Research in the Middle Euphrates, Syria.* Tokyo: Kokushikan University.

Olsen, J., N. J. Anderson, and M. F. Knudsen. 2012. Variability of the North Atlantic Oscillation over the past 5,200 years. *Nature Geoscience* 5: 808–812.

Pappi, C. 2006. The Jebel Bishri in the physical and cultural landscape of the ancient Near East. *KASKAL* 3: 241–256.

Parker, A. G., and A. S. Goudie. 2008. Geomorphological and palaeoenvironmental investigations in the southeastern Arabian Gulf region and the implication for the archaeology of the region. *Geomorphology* 101: 458–470.

Parker, A. G., A. S. Goudie, S. Stokes, K. White; M. J. Hodson, M. Manning, and D. Kennet. 2006. A record of Holocene climate change from lake geochemical analyses in southeastern Arabia. *Quaternary Research* 66: 465–476.

Peck, V. L., C. S. Allen, S. Kender, E. L. McClymont, and D. A. Hodgson. 2015. Oceanographic variability on the West Antarctic Peninsula during the Holocene and the influence of upper circumpolar deep water. *Quaternary Science Reviews* 119: 54–65.

Petrie, C. A. 2014. Iran and Uruk Mesopotamia: Chronologies and Connections in the Fourth Millennium. In *Preludes to Urbanism: The Late Chalcolithic of Mesopotamia*, ed. A. McMahon and H. Crawford. Cambridge: McDonald Institute of Archaeological Research, University of Cambridge, 157–172.

Petrie, C. A., J. Bates, T. Higham & R. N. Singh. 2016. Feeding ancient cities in South Asia: dating the adoption of rice, millet and tropical pulses in the Indus civilization. *Antiquity* 90: 1489–1504. doi:10.15184/aqy.2016.210

Peyron, O., M. Magny, S. Goring, S. Joannin, J.-L. de Beaulieu, E. Brugiapaglia, L. Sadori, G. Garfi, K. Kouli, C. Ioakim, and N. Combourieu-Nebout. 2013. Contrasting patterns of climatic changes during the Holocene across the Italian peninsula reconstructed from pollen data. *Climate of the Past* 9: 1233–1252. doi: 10.5194/cp-9-1233-2013

Pfälzner, P. 2012. Household dynamics in late Third Millennium northern Mesopotamia. In *Seven Generations Since the Fall of Akkad,* ed. H. Weiss. Studia Chaburensia, vol. 3. Wiesbaden: Harrassowitz, 145–162.

Ponton, C., L. Giosan, T. I. Eglinton, D. Q. Fuller, J. E. Johnson, P. Kumar, and T. S. Collett. 2012. Holocene aridification of India. *Geophysical Research Letters* 39 (3): L03704. doi:10.1029/2011GL050722

Postgate, J. N. 1984. The problem of yields in Sumerian texts. *Bulletin on Sumerian Agriculture* 1: 97–102.

Postgate, J. N. 1986. The transition from Uruk to Early Dynastic: continuities and discontinuities in the record of settlement. In *Ğamdat Nasr: period or regional style?* ed. U. Finkbeiner and W. Rollig. Wiesbaden: Ludwig Reichert, 90–106.

Postgate, J. N.1992. The land of Assur and the yoke of Assur. *World Archaeology* 23: 247–263. doi: 10.1080/00438243.1992.9980178

Potts, D. T. 2007. Babylonian sources of exotic raw materials. In *The Babylonian World*, ed. G. Leick. London: Routledge, 124–140.

Powell, M. A. 1990. Masse und Gewichte. *Reallexikon der Assyriologie und Vorderasiatischen Archäologie* 7: 457–517.

Prasad, S., A. Anoop, N. Riedel, S. Sarkar, P. Menzel, N. Basavaiah, R. Krishnan, D. Fuller, et al. 2014. Prolonged monsoon droughts and links to Indo-Pacific warm-pool: A Holocene record from Lonar Lake, central India. *Earth and Planetary Science Letters* 391: 171–182.

Pustovoytov, K., K. Schmidt, and H. Taubald. 2007. Evidence for Holocene environmental changes in the northern Fertile Crescent provided by pedogenic carbonate coatings. *Quaternary Research* 67: 315–327.

P'yankova, L. 1994. Central Asia in the Bronze Age: sedentary and nomadic cultures. *Antiquity* 68: 355–372.

Railsback, L. B., F. Liang, J. R. V. Romaní, A. Grandal-d'Anglade, M. V. Rodríguez, L. S. Fidalgo, D. F. Mosquera, H. Cheng, and R. L. Edwards. 2011. Petrographic and isotopic evidence for Holocene long-term climate change and shorter-term environmental shifts from a stalagmite from the Serra do Courel of northwestern Spain, and implications for climatic history across Europe and the Mediterranean. *Palaeogeography, Palaeoclimatology, Palaeoecology* 305: 172–184.

Ramsey, C. B., M. W. Dee, J. M. Rowland, T. F. G. Higham, S. A. Harris, F. Brock, A. Quiles, E. M. Wild, E. S. Marcus, and A. J. Shortland. 2010. Radiocarbon-based chronology for dynastic Egypt. *Science* 328: 1554–1557.

Reed, J. M., N. Roberts, and M. J. Leng. 1999. An evaluation of the diatom response to Late Quaternary environmental change in two lakes in the Konya Basin, Turkey, by comparison with stable isotope data. *Quaternary Science Reviews* 18: 631–646.

Regattieri, E., G. Zanchetta, R. N. Drysdale, I. Isola, J. C. Hellstrom, and L. Dallai. 2014. Lateglacial to Holocene trace element record (Ba, Mg, Sr) from Corchia Cave (Apuan Alps, central Italy): paleoenvironmental implications. *Journal of Quaternary Science* 29: 381–392.

Revel, M., C. Colin, S. Bernasconi, N. Combourieu-Nebout, E. Ducassou, F. E. Grousset, Y. Rolland, S. Migeon, D. Bosch, P. Brunet, Y. Zhao, and J. Mascle. 2014. 21,000 years of Ethiopian African monsoon variability recorded in sediments of the western Nile deep-sea fan. *Regional Environmental Change* 14: 1685–1696.

Ristvet, L.. 2005. "Settlement, economy, and society in the Tell Leilan Region, Syria, 3000–1000 BC." Ph.D. diss. University of Cambridge.

Ristvet, L. 2008. Legal and archaeological territories of the Second Millennium BC in northern Mesopotamia. *Antiquity* 82: 585–599.

Ristvet, L. 2012. The development of underdevelopment? Imperialism, economic exploitation and settlement dynamics on the Khabur Plains, ca. 2300–2200 BC. In *Seven Generations Since the Fall of Akkad*, ed. H. Weiss. Studia Chaburensia, vol. 3. Wiesbaden: Harrassowitz, 241–261.

Ristvet, L., and H. Weiss. 2013. The Ḫābūr region in the Old Babylonian period. In *Archéologie et Histoire de la Syrie I. La Syrie de l'époque néolithique à l'âge du fer*, ed. W. Orthmann, P. Matthiae, and M. al-Maqdissi. Schriften zur Vorderasiatischen Archäologie 1/1. Wiesbaden: Harrassowitz, 257–272.

Roberts, N., J. M. Reed, M. J. Leng, C. Kuzucuoğlu, M. Fontugne, J. Bertaux, H. Woldring, S. Bottema, et al. 2001. The tempo of Holocene climatic change in the eastern Mediterranean region: new high-resolution crater-lake sediment data from central Turkey. *The Holocene* 11: 721–736.

Roberts, N. W. Eastwood, C. Kuzucuoğlu, G. Fiorentino, and V. Carracuta. 2011. Climate, vegetation and cultural change in the eastern Mediterranean during the mid-Holocene environmental transition. *The Holocene* 21: 147–162.

Rosenthal, Y., B. K. Linsley, and D. W. Oppo. 2013. Pacific Ocean Heat Content During the Past 10,000 Years. *Science* 342: 617–621.

Rothman, M. S. (Ed.). 2001. *Uruk Mesopotamia & Its Neighbors: Cross-Cultural Interactions in the Era of State Formation.* Santa Fe, NM: School of American Research.

Rova, E. and H. Weiss. (Eds.) *The Origins of North Mesopotamian Civilization: Ninevite 5 Chronology, Economy, Society.* Subartu IX. Turnhout, Belgium: Brepols.

Sadori, L., E. Ortu, O. Peyron, G. Zanchetta, B. Vannière, M. Desmet, and M. Magny. 2013. The last 7 millennia of vegetation and climate changes at Lago di Pergusa (central Sicily, Italy). *Climate of the Past* 9: 1969–1984.

Sadori, L., M. Giardini, E. Gliozzi, I. Mazzini, R. Sulpizio, A. van Welden, and G. Zanchetta. 2015. Vegetation, climate and environmental history of the last 4500 years at Lake Shkodra (Albania/Montenegro). *The Holocene* 25: 435–444.

Sallaberger, W. 2009. Die Amurriter-Mauer in Mesopotamien: der älteste historische Grenzwall gegen Nomaden vor 4000 Jahren. In *Mauern als Grenzen,* ed. A. Nunn. Mainz: Phillipp von Zabern, 27–38.

Sallaberger, W., and A. Pruss. 2015. Home and work in Early Bronze Age Mesopotamia: "ration lists" and "private houses" at Tell Beydar/Nabada. In *Labor in the Ancient World,* ed. P. Steinkeller and M. Hudson. International Scholars Conferences on Ancient Near Eastern Economics 5 Dresden: Islet, 69–136.

Sallaberger, W., and I. Schrakamp. 2015. Philological data for a historical chronology of Mesopotamia in the 3rd Millennium. In *History and Philology,* ed. W. Sallaberger and I. Schrakamp. ARCANE, vol. 3. Turnhout, Belgium: Brepols, 1–136.

Salzer, M. W., A. G. Bunn, N. E. Graham, and M. K. Hughes. 2014. Five millennia of paleotemperature from tree-rings in the Great Basin, USA. *Climate Dynamics* 42: 1517–1526.

Sandweiss, D. H., R. S. Solís, M. E. Moseley, D. K. Keefer, and C. R. Ortloff. 2009. Environmental change and economic development in coastal Peru between 5,800 and 3,600 years ago. *Proceedings of the National Academy of Sciences* 106: 1359–1363. doi: 10.1073/pnas.0812645106

Schefuss, E., H. Kuhlmann, G. Mollenhauer, M. Prange, and J. Pätzold. 2011. Forcing of wet phases in southeast Africa over the past 17,000 years. *Nature* 480: 509–512. doi: 10.1038/nature10685

Schittek, K., M. Forbriger, B. Mächtle, F. Schäbitz, V. Wennrich, M. Reindel, and B. Eitel. 2015. Holocene envirionmental changes in the highlands of the southern Peruvian Andes (14° S) and their impact on pre-Columbian cultures. *Climate of the Past* 11: 27–44. doi: 10.5194/cp-11/27-2015

Schmidt, A., M. Quigley, M. Fattahi, G. Azizi, M. Maghsoudi, and H. Fazeli. 2011. Holocene settlement shifts and palaeoenvironments on the Central Iranian Plateau: investigating linked systems. *The Holocene* 21: 583–595.

Schmidt, R., J. Müller, R. Drescher-Schneider, R. Krisai, K. Szeroczyńska, and A. Barić. 2000. Changes in lake level and trophy at Lake Vrana, a large karstic lake on the Island of Cres (Croatia), with respect to palaeoclimate and anthropogenic impacts during the last approx. 16,000 years. *Journal of Limnology* 59 (2): 113–130.

Schwartz, G. M. 2007. Taking the long view on collapse: a Syrian perspective. In *Sociétés humaines et changement climatique à la fin du troisième millénaire: une crise à-t-elle eu lieu en Haute Mésopotamie?* ed. C. Kuzucuoğlu and C. Marro. Institut français d'études anatoliennes GeorgesDumézil—Istanbul. Varia Anatolica, vol. 19. Paris: de Boccard, 45–67.

Schwartz, G. M., H. H. Curvers, S. S. Dunham, and J. A. Weber. 2012. From urban origins to imperial integration in western Syria: Umm el-Marra 2006, 2008. *American Journal of Archaeology* 116: 157–193.

Scussolini, P., T. Vegas-Vilarrúbia, V. Rull, J. P. Corella, B. Valero-Garcés, and J. Gomà. 2011. Middle and late Holocene climate change and human impact inferred from diatoms, algae and aquatic macrophyte pollen in sediments from Lake Montcortès (NE Iberian Peninsula). *Journal of Paleolimnology* 46: 369–385.

Sharlach, T. M. 2004. *Provincial Taxation and the Ur III State*. Leiden: Brill.

Sommerfeld, W., A. Archi, and H. Weiss. 2004. Why "Dada Measured 40,000 liters of barley from Nagar for Sippar." 4ICAANE Berlin, March 29–April 3, 2004. Tell Leilan Project Poster Presentations. (See Yale Tell Leilan Project website.)

Staubwasser, M., and H. Weiss. 2006. Holocene climate and cultural evolution in late prehistoric-early historic West Asia. *Quaternary Research* 66: 372–387.

Stevens, L. R., E. Ito, A. Schwalb, and H. E. Wright Jr. 2006. Timing of atmospheric precipitation in the Zagros Mountains inferred from a multi-proxy record from Lake Mirabad, Iran. *Quaternary Research* 66: 494–500.

Stevens, L. R., H. E. Wright, and E. Ito. 2001. Proposed changes in seasonality of climate during the Lateglacial and Holocene at Lake Zeribar, Iran. *The Holocene* 11: 747–755.

Styring, A. K., M. Charles, F. Fantone, M. M. Hald, A. McMahon, R. H. Meadow, G. K. Nicholls, A. K. Patel, M. C. Pitre, A. Smith, A. Sołtysiak, G. Stein, J. A., Weber, H. Weiss, and A. Bogaard. 2017. Isotope evidence for agricultural extensification reveals how the world's first cities were fed. *Nature Plants* 3(17076): 1–11.

Thomas, R. J., H. El-Dessougi, and A. Tubeileh. 2006. Soil system management under arid and semi-arid conditions. In *Biological Approaches to Sustainable Soil Systems*, ed. N. Uphoff. Boca Raton: CRC Press. doi:10.1201/9781420017113. ch4

Torrescano-Valle, N., and G. A. Islebe. 2015. Holocene paleoecology, climate history and human influence in the southwestern Yucatan Peninsula. *Review of Palaeobotany and Palynology* 217: 1–8.

Triantaphyllou, M. V., A. Gogou, I. Bouloubassi, M. Dimiza, K. Kouli, G. Rousakis, U. Kotthoff, K.-C. Emeis, M. Papanikolaou, M. Athanasiou, C. Parinos, C. Ioakim, and V. Lykousis. 2014. Evidence for a warm and humid mid-Holocene episode in the Aegean and northern Levantine seas (Greece, NE Mediterranean). *Regional Environmental Change* 14: 1697–1712.

Turner, R., N. Roberts, and M. D. Jones. 2008. Climatic pacing of Mediterranean fire histories from lake sedimentary microcharcoal. *Global and Planetary Change* 63: 317–324.

Ur, J. 2015. Urban adaptations to climate change in northern Mesopotamia. In *Climate and Ancient Societies*, ed. S. Kerner, R. J. Dann, and P. Bangsgaard. Copenhagen: Museum, Tuscalanum Press

Ülgen, U. B., S. O. Franz, D. Biltekin, M. N. Çagatay, P. A. Roeser, L. Doner, and J. Thein. 2012. Climatic and environmental evolution of Lake Iznik (NW Turkey) over the last ~4700 years. *Quaternary International* 274: 88–101.

Valladas, H., J. Evin, and M. Arnold. 1996. Datation par la methode du carbone 14 des couches Obeid 0 et 1 de Tel Oueilli (Iraq). In *'Oueilli: Travaux de 1987 et 1989*, ed. J.-J. Huot. Paris: Editions Recherche sur les Civilisations, 381–384.

Van de Mieroop, M. 1999. *Cuneiform Texts and the Writing of History*. New York: Routledge.

van der Pflicht, J., P. M. M. G. Akkermans, O. Nieuwenhuyse, A. Kaneda, and A. Russell. 2011. Tell Sabi Abyad, Syria: radiocarbon chronology, cultural change, and the 8.2 ka event. *Radiocarbon* 53: 229–243.

Verheyden, S., F. H. Nader, H. J. Cheng, L. R. Edwards, and R. Swennen. 2008. Paleoclimate reconstruction in the Levant region from the geochemistry of a Holocene stalagmite from the Jeita Cave, Lebanon. *Quaternary Research* 70: 368–381.

Visicato, G. 1999. The Sargonic archive at Tell Suleimah. *Journal of Cuneiform Studies* 51: 17–30. doi: 10.2307/1359727

Visicato, G. 2000. *The Power and the Writing: The Sarly scribes of Mesopotamia*. Bethesda: CDL.

Voûte, C. 1961. A comparison between some hydrological observations made in the Jurassic and Cenomian limestone mountains situated to the west and to the east of the Ghab Graben (U.A.R., Syria). In *Eaux souterraines dans les zones arides: colloque d'Athènes, 10–9–18–9*. Gentbrugge, Belgium: Association international d'hydrologie scientifique, 160–66.

Wagner, B., A. F. Lotter, N. Nowaczyk, J. M. Reed, A. Schwalb, R. Sulpizio, V. Valsecchi, M. Wessels, and G. Zanchetta. 2009. A 40,000-year record of environmental change from ancient Lake Ohrid (Albania and Macedonia). *Journal of Paleolimnology* 41: 407–430.

Walker, M. J. C., M. Berkelhammer, S. Björck, L. C. Cwynar, D. A. Fisher, A. J. Long, J. J. Lowe, R. M. Newnham, S. O. Rasmussen, and H. Weiss. 2012. Formal subdivision of the Holocene Series/Epoch: a discussion paper by a working group of INTIMATE (integration of ice-core, marine and terrestrial records) and the Subcommission on Quaternary Stratigraphy (International Commission on Stratigraphy). *Journal of Quaternary Science* 27: 649–659.

Wang, J., L. Zhu, Y. Wang, P. Peng, Q. Ma, T. Haberzetti, T. Kasper, T. Matsunaka, and T. Nakamura. 2015. Variability of the ^{14}C reservoir effects in Lake Tangra Yumco, Central Tibet (China), determined from recent sedimentation rates and dating of plant fossils. *Quaternary International*. doi: 10.1016/j.quaint.2015.10.084

Weiberg, E., and M. Finné. 2013. Mind or matter? People-environment interactions and the demise of early Helladic II society in the northeastern Peloponnese. *American Journal of Archaeology* 117: 1–31.

Weiss, H. 1986. The origins of Tell Leilan and the conquest of space in third millennium Mesopotamia. In *The Origins of Cities in Dry-Farming Syria and Mesopotamia in the Third Millennium B.C.*, ed. H. Weiss. Guilford, CT: Four Quarters, 71–108.

————. 1990. Tell Leilan 1989: New data for mid-third millennium urbanization and state formation. *Mitteilungen der Deutschen Orient-Gesellschaft zu Berlin* 122: 193–218.

————. 2000. Beyond the Younger Dryas: collapse as adaptation to abrupt climate change in ancient West Asia and the eastern Mediterranean. In

Environmental Disaster and the Archaeology of Human Response, ed. G. Bawden and R. M. Reycraft. Anthropological Papers 7. Albuquerque: Maxwell Museum of Anthropology, 75–98.

_____. 2003. Ninevite 5 periods and processes. In *The Origins of North Mesopotamian Civilization: Ninevite 5 Chronology, Economy, Society*, ed. E. Rova and H. Weiss. Subartu IX. Turnhout, Belgium: Brepols, 593–624.

_____. 2012. Quantifying collapse. In *Seven Generations Since the Fall of Akkad*, ed. H. Weiss. Studia Chaburensia, vol. 3. Wiesbaden: Harrassowitz, 1–24.

_____. 2014. The northern Levant during the Intermediate Bronze Age: altered trajectories. In *The Oxford Handbook of the Archaeology of the Levant: c. 8000–332 BCE*, ed. Margreet L. Steiner and Ann E. Killebrew. Oxford: Oxford University Press, 367–387.

Weiss, H., M.-A. Courty, W. Wetterstrom, F. Guichard, L. Senior, R. Meadow, and A. Curnow. 1993. The genesis and collapse of third millennium north Mesopotamian civilization. *Science* 261: 995–1004.

Weiss, H., S. W. Manning, L. Ristvet, L. Mori, M. Besonen, A. McCarthy, P. Quenet, A. Smith, and Z. Bahrani. 2012. Tell Leilan Akkadian Imperialization, Collapse and Short-Lived Reoccupation Defined by High-Resolution Radiocarbon Dating. In *Seven Generations Since the Fall of Akkad*, ed. H. Weiss. Studia Chaburensia, vol. 3. Wiesbaden: Harrassowitz, 163–192.

Welc, F., and L. Marks. 2014. Climate change at the end of the Old Kingdom in Egypt around 4200 BP: New geoarchaeological evidence. *Quaternary International* 324: 124–133.

Weldeab, S. W., R. R. Schneider, M. Kölling, and G. Wefer. 2005. Holocene African droughts relate to eastern equatorial Atlantic cooling. *Geology* 33: 981–984.

Westenholz, A. 1987. *Old Sumerian and Old Akkadian Texts in Philadelphia, Chiefly from Nippur, Part Two: The "Akkadian" Texts, the Enlilemaba Texts, and the Onion Archive*. Carsten Niebuhr Institute. Copenhagen: Museum Tusculanum.

Westenholz, A. 1999. The Old Akkadian Period: History and Culture. In *Mesopotamien: Akkade-Zeit und Ur III-Zeit*, ed. P. Attinger and M. Wäfler. Orbis Biblicus et Orientalis 160(3). Göttingen: Vandenhoeck & Ruprecht, 17–117.

Westenholz, J. G. 1998. Relations between Mesopotamia and Anatolia in the age of the Sargonic kings. In *XXXIV International Assyriology Congress*, ed. H. Erkanal, V. Donbaz, and A. Uğuroğlu. Ankara: Türk Tarih Kurumu, 5–22.

White, P. F., T. T. Teacher, A. Termanini. 1997. Nitrogen cycling in semi-arid Mediterranean zones: removal and return of nitrogen to pastures by grazing sheep. *Australian Journal of Agricultural Research* 48: 317–322

Wick, L., G. Lemcke, and M. Sturm. 2003. Evidence of Lateglacial and Holocene climatic change and human impact in eastern Anatolia: high-resolution pollen, charcoal, isotopic and geochemical records from the laminated sediments of Lake Van, Turkey. *The Holocene* 13: 665–675.

Wilkinson, T. J. 1982. The Definition of Ancient Manured Zones by Means of Extensive Sherd-Sampling Techniques. Journal of Field Archaeology 9: 323–333.

Wilkinson, T. J. 1990.Town & Country in Southeastern Anatolia: Settlement and landuse at Kurban Hoyuk and other sites in the Lower Karababa Basin. Chicago: Oriental Institute of the University of Chicago.

Wilkinson, T. J. 2003. *Archaeological Landscapes of the Near East*. Tucson: University of Arizona.

Wilkinson, T. J., J. H. Christiansen, J. Ur, M. Widell, and M. al-Taweel. 2007. Urbanization within a dynamic environment: modeling Bronze Age communities in Upper Mesopotamia. *American Anthropologist* 109: 52–68.

Wirth, E. 1971. *Syrien: eine geographische Landeskunde*. Darmstadt: Wissenschaftliche Buchgesellschaft.

Yener, K. A. 2005. Conclusions. In *The Amuq Valley Regional Projects, Vol. 1: Surveys in the Plain of Antioch and Orontes Delta, Turkey, 1995–2002*, ed. K. A. Yener. Chicago: Oriental Institute Press of the University of Chicago, 193–202.

Zanchetta, G., A. Van Welden, I. Baneschi, R. Drysdale, L. Sadori, N. Roberts, M. Giardini, C. Beck, V. Pascucci, and R. Sulpizio. 2012. Multiproxy record for the last 4500 years from Lake Shkodra (Albania/Montenegro). *Journal of Quaternary Science* 27: 780–789.

Zanchetta, G., E. Regattieri, I. Isola, R. N. Drysdale, M. Bini, I Baneschi, and J. C. Hellstrom. 2016. The so-called "4.2 event" in the central Mediterranean and its climatic teleconnections. *Alpine and Mediterranean Quaternary* 29: 5–17.

Zangger, E., M. E. Timpson, S. B. Yazvenko, F. Kuhnke, and J. Knauss. 1997. The Pylos regional archaeological project, part II: landscape evolution and site preservation. *Hesperia* 66 (4): 549–641.

Zhang, Q.-B., and R. J. Hebda. 2005. Abrupt climate change and variability in the past four millennia of the southern Vancouver Island, Canada. *Geophysical Research Letters* 32: L16708. doi: 10.1029/2005GL022913

Appendix 3.1 Proxymap Sites and Sources

SITE #	SITE/RECORD NAME	BIBLIOGRAPHIC REFERENCE
1	Crag Cave	McDermott, F., D.P. Mattey, and C. Hawkesworth. 2001. Centennial-Scale Holocene Climate Variability Revealed by a High-Resolution Speleothem $\delta^{18}O$ Record from SW Ireland. *Science* 294 (5545): 1328-1331. doi 10.1126/science.1063678.
2	Kharabuluk Mire	Kremenetski, C.V., O.A. Chichagova, and N.I. Shishlina. 1999. Palaeoecological evidence for Holocene vegetation, climate and land-use change in the low Don basin and Kalmuk area, southern Russia. *Vegetation History and Archaeobotany* 8 (4): 233-246. doi 10.1007/BF01291776.
3	Razdorskaya	Kremenetski, C.V. 1991. Palaeoecology of the Earliest Farmers and Herders of the Russian Plain [Paleoekologiya Drevneishikh Zemledeltsev i Skotovodov Russkoi Ravniny], Thesis, 193pp. Moscow: Institute of Geography.
4	Spannagel Cave	Vollweiler, N., D. Scholz, C. Mühlinghaus, A. Mangini, and C. Spötl. 2006. A precisely dated climate record for the last 9 kyr from three high alpine stalagmites, Spannagel Cave, Austria. *Geophysical Research Letters* 33 (20): L20703. doi 10.1029/2006GL027662.
5	Ursilor Cave	Onac, B.P., S. Constantin, J. Lundberg, and S.-E. Lauritzen. 2002. Isotopic climate record in a Holocene stalagmite from Ursilor Cave (Romania). *Journal of Quaternary Science* 17 (4): 319–327. doi 10.1002/jqs.685.
6	Kardashinski Swamp	Kremenetski, C.V. 1995. Holocene vegetation and climate history of southwestern Ukraine. *Review of Palaeobotany and Palynology* 85 (3–4): 289-301. doi 10.1016/0034-6667(94)00123-2.
7	Lake Vrana	Schmidt et al. 2000 Schneider, R. Krisai, K. Szeroczyńska, and A. Barić. 2000. Changes in lake level and trophy at Lake Vrana, a large karstic lake on the Island of Cres (Croatia), with respect to palaeoclimate and anthropogenic impacts during the last approx. 16,000 years. *Journal of Limnology* 59 (2): 113-130. doi 10.4081/jlimnol.2000.113.

SITE #	SITE/RECORD NAME	BIBLIOGRAPHIC REFERENCE
8	Heraklean Peninsula and Chyornaya Valley	Cordova, C.E., and P.H. Lehman. 2005. Holocene environmental change in southwestern Crimea (Ukraine) in pollen and soil records. *The Holocene* 15 (2): 263-277. doi 10.1191/0959683605hl791rp.
9	Lake Petit	Brisset et al. 2013 Brisset, E., C. Miramont, F. Guiter, E.J. Anthony, K. Tachikawa, J. Poulenard, F. Arnaud, C. Delhon, J.-D. Meunier, E. Bard, and F. Suméra. 2013. Non-reversible geosystem destabilisation at 4200 cal. BP: Sedimentological, geochemical and botanical markers of soil erosion recorded in a Mediterranean alpine lake. *The Holocene* 23 (12): 1863-1874. doi 10.1177/0959683613508158.
10	Buca della Renella	Drysdale, R., G. Zanchetta, J. Hellstrom, R. Maas, A. Fallick, M. Pickett, I. Cartwright, and L. Piccini. 2006. Late Holocene drought responsible for the collapse of Old World civilizations is recorded in an Italian cave flowstone. *Geology* 34 (2): 101-104. doi 10.1130/g22103.1.
11	Corchia Cave	Regattieri, E., G. Zanchetta, R.N. Drysdale, I. Isola, J.C. Hellstrom, and L. Dallai. 2014. Lateglacial to Holocene trace element record (Ba, Mg, Sr) from Corchia Cave (Apuan Alps, central Italy): paleoenvironmental implications. *Journal of Quaternary Science* 29 (4): 381-392. doi 10.1002/jqs.2712.
		Zanchetta, G., M. Bar-Matthews, R.N. Drysdale, P. Lionello, A. Ayalon, J.C. Hellstrom, I. Isola, and E. Regattieri. IN PRESS. Coeval dry events in the central and eastern Mediterranean basin at 5.2 and 5.6 ka recorded in Corchia (Italy) and Soreq Cave (Israel) speleothems. *Global and Planetary Change*. doi 10.1016/j.gloplacha.2014.07.013.
12	Arkhyz	Kvavadze, E.V., and Y.V. Efremov. 1996. Palynological studies of lake and lake-swamp sediments of the Holocene in the high mountains of Arkhyz (Western Caucasus). *Acta Palaeobotanica* 36 (1): 107-119.
13	Lake Accesa	Magny, M., J.-L. de Beaulieu, R. Drescher-Schneider, B. Vannière, A.-V. Walter-Simonnet, Y. Miras, L. Millet, G. Bossuet, O. Peyron, E. Brugiapaglia, and A. Leroux. 2007. Holocene climate changes in the central Mediterranean

as recorded by lake-level fluctuations at Lake Accesa (Tuscany, Italy). *Quaternary Science Reviews* 26 (13–14): 1736-1758. doi 10.1016/ j.quascirev.2007.04.014.

Peyron, O., M. Magny, S. Goring, S. Joannin, J.L. de Beaulieu, E. Brugiapaglia, L. Sadori, G. Garfi, K. Kouli, C. Ioakim, and N. Combourieu-Nebout. 2013. Contrasting patterns of climatic changes during the Holocene across the Italian Peninsula reconstructed from pollen data. *Climate of the Past* 9 (3): 1233-1252. doi 10.5194/cp-9-1233-2013.

Zanchetta, G., E. Regattieri, I. Isola, R. Drysdale, M. Bini, I. Baneschi, J. C. Hellstrom. 2016. The so-called "4.2 event" in the central Mediterranean and its climatic teleconnections. *Alpine and Mediterranean Quaternary* 29: 5–17.

SITE #	SITE/RECORD NAME	BIBLIOGRAPHIC REFERENCE
14	Straldzha Mire	Connor, S.E., S.A. Ross, A. Sobotkova, A.I.R. Herries, S.D. Mooney, C. Longford, and I. Iliev. 2013. Environmental conditions in the SE Balkans since the Last Glacial Maximum and their influence on the spread of agriculture into Europe. *Quaternary Science Reviews* 68: 200-215. doi 10.1016/j.quascirev.2013.02.011.
15	Cova da Arcoia	Railsback, L.B., F. Liang, J.R. Vidal Romaní, A. Grandal-d'Anglade, M. Vaqueiro Rodríguez, L. Santos Fidalgo, D. Fernández Mosquera, H. Cheng, and R.L. Edwards. 2011. Petrographic and isotopic evidence for Holocene long-term climate change and shorter-term environmental shifts from a stalagmite from the Serra do Courel of northwestern Spain, and implications for climatic history across Europe and the Mediterranean. *Palaeogeography, Palaeoclimatology, Palaeoecology* 305 (1–4): 172-184. doi 10.1016/j.palaeo.2011.02.030.
16	Lake Montcortes	Scussolini, P., T. Vegas-Vilarrúbia, V. Rull, J. Corella, B. Valero-Garcés, and J. Gomà. 2011. Middle and late Holocene climate change and human impact inferred from diatoms, algae and aquatic macrophyte pollen in sediments from Lake Montcortès (NE Iberian Peninsula). *Journal of Paleolimnology* 46 (3): 369-385. doi 10.1007/s10933-011-9524-y.

17	Central Italy Maar Lakes	Magri, D., and I. Parra. 2002. Late Quaternary western Mediterranean pollen records and African winds. *Earth and Planetary Science Letters* 200 (3–4): 401-408. doi 10.1016/ S0012-821X(02)00619-2.
		Magri, D. 1997. "Middle and Late Holocene Vegetation and Climate Changes in Peninsular Italy." In *Third Millennium BC Climate Change and Old World Collapse*, edited by H. Nüzhet Dalfes, George Kukla and Harvey Weiss, 517-530. Springer Berlin Heidelberg.
		Alessio, M., L. Allegri, F. Bella, G. Calderoni, C. Cortesi, G. Dai Pra, D. De Rita, D. Esu, M. Follieri, S. Improta, D. Magri, B. Narcisi, V. Petrone, and L. Sadori. 1986. ^{14}C dating, geochemical features, faunistic and pollen analyses of the uppermost 10 m core from Valle di Castiglione (Rome, Italy). *Geologica Romana* 25: 287-308.
		Magri, D. 1999. Late Quaternary vegetation history at Lagaccione near Lago di Bolsena (central Italy). *Review of Palaeobotany and Palynology* 106 (3–4): 171-208. doi 10.1016/ S0034-6667(99)00006-8.
		Magri, D., and L. Sadori. 1999. Late Pleistocene and Holocene pollen stratigraphy at Lago di Vico, central Italy. *Vegetation History and Archaeobotany* 8 (4): 247-260. doi 10.1007/BF01291777.
18	Lake Shkodra	Zanchetta, G., A. Van Welden, I. Baneschi, R. Drysdale, L. Sadori, N. Roberts, M. Giardini, C. Beck, V. Pascucci, and R. Sulpizio. 2012. Multiproxy record for the last 4500 years from Lake Shkodra (Albania/Montenegro). *Journal of Quaternary Science* 27 (8): 780-789. doi 10.1002/ jqs.2563.
		Mazzini, I., E. Gliozzi, R. Koci, I. Soulie-Märsche, G. Zanchetta, I. Baneschi, L. Sadori, M. Giardini, A. Van Welden, and S. Bushati. 2015. Historical evolution and middle to late Holocene environmental changes in Lake Shkodra (Albania): new evidences from micropaleontological analysis. *Palaeogeography, Palaeolimnology, Palaeoclimatology* 419: 47–59.

SITE #	SITE/RECORD NAME	BIBLIOGRAPHIC REFERENCE
19	Lake Dojran	Francke, A., B. Wagner, M.J. Leng, and J. Rethemeyer. 2013. A Late Glacial to Holocene record of environmental change from Lake Dojran (Macedonia, Greece). *Climate of the Past* 9 (1): 481-498. doi 10.5194/cp-9-481-2013.
20	Central Caspian Sea	Leroy, S.A.G., F. Marret, E. Gibert, F. Chalié, J.L. Reyss, and K. Arpe. 2007. River inflow and salinity changes in the Caspian Sea during the last 5500 years. *Quaternary Science Reviews* 26 (25–28): 3359-3383. doi 10.1016/j.quascirev.2007.09.012.
		Leroy, S.A.G., L. López-Merino, A. Tudryn, F. Chalié, and F. Gasse. 2014. Late Pleistocene and Holocene palaeoenvironments in and around the middle Caspian basin as reconstructed from a deep-sea core. *Quaternary Science Reviews* 101: 91-110. doi 10.1016/j.quascirev.2014.07.011.
21	Lake Ohrid	Wagner, B., A. Lotter, N. Nowaczyk, J. Reed, A. Schwalb, R. Sulpizio, V. Valsecchi, M. Wessels, and G. Zanchetta. 2009. A 40,000-year record of environmental change from ancient Lake Ohrid (Albania and Macedonia). *Journal of Paleolimnology* 41 (3): 407-430. doi 10.1007/s10933-008-9234-2.
22	Yeniçaga Golu	Bottema, S. 1997. "Third Millennium Climate in the Near East Based upon Pollen Evidence." In *Third Millennium BC Climate Change and Old World Collapse*, edited by H. Nüzhet Dalfes, George Kukla and Harvey Weiss, 489-515. Springer Berlin Heidelberg.
		van Zeist, W., and S. Bottema. 1991. *Late Quaternary Vegetation of the Near East, Tübinger Atlas des Vorderen Orients (TAVO)*. Wiesbaden: Dr. Ludwig Reichert Verlag.
23	Abant Golu	Bottema, S. 1997. "Third Millennium Climate in the Near East Based upon Pollen Evidence." In *Third Millennium BC Climate Change and Old World Collapse*, edited by H. Nüzhet Dalfes, George Kukla and Harvey Weiss, 489-515. Springer Berlin Heidelberg.

SITE #	SITE/RECORD NAME	BIBLIOGRAPHIC REFERENCE
24	Lake Iznik	Ülgen, U.B., S.O. Franz, D. Biltekin, M.N. Çagatay, P.A. Roeser, L. Doner, and J. Thein. 2012. Climatic and environmental evolution of Lake Iznik (NW Turkey) over the last ~4700 years. *Quaternary International* 274 (0): 88-101. doi 10.1016/j.quaint.2012.06.016.
25	Kaz Gölü	Bottema, S. 1997. "Third Millennium Climate in the Near East Based upon Pollen Evidence." In *Third Millennium BC Climate Change and Old World Collapse*, edited by H. Nüzhet Dalfes, George Kukla and Harvey Weiss, 489-515. Springer Berlin Heidelberg.
26	Yenişehir	Bottema, S., H. Woldring, and I. Kayan. 2001. "The late Quaternary vegetation history of western Turkey." In *The Ilıpınar Excavations II*, edited by J.J. Roodenberg and L.C. Thissen, 327-354. Leiden: Nederlands Instituut voor het Nabije Oosten.
27	Lago Alimini Piccolo	Di Rita, F., and D. Magri. 2009. Holocene drought, deforestation and evergreen vegetation development in the central Mediterranean: a 5500 year record from Lago Alimini Piccolo, Apulia, southeast Italy. *The Holocene* 19 (2): 295-306. doi 10.1177/0959683608100574.
28	Rezina Marsh	Willis, K.J. 1992b. The late Quaternary vegetational history of northwest Greece, II. Rezina marsh. *New Phytologist* 121 (1): 119-138. doi 10.1111/j.1469-8137.1992.tb01098.x.
29	Gramousti Lake	Willis, K.J. 1992a. The late Quaternary vegetational history of northwest Greece, I. Lake Gramousti. *New Phytologist* 121 (1): 101-117. doi 10.1111/j.1469-8137.1992.tb01097.x.
30	Tecer Lake	Kuzucuoğlu, C., W. Dörfler, S. Kunesch, and F. Goupille. 2011. Mid- to late-Holocene climate change in central Turkey: The Tecer Lake record. *The Holocene* 21 (1): 173-188. doi 10.1177/0959683610384163.
31	Söğûtlû Marsh	Bottema, S. 1995. Holocene vegetation of the Van area: palynological and chronological evidence from Söğütlü, Turkey. *Vegetation History and Archaeobotany* 4 (3): 187-193. doi 10.1007/BF00203937.

SITE #	SITE/RECORD NAME	BIBLIOGRAPHIC REFERENCE
		Bottema, S. 1997. "Third Millennium Climate in the Near East Based upon Pollen Evidence." In *Third Millennium BC Climate Change and Old World Collapse*, edited by H. Nüzhet Dalfes, George Kukla and Harvey Weiss, 489-515. Springer Berlin Heidelberg.
32	Lake Van	Wick, L., G. Lemcke, and M. Sturm. 2003. Evidence of Lateglacial and Holocene climatic change and human impact in eastern Anatolia: high-resolution pollen, charcoal, isotopic and geochemical records from the laminated sediments of Lake Van, Turkey. *The Holocene* 13 (5): 665-675. doi 10.1191/0959683603hl653rp.
		Lemcke, G., and M. Sturm. 1997. "$\delta^{18}O$ and Trace Element Measurements as Proxy for the Reconstruction of Climate Changes at Lake Van (Turkey): Preliminary Results." In *Third Millennium BC Climate Change and Old World Collapse*, edited by H. Nüzhet Dalfes, George Kukla and Harvey Weiss, 653-678. Springer Berlin Heidelberg.
33	Eski Açigol	Roberts, N., J.M. Reed, M.J. Leng, C. Kuzucuoğlu, M. Fontugne, J. Bertaux, H. Woldring, S. Bottema, S. Black, E. Hunt, and M. Karabiyikoğlu. 2001. The tempo of Holocene climatic change in the eastern Mediterranean region: new high-resolution crater-lake sediment data from central Turkey. *The Holocene* 11 (6): 721-736. doi 10.1191/095968301195744.
34	Arslantepe	Masi, A., L. Sadori, G. Zanchetta, I. Baneschi, and M. Giardini. 2013. Climatic interpretation of carbon isotope content of mid-Holocene archaeological charcoals from eastern Anatolia. *Quaternary International* 303 (0): 64-72. doi 10.1016/j.quaint.2012.11.010.
35	Northern and Southern Aegean Sea	Ehrmann, W., G. Schmiedl, Y. Hamann, T. Kuhnt, C. Hemleben, and W. Siebel. 2007. Clay minerals in late glacial and Holocene sediments of the northern and southern Aegean Sea. *Palaeogeography, Palaeoclimatology, Palaeoecology* 249 (1–2): 36-57. doi 10.1016/j.palaeo.2007.01.004.

SITE #	SITE/RECORD NAME	BIBLIOGRAPHIC REFERENCE
		Kuhnt, T., G. Schmiedl, W. Ehrmann, Y. Hamann, and N. Andersen. 2008. Stable isotopic composition of Holocene benthic foraminifers from the Eastern Mediterranean Sea: Past changes in productivity and deep water oxygenation. *Palaeogeography, Palaeoclimatology, Palaeoecology* 268 (1–2): 106-115. doi 10.1016/ j.palaeo.2008.07.010.
36	Lago Preola	Magny, M., B. Vannière, C. Calo, L. Millet, A. Leroux, O. Peyron, G. Zanchetta, T. La Mantia, and W. Tinner. 2011. Holocene hydrological changes in south-western Mediterranean as recorded by lake-level fluctuations at Lago Preola, a coastal lake in southern Sicily, Italy. *Quaternary Science Reviews* 30 (19–20): 2459-2475. doi 10.1016/j.quascirev.2011.05.018.
37	Lake Lerna/Argive Plain	Jahns, S. 1993. On the Holocene vegetation history of the Argive Plain (Peloponnese, southern Greece). *Vegetation History and Archaeobotany* 2 (4): 187-203. doi 10.1007/ BF00198161.
38	Puerto de Mazarrón	Navarro-Hervás, F., M.-M. Ros-Salas, T. Rodríguez-Estrella, E. Fierro-Enrique, J.-S. Carrión, J. García-Veigas, J.-A. Flores, M.Á. Bárcena, and M.S. García. 2014. Evaporite evidence of a mid-Holocene (c. 4550–4400 cal. yr BP) aridity crisis in southwestern Europe and palaeoenvironmental consequences. *The Holocene* 24 (4): 489-502. doi 10.1177/ 0959683613520260.
39	Lago di Pergusa	Sadori, L., and B. Narcisi. 2001. The Postglacial record of environmental history from Lago di Pergusa, Sicily. *The Holocene* 11 (6): 655-671. doi 10.1191/095968301195681.
		Sadori, L., E. Ortu, O. Peyron, G. Zanchetta, B. Vannière, M. Desmet, and M. Magny. 2013. The last 7 millennia of vegetation and climate changes at Lago di Pergusa (central Sicily, Italy). *Climate of the Past* 9 (4): 1969-1984. doi 10.5194/ cp-9-1969-2013.

SITE #	SITE/RECORD NAME	BIBLIOGRAPHIC REFERENCE
		Peyron, O., M. Magny, S. Goring, S. Joannin, J.L. de Beaulieu, E. Brugiapaglia, L. Sadori, G. Garfi, K. Kouli, C. Ioakim, and N. Combourieu-Nebout. 2013. Contrasting patterns of climatic changes during the Holocene across the Italian Peninsula reconstructed from pollen data. *Climate of the Past* 9 (3): 1233-1252. doi 10.5194/cp-9-1233-2013.
40	Konya Basin Lakes	Roberts, N., S. Black, P. Boyer, W.J. Eastwood, H.I. Griffiths, H.F. Lamb, M.J. Leng, R. Parish, J.M. Reed, D. Twigg, and H. Yiğitbaşioğlu. 1999. Chronology and stratigraphy of Late Quaternary sediments in the Konya Basin, Turkey: Results from the KOPAL Project. *Quaternary Science Reviews* 18 (4–5): 611-630. doi 10.1016/S0277-3791(98)00100-0.
		Leng, M.J., N. Roberts, J.M. Reed, and H.J. Sloane. 1999. Late Quaternary palaeohydrology of the Konya Basin, Turkey, based on isotope studies of modern hydrology and lacustrine carbonates. *Journal of Paleolimnology* 22 (2): 187-204. doi 10.1023/A:1008024127346.
		Reed, J.M., N. Roberts, and M.J. Leng. 1999. An evaluation of the diatom response to Late Quaternary environmental change in two lakes in the Konya Basin, Turkey, by comparison with stable isotope data. *Quaternary Science Reviews* 18 (4–5): 631-646. doi 10.1016/S0277-3791(98)00101-2.
		Boyer, P., N. Roberts, and D. Baird. 2006. Holocene environment and settlement on the Çarşamba alluvial fan, south-central Turkey: Integrating geoarchaeology and archaeological field survey. *Geoarchaeology* 21 (7): 675-698. doi 10.1002/gea.20133.
41	Göbekli Tepe	Pustovoytov, K., K. Schmidt, and H. Taubald. 2007. Evidence for Holocene environmental changes in the northern Fertile Crescent provided by pedogenic carbonate coatings. *Quaternary Research* 67 (3): 315-327. doi 10.1016/j.yqres.2007.01.002.

SITE #	SITE/RECORD NAME	BIBLIOGRAPHIC REFERENCE
42	Southeastern Caspian Sea	Leroy, S.A.G., A.A. Kakroodi, S. Kroonenberg, H.K. Lahijani, H. Alimohammadian, and A. Nigarov. 2013. Holocene vegetation history and sea level changes in the SE corner of the Caspian Sea: relevance to SW Asia climate. *Quaternary Science Reviews* 70: 28-47. doi 10.1016/j.quascirev.2013.03.004.
43	Golhisar Gölü	Eastwood, W.J., N. Roberts, H.F. Lamb, and J.C. Tibby. 1999. Holocene environmental change in southwest Turkey: a palaeoecological record of lake and catchment-related changes. *Quaternary Science Reviews* 18 (4–5): 671-695. doi 10.1016/ S0277-3791(98)00104-8.
		Eastwood, W.J., M.J. Leng, N. Roberts, and B. Davis. 2007. Holocene climate change in the eastern Mediterranean region: a comparison of stable isotope and pollen data from Lake Gölhisar, southwest Turkey. *Journal of Quaternary Science* 22 (4): 327-341. doi 10.1002/jqs.1062. Leng, M.J., M.D. Jones, M.R. Frogley, W.J. Eastwood, C.P. Kendrick, and C.N. Roberts. 2010. Detrital carbonate influences on bulk oxygen and carbon isotope composition of lacustrine sediments from the Mediterranean. *Global and Planetary Change* 71 (3–4): 175-182. doi 10.1016/j.gloplacha.2009.05.005.
44	Borreguiles de la Virgen Bog	Jiménez-Moreno, G., and R.S. Anderson. 2012. Holocene vegetation and climate change recorded in alpine bog sediments from the Borreguiles de la Virgen, Sierra Nevada, southern Spain. *Quaternary Research* 77 (1): 44-53. doi 10.1016/j.yqres.2011.09.006.
45	Osmanaga Lagoon (Pylos)	Zangger, E., M.E. Timpson, S.B. Yazvenko, F. Kuhnke, and J. Knauss. 1997. The Pylos Regional Archaeological Project, Part II: Landscape Evolution and Site Preservation. *Hesperia* 66 (4): 549-641.
46	Tell Leilan	Weiss, H., M.-A. Courty, W. Wetterstrom, F. Guichard, L. Senior, R. Meadow, and A. Curnow. 1993. The Genesis and Collapse of Third Millennium North Mesopotamian Civilization. *Science* 261 (5124): 995-1004. doi 10.1126/science.261.5124.995.

SITE #	SITE/RECORD NAME	BIBLIOGRAPHIC REFERENCE
47	Sierra de Gador	Carrión, J.S., P. Sánchez-Gómez, J.F. Mota, R. Yll, and C. Chaín. 2003. Holocene vegetation dynamics, fire and grazing in the Sierra de Gádor, southern Spain. *The Holocene* 13 (6): 839-849. doi 10.1191/0959683603hl662rp.
48	Kos Basin, Southern Aegean Sea	Triantaphyllou, M.V., P. Ziveri, A. Gogou, G. Marino, V. Lykousis, I. Bouloubassi, K.C. Emeis, K. Kouli, M. Dimiza, A. Rosell-Melé, M. Papanikolaou, G. Katsouras, and N. Nunez. 2009. Late Glacial–Holocene climate variability at the south-eastern margin of the Aegean Sea. *Marine Geology* 266 (1–4): 182-197. doi 10.1016/j.margeo.2009.08.005.
49	Tell Mardikh (Ebla)	Fiorentino, G., V. Caracuta, L. Calcagnile, M. D'Elia, P. Matthiae, F. Mavelli, and G. Quarta. 2008. Third millennium B.C. climate change in Syria highlighted by Carbon stable isotope analysis of ^{14}C-AMS dated plant remains from Ebla. *Palaeogeography, Palaeoclimatology, Palaeoecology* 266 (1–2): 51-58. doi 10.1016/j.palaeo.2008.03.034.
		Gallet, Y., M. D'Andrea, A. Genevey, F. Pinnock, M. Le Goff, and P. Matthiae. 2014. Archaeomagnetism at Ebla (Tell Mardikh, Syria). New data on geomagnetic field intensity variations in the Near East during the Bronze Age. *Journal of Archaeological Science* 42 (0): 295-304. doi 10.1016/j.jas.2013.11.007.
50	Lake Zeribar	Stevens, L.R., H.E. Wright, and E. Ito. 2001. Proposed changes in seasonality of climate during the Lateglacial and Holocene at Lake Zeribar, Iran. *The Holocene* 11 (6): 747-755. doi 10.1191/095968301195762.
		Schmidt, A., M. Quigley, M. Fattahi, G. Azizi, M. Maghsoudi, and H. Fazeli. 2011. Holocene settlement shifts and palaeoenvironments on the Central Iranian Plateau: Investigating linked systems. *The Holocene* 21 (4): 583-595. doi 10.1177/0959683610385961.

SITE #	SITE/RECORD NAME	BIBLIOGRAPHIC REFERENCE
51	Ghab Valley	Bottema, S., and H. Woldring. 1990. "Anthropogenic indicators in the pollen record of the Eastern Mediterranean." In *Man's Role in the Shaping of the Eastern Mediterranean Landscape. Proceedings of the INQUA/BAI Symposium on the Impact of Ancient Man on the Landscape of the Eastern Mediterranean Regiona and the Near East, Groningen, Netherlands, 6-9 March 1989*, edited by Sytze Bottema, G. Entjes-Nieborg and W. van Zeist, 231-264. Rotterdam: A.A. Balkema.
		van Zeist, W., and H. Woldring. 1980. Holocene vegetation and climate of northwestern Syria. *Palaeohistoria* 22: 111-125.
		van Zeist, W., and S. Bottema. 1991. *Late Quaternary Vegetation of the Near East, Tübinger Atlas des Vorderen Orients (TAVO)*. Wiesbaden: Dr. Ludwig Reichert Verlag.
52	Tell Tweini	Kaniewski et al. 2008 Kaniewski, D., E. Paulissen, E. Van Campo, M. Al-Maqdissi, J. Bretschneider, and K. Van Lerberghe. 2008. Middle East coastal ecosystem response to middle-to-late Holocene abrupt climate changes. *Proceedings of the National Academy of Sciences* 105 (37): 13941-13946. doi 10.1073/pnas.0803533105.
53	Bouara Saltflats	Bottema, S. 1997. "Third Millennium Climate in the Near East Based upon Pollen Evidence." In *Third Millennium BC Climate Change and Old World Collapse*, edited by H. Nüzhet Dalfes, George Kukla and Harvey Weiss, 489-515. Springer Berlin Heidelberg.
		Gremmen, W.H.E., and S. Bottema. 1991. "Palynological Investigations in the Syrian Ğazīra." In *Die Rezente Umwelt von Tall Šēḫ Ḥamad und Daten zur Umweltrekonstruktion der Assyrischen Stadt Dūr-Katlimmu*, edited by Hartmut Kühne, 105–116. Berlin: Dietrich Reimer Verlag.
54	Jeita Cave	Verheyden, S., F.H. Nader, H.J. Cheng, L.R. Edwards, and R. Swennen. 2008. Paleoclimate reconstruction in the Levant region from the geochemistry of a Holocene stalagmite from the Jeita cave, Lebanon. *Quaternary Research* 70 (3): 368-381. doi 10.1016/j.yqres.2008.05.004.

SITE #	SITE/RECORD NAME	BIBLIOGRAPHIC REFERENCE
55	Lake Hula	Baruch, U., and S. Bottema. 1999. "A new pollen diagram from Lake Hula: vegetational, climatic and anthropogenic implications." In *Ancient Lakes: Their Cultural and Biological Diversity*, edited by Hiroya Kawanabe, George W. Coulter and Anna C. Roosevelt, 75-86. Ghent, Belgium: Kenobi Productions.
56	Lake Mirabad	Griffiths, H.I., A. Schwalb, and L.R. Stevens. 2001. Evironmental change in southwestern Iran: the Holocene ostracod fauna of Lake Mirabad. *The Holocene* 11 (6): 757-764. doi 10.1191/095968301957771.
		Stevens, L.R., E. Ito, A. Schwalb, and H.E. Wright Jr. 2006. Timing of atmospheric precipitation in the Zagros Mountains inferred from a multi-proxy record from Lake Mirabad, Iran. *Quaternary Research* 66 (3): 494-500. doi 10.1016/j.yqres.2006.06.008.
		Schmidt, A., M. Quigley, M. Fattahi, G. Azizi, M. Maghsoudi, and H. Fazeli. 2011. Holocene settlement shifts and palaeoenvironments on the Central Iranian Plateau: Investigating linked systems. *The Holocene* 21 (4): 583-595. doi 10.1177/0959683610385961.
57	Tso Moriri Lake	Leipe, C., D. Demske, and P.E. Tarasov. IN PRESS. A Holocene pollen record from the northwestern Himalayan lake Tso Moriri: Implications for palaeoclimatic and archaeological research. *Quaternary International*. doi 10.1016/j.quaint.2013.05.005.
58	Soreq Cave	Bar-Matthews, M., A. Ayalon, A. Kaufman, and G.J. Wasserburg. 1999. The Eastern Mediterranean paleoclimate as a reflection of regional events: Soreq cave, Israel. *Earth and Planetary Science Letters* 166 (1–2): 85-95. doi 10.1016/S0012-821X(98)00275-1.
		Zanchetta, G., M. Bar-Matthews, R.N. Drysdale, P. Lionello, A. Ayalon, J.C. Hellstrom, I. Isola, and E. Regattieri. IN PRESS. Coeval dry events in the central and eastern Mediterranean basin at 5.2 and 5.6 ka recorded in Corchia (Italy) and Soreq Cave (Israel) speleothems. *Global and Planetary Change*. doi 10.1016/j.gloplacha.2014.07.013.

SITE #	SITE/RECORD NAME	BIBLIOGRAPHIC REFERENCE
59	Nile Deep-Sea Fan	Blanchet, C.L., R. Tjallingii, M. Frank, J. Lorenzen, A. Reitz, K. Brown, T. Feseker, and W. Brückmann. 2013. High- and low-latitude forcing of the Nile River regime during the Holocene inferred from laminated sediments of the Nile deep-sea fan. *Earth and Planetary Science Letters* 364: 98-110. doi 10.1016/j.epsl.2013.01.009.
60	Dead Sea	Frumkin, A., I. Carmi, I. Zak, and M. Magaritz. 1994. "Middle Holocene environmental change determined from the salt caves of Mount Sedom, Israel." In *Late Quaternary Chronology and Paleoclimates of the Eastern Mediterranean. Papers from the 14th International Radiocarbon Conference, Tuscon, Arizona, 20-24 May 1991*, edited by Ofer Bar-Yosef and Renée S. Kra, 315–322. Tucson (Arizona): University of Arizona Press.
		Migowski, C., M. Stein, S. Prasad, J.F.W. Negendank, and A. Agnon. 2006. Holocene climate variability and cultural evolution in the Near East from the Dead Sea sedimentary record. *Quaternary Research* 66 (3): 421-431. doi 10.1016/j.yqres.2006.06.010.
		Frumkin, A. 2009. Stable isotopes of a subfossil Tamarix tree from the Dead Sea region, Israel, and their implications for the Intermediate Bronze Age environmental crisis. *Quaternary Research* 71 (3): 319-328. doi 10.1016/j.yqres.2009.01.009.
61	Burullus Lagoon	Bernhardt, C.E., B.P. Horton, and J.-D. Stanley. 2012. Nile Delta vegetation response to Holocene climate variability. *Geology* 40 (7): 615-618. doi 10.1130/g33012.1.
62	Manzala Lagoon	Goodfriend, G.A., and D.J. Stanley. 1999. Rapid strand-plain accretion in the northeastern Nile Delta in the 9th century A.D. and the demise of the port of Pelusium. *Geology* 27 (2): 147-150. doi 10.1130/0091-7613(1999)027<0147:rspait>2.3.co;2.
		Krom, M.D., J.D. Stanley, R.A. Cliff, and J.C. Woodward. 2002. Nile River sediment fluctuations over the past 7000 yr and their key role in sapropel development. *Geology* 30 (1): 71-74. doi 10.1130/0091-7613(2002)030<0071:nrsfot> 2.0.co;2.

SITE #	SITE/RECORD NAME	BIBLIOGRAPHIC REFERENCE
		Stanley, J.-D., M.D. Krom, R.A. Cliff, and J.C. Woodward. 2003. Short contribution: Nile flow failure at the end of the Old Kingdom, Egypt: Strontium isotopic and petrologic evidence. *Geoarchaeology* 18 (3): 395-402. doi 10.1002/gea.10065.
63	Lake Maharlou	Djamali, M., J.-L. De Beaulieu, N. Miller, V. Andrieu-Ponel, P. Ponel, R. Lak, N. Sadeddin, H. Akhani, and H. Fazeli. 2009. Vegetation history of the SE section of the Zagros Mountains during the last five millennia; a pollen record from the Maharlou Lake, Fars Province, Iran. *Vegetation History and Archaeobotany* 18 (2): 123-136. doi 10.1007/s00334-008-0178-2.
64	Faiyum Depression	Hassan, F.A. 1986. Holocene lakes and prehistoric settlements of the Western Faiyum, Egypt. *Journal of Archaeological Science* 13 (5): 483-501. doi 10.1016/0305-4403(86)90018-X.
		Hassan, F.A., G.J. Tassie, R. Flower, M. Hughes, and M. Hamden. 2006. Modelling environmental and settlement change in the Fayum. *Egyptian Archaeology* 29: 37-40.
		Davoli, P. 2013. "Fayyum." In *The Encyclopedia of Ancient History*, edited by Roger S. Bagnall, Kai Brodersen, Craige B. Champion, Andrew Erskine and Sabine R. Huebner, 2649–2652. John Wiley & Sons, Inc.
65	Northern Red Sea and Gulf of Aqaba	Arz, H.W., F. Lamy, J. Pätzold, P.J. Müller, and M. Prins. 2003. Mediterranean Moisture Source for an Early-Holocene Humid Period in the Northern Red Sea. *Science* 300 (5616): 118-121. doi 10.1126/science.1080325.
66	Paleolake Kotla Dahar	Dixit, Y., D.A. Hodell, and C.A. Petrie. 2014. Abrupt weakening of the summer monsoon in northwest India ~4100 yr ago. *Geology* 42 (4): 339-342. doi 10.1130/g35236.1.
67	Thar Desert Salt Lakes	Enzel, Y., L.L. Ely, S. Mishra, R. Ramesh, R. Amit, B. Lazar, S.N. Rajaguru, V.R. Baker, and A. Sandler. 1999. High-Resolution Holocene Environmental Changes in the Thar Desert, Northwestern India. *Science* 284 (5411): 125-128. doi 10.1126/science.284.5411.125.

SITE #	SITE/RECORD NAME	BIBLIOGRAPHIC REFERENCE
		Kajale, M.D., and B.C. Deotare. 1997. Late Quaternary environmental studies on salt lakes in western Rajasthan, India: a summarised view. *Journal of Quaternary Science* 12 (5): 405-412. doi 10.1002/(SICI)1099-1417(199709/10)12:5<405::AID-JQS323>3.0.CO;2-N.
68	Shaban Deep (Northern Red Sea)	Arz, H.W., F. Lamy, and J. Pätzold. 2006. A pronounced dry event recorded around 4.2 ka in brine sediments from the northern Red Sea. *Quaternary Research* 66 (3): 432-441. doi 10.1016/j.yqres.2006.05.006.
69	Awafi	Parker, A.G., L. Eckersley, M.M. Smith, A.S. Goudie, S. Stokes, S. Ward, K. White, and M.J. Hodson. 2004. Holocene vegetation dynamics in the northeastern Rub' al-Khali desert, Arabian Peninsula: a phytolith, pollen and carbon isotope study. *Journal of Quaternary Science* 19 (7): 665-676. doi 10.1002/jqs.880.
		Parker, A.G., A.S. Goudie, S. Stokes, K. White, M.J. Hodson, M. Manning, and D. Kennet. 2006. A record of Holocene climate change from lake geochemical analyses in southeastern Arabia. *Quaternary Research* 66 (3): 465-476. doi 10.1016/j.yqres.2006.07.001.
		Parker, A.G., and A.S. Goudie. 2008. Geomorphological and palaeoenvironmental investigations in the southeastern Arabian Gulf region and the implication for the archaeology of the region. *Geomorphology* 101 (3): 458-470. doi 10.1016/j.geomorph.2007.04.028.
70	Mawmluh Cave	Berkelhammer, M., A. Sinha, L. Stott, H. Cheng, F.S.R. Pausata, and K. Yoshimura. 2013. "An Abrupt Shift in the Indian Monsoon 4000 Years Ago." In *Climates, Landscapes, and Civilizations*, edited by Liviu Giosan, Dorian Q. Fuller, Kathleen Nicoll, Rowan K. Flad and Peter D. Clift, 75-88. Washington, DC: American Geophysical Union.
71	Southern Pakistan	von Rad, U., M. Schaaf, K.H. Michels, H. Schulz, W.H. Berger, and F. Sirocko. 1999. A 5000-yr Record of Climate Change in Varved Sediments from the Oxygen Minimum Zone off Pakistan, Northeastern Arabian Sea. *Quaternary Research* 51 (1): 39-53. doi 10.1006/qres.1998.2016.

SITE #	SITE/RECORD NAME	BIBLIOGRAPHIC REFERENCE
		Staubwasser, M., F. Sirocko, P.M. Grootes, and M. Segl. 2003. Climate change at the 4.2 ka BP termination of the Indus valley civilization and Holocene south Asian monsoon variability. *Geophysical Research Letters* 30 (8): 1425. doi 10.1029/2002GL016822.
72	Gulf of Oman	Cullen, H.M., P.B. deMenocal, S. Hemming, G. Hemming, F.H. Brown, T. Guilderson, and F. Sirocko. 2000. Climate change and the collapse of the Akkadian empire: Evidence from the deep sea. *Geology* 28 (4): 379-382. doi 10.1130/0091-7613(2000)28<379:ccatco>2.0.co;2.
73	Nal Sarovar	Prasad, S., S. Kusumgar, and S.K. Gupta. 1997. A mid to late Holocene record of palaeoclimatic changes from Nal Sarovar: a palaeodesert margin lake in western India. *Journal of Quaternary Science* 12 (2): 153-159. doi 10.1002/(SICI)1099-1417(199703/04)12:2<153::AID-JQS300>3.0.CO;2-X.
		Prasad, S., and Y. Enzel. 2006. Holocene paleoclimates of India. *Quaternary Research* 66 (3): 442-453. doi 10.1016/j.yqres.2006.05.008.
74	Wadhwana Lake	Prasad, V., A. Farooqui, A. Sharma, B. Phartiyal, S. Chakraborty, S. Bhandari, R. Raj, and A. Singh. 2014. Mid–late Holocene monsoonal variations from mainland Gujarat, India: A multi-proxy study for evaluating climate culture relationship. *Palaeogeography, Palaeoclimatology, Palaeoecology* 397 (0): 38-51. doi 10.1016/j.palaeo.2013.05.025.
75	Central Red Sea	Edelman-Furstenberg, Y., A. Almogi-Labin, and C. Hemleben. 2009. Palaeoceanographic evolution of the central Red Sea during the late Holocene. *The Holocene* 19 (1): 117-127. doi 10.1177/0959683608098955.
76	Lake Yoa	Kröpelin, S., D. Verschuren, A.-M. Lézine, H. Eggermont, C. Cocquyt, P. Francus, J.-P. Cazet, M. Fagot, B. Rumes, J.M. Russell, F. Darius, D.J. Conley, M. Schuster, H. von Suchodoletz, and D.R. Engstrom. 2008. Climate-Driven Ecosystem Succession in the Sahara: The Past 6000 Years. *Science* 320 (5877): 765-768. doi 10.1126/science.1154913.

SITE #	SITE/RECORD NAME	BIBLIOGRAPHIC REFERENCE
77	Qunf Cave	Fleitmann, D., S.J. Burns, M. Mudelsee, U. Neff, J. Kramers, A. Mangini, and A. Matter. 2003. Holocene Forcing of the Indian Monsoon Recorded in a Stalagmite from Southern Oman. *Science* 300 (5626): 1737-1739. doi 10.1126/science.1083130.
78	Western Ghats	Bentaleb, I., C. Caratini, M. Fontugne, M.T. Morzadec-Kerfourn, J.P. Pascal, and C. Tissot. 1997. "Monsoon Regime Variations During the Late Holocene in the Southwestern India." In *Third Millennium BC Climate Change and Old World Collapse*, edited by H. Nüzhet Dalfes, George Kukla and Harvey Weiss, 475-488. Springer Berlin Heidelberg.
79	Chad Basin Lakes	Gasse, F. 2000. Hydrological changes in the African tropics since the Last Glacial Maximum. *Quaternary Science Reviews* 19 (1–5): 189-211. doi 10.1016/S0277-3791(99)00061-X.
		Servant, M., and S. Servant. 1970. Les formations lacustres et les diatomées du quaternaire récent du fond de la cuvette tchadienne. *Revue de Géographie Physique et de Géologie Dynamique* 12 (1): 63-76.
80	Tigray Plateau	Terwilliger, V.J., Z. Eshetu, J.-R. Disnar, J. Jacob, W. Paul Adderley, Y. Huang, M. Alexandre, and M.L. Fogel. 2013. Environmental changes and the rise and fall of civilizations in the northern Horn of Africa: An approach combining δD analyses of land-plant derived fatty acids with multiple proxies in soil. *Geochimica et Cosmochimica Acta* 111: 140-161. doi 10.1016/j.gca.2012.10.040.
81	Kajemarum Oasis/ Jikariya Lake	Holmes, J.A., F.A. Street-Perrott, M.J. Allen, P.A. Fothergill, D.D. Harkness, D. Kroon, and R.A. Perrott. 1997. Holocene palaeolimnology of Kajemamm Oasis, Northern Nigeria: an isotopic study of ostracodes, bulk carbonate and organic carbon. *Journal of the Geological Society* 154 (2): 311-319. doi 10.1144/gsjgs.154.2.0311.
		Street-Perrott, F.A., J.A. Holmes, M.P. Waller, M.J. Allen, N.G.H. Barber, P.A. Fothergill, D.D. Harkness, M. Ivanovich, D. Kroon, and R.A. Perrott. 2000. Drought and dust deposition in the West African Sahel: a 5500-year record from Kajemarum Oasis, northeastern Nigeria. *The Holocene* 10 (3): 293-302. doi 10.1191/095968300678141274.

SITE #	SITE/RECORD NAME	BIBLIOGRAPHIC REFERENCE
		Wang, H., J. A. Holmes, F. A. Street-Perrott, M. P. Waller, and R. A. Perrott. 2008. Holocene environmental change in the West African Sahel: sedimentological and mineral-magnetic analyses of lake sediments from Jikariya Lake, northeastern Nigeria. *Journal of Quaternary Science* 23:449–460.
		Cockerton, H.E., J.A. Holmes, F.A. Street-Perrott, and K.J. Ficken. 2014. Holocene dust records from the West African Sahel and their implications for changes in climate and land surface conditions. *Journal of Geophysical Research: Atmospheres* 119 (14): 2013JD021283. doi 10.1002/2013JD021283.
82	Gulf of Aden/ Horn of Africa	Tierney, J.E., and P.B. deMenocal. 2013. Abrupt Shifts in Horn of Africa Hydroclimate Since the Last Glacial Maximum. *Science* 342 (6160): 843-846. doi 10.1126/science.1240411.
83	Lake Abhe	Gasse, F. 1977. Evolution of Lake Abhe (Ethiopia and TFAI), from 70,000 b.p. *Nature* 265 (5589): 42-45. doi 10.1038/265042a0.
84	Ziway-Shala Lake System	Gasse, F. 2000. Hydrological changes in the African tropics since the Last Glacial Maximum. *Quaternary Science Reviews* 19 (1–5): 189-211. doi 10.1016/S0277-3791(99)00061-X.
85	Lake Tilo	Lamb, A.L., M.J. Leng, H.F. Lamb, and M.U. Mohammed. 2000. A 9000-year oxygen and carbon isotope record of hydrological change in a small Ethiopian crater lake. *The Holocene* 10 (2): 167-177. doi 10.1191/095968300677444611.
86	Lake Bosumtwi	Marret, F., J. Maley, and J. Scourse. 2006. Climatic instability in west equatorial Africa during the Mid- and Late Holocene. *Quaternary International* 150 (1): 71-81. doi 10.1016/j.quaint.2006.01.008.
		Shanahan, T.M., J.T. Overpeck, W.E. Sharp, C.A. Scholz, and J.A. Arko. 2007. Simulating the response of a closed-basin lake to recent climate changes in tropical West Africa (Lake Bosumtwi, Ghana). *Hydrological Processes* 21 (13): 1678-1691. doi 10.1002/hyp.6359.

SITE #	SITE/RECORD NAME	BIBLIOGRAPHIC REFERENCE
87	Chew Bahir Basin	Foerster et al. 2012 Foerster, V., A. Junginger, O. Langkamp, T. Gebru, A. Asrat, M. Umer, H.F. Lamb, V. Wennrich, J. Rethemeyer, N. Nowaczyk, M.H. Trauth, and F. Schaebitz. 2012. Climatic change recorded in the sediments of the Chew Bahir basin, southern Ethiopia, during the last 45,000 years. *Quaternary International* 274 (0): 25-37. doi 10.1016/j.quaint.2012.06.028.
88	Lake Ossa	Nguetsop, V.F., S. Servant-Vildary, and M. Servant. 2004. Late Holocene climatic changes in west Africa, a high resolution diatom record from equatorial Cameroon. *Quaternary Science Reviews* 23 (5–6): 591-609. doi 10.1016/j.quascirev.2003.10.007.
89	Lake Turkana (Kenya)	Johnson, T.C., and E.O. Odada, eds. 1996. *The Limnology, Climatology and Paleoclimatology of the East African Lakes.* Amsterdam: Gordon and Breach Publishers.
90	Congo Basin/ Equatorial Africa	Maley, J. 1997. "Middle to Late Holocene Changes in Tropical Africa and Other Continents: Paleomonsoon and Sea Surface Temperature Variations." In *Third Millennium BC Climate Change and Old World Collapse*, edited by H. Nüzhet Dalfes, George Kukla and Harvey Weiss, 611-640. Springer Berlin Heidelberg. Marret, F., J. Maley, and J. Scourse. 2006. Climatic instability in west equatorial Africa during the Mid- and Late Holocene. *Quaternary International* 150 (1): 71-81. doi 10.1016/j.quaint.2006.01.008.
91	Lake Edward	Russell, J.M., and T.C. Johnson. 2005. A high-resolution geochemical record from Lake Edward, Uganda Congo and the timing and causes of tropical African drought during the late Holocene. *Quaternary Science Reviews* 24 (12–13): 1375-1389. doi 10.1016/j.quascirev.2004.10.003.

SITE #	SITE/RECORD NAME	BIBLIOGRAPHIC REFERENCE
92	Kilimanjaro	Thompson, L.G., E. Mosley-Thompson, M.E. Davis, K.A. Henderson, H.H. Brecher, V.S. Zagorodnov, T.A. Mashiotta, P.-N. Lin, V.N. Mikhalenko, D.R. Hardy, and J. Beer. 2002. Kilimanjaro Ice Core Records: Evidence of Holocene Climate Change in Tropical Africa. *Science* 298 (5593): 589-593. doi 10.1126/science.1073198.
93	Lake Challa	Tierney, J.E., J.M. Russell, J.S. Sinninghe Damsté, Y. Huang, and D. Verschuren. 2011. Late Quaternary behavior of the East African monsoon and the importance of the Congo Air Boundary. *Quaternary Science Reviews* 30 (7–8): 798-807. doi 10.1016/j.quascirev.2011.01.017.
94	Tatos Basin (Mauritian Lowlands)	de Boer, E.J., R. Tjallingii, M.I. Vélez, K.F. Rijsdijk, A. Vlug, G.-J. Reichart, A.L. Prendergast, P.G.B. de Louw, F.B.V. Florens, C. Baider, and H. Hooghiemstra. 2014. Climate variability in the SW Indian Ocean from an 8000-yr long multi-proxy record in the Mauritian lowlands shows a middle to late Holocene shift from negative IOD-state to ENSO-state. *Quaternary Science Reviews* 86: 175–189. doi 10.1016/j.quascirev.2013.12.026.
95	Gueldaman GLD1 Cave	Ruan, J., F. Kherbouche, D. Genty, D. Blamart, H. Cheng, F. Dewilde, S. Hachi, L.R. Edwards, E. Régnier, and J.-L. Michelot. 2015. Evidence of a prolonged drought ca. 4200 yr BP correlated with prehistoric settlement abandonment from the Gueldaman GLD1 Cave, N-Algeria. *Climate of the Past Discussions* 11: 2729-2762. doi 10.5194/cpd-11-2729-2015.
96	Gulf of Guinea	Weldeab, S., D.W. Lea, R.R. Schneider, and N. Andersen. 2007. Centennial scale climate instabilities in a wet early Holocene West African monsoon. *Geophysical Research Letters* 34: L24702. doi 10.1029/2007GL031898.

SITE #	SITE/RECORD NAME	BIBLIOGRAPHIC REFERENCE
97	Doñana National Park	Jiménez-Moreno, G., A. Rodríguez-Ramírez, J.N. Pérez-Asensio, J.S. Carrión, J.A. López-Sáez, J.J. Villarías-Robles, S. Celestino-Pérez, E. Cerrillo-Cuenca, Á. León, and C. Contreras. 2015. Impact of late-Holocene aridification trend, climate variability and geodynamic control on the environment from a coastal area in SW Spain. *The Holocene* 25 (4): 607-617. doi 10.1177/0959683614565955.
98	Lake Prespa	Leng, M.J., I. Baneschi, G. Zanchetta, C.N. Jex, B. Wagner, and H. Vogel. 2010. Late Quaternary palaeoenvironmental reconstruction from Lakes Ohrid and Prespa (Macedonia/Albania border) using stable isotopes. *Biogeosciences* 7 (10): 3109-3122. doi 10.5194/bg-7-3109-2010.
99	Kotychi Lagoon	Haenssler, E., I. Unkel, W. Dörfler, and M.-J. Nadeau. 2014. Driving mechanisms of Holocene lagoon development and barrier accretion in Northern Elis, Peloponnese, inferred from the sedimentary record of the Kotychi Lagoon. *E&G Quaternary Science Journal* 63 (1): 60-77. doi 10.3285/eg.63.1.04.
100	Acre, Israel	Kaniewski, D., E. Van Campo, C. Morhange, J. Guiot, D. Zviely, I. Shaked, T. Otto, and M. Artzy. 2013. Early urban impact on Mediterranean coastal environments. *Scientific Reports* 3: 3540. doi 10.1038/srep03540.
101	Koçain Cave	Göktürk, O.M. 2011. "Climate in the Eastern Mediterranean through the Holocene inferred from Turkish stalagmites." Ph.D. Dissertation, 113pp. Universität Bern.
102	Gulf of Lions	Jalali, B., M.A. Sicre, M.A. Bassetti, and N. Kallel. 2016. Holocene climate variability in the North-Western Mediterranean Sea (Gulf of Lions). *Climate of the Past* 12 (1): 91-101. doi 10.5194/cp-12-91-2016.
103	Lake Neor	Sharifi, A., A. Pourmand, E.A. Canuel, E. Ferer-Tyler, L.C. Peterson, B. Aichner, S.J. Feakins, T. Daryaee, M. Djamali, A. Naderi Beni, H.A.K. Lahijani, P.K. Swart. 2015. Abrupt climate variability since the last deglaciation based on a high-resolution, multi-proxy peat record from NW Iran: The hand that rocked the Cradle of Civilization? *Quaternary Science Reviews* 23: 215–230. doi 10.1016/j.quascirev.2015.07.006.

SITE #	SITE/RECORD NAME	BIBLIOGRAPHIC REFERENCE
104	Gulf of Gaeta	Margaritelli, G., M. Vallefuoco, F. Di Rita, L. Capotondi, L.G. Bellucci, D.D. Insinga, P. Petrosino, S. Bonomo, I. Cacho, A. Cascella, L. Ferraro, F. Florindo, C. Lubritto, P.C. Lurcock, D. Magric, N. Pelosi, R. Rettori, F. Lirer. 2016. Marine response to climate changes during the last five millennia in the central Mediterranean Sea. *Global and Planetary Change* 142: 53–72. doi 10.1016/j.gloplacha.2016.04.007.
105	Lazaun Glacier	Krainer, K., D. Bressan, B. Dietre, J.N. Haas, I. Hajdas, K. Lang, V. Mair, U. Nickus, D. Reidl, H. Thies, D. Tonidandel. 2014. A 10,300-year-old permafrost core from the active rock glacier Lazaun, southern Ötztal Alps (South Tyrol, northern Italy). *Quaternary Research* 397: 349–359. doi 10.1016/j.yqres.2014.12.005.
106	Lake Rara	Nakamura, A., Y. Yokoyama, H. Maemoku, H. Yagi, M. Okamura, H. Matsuoka, N. Miyake, T. Osada, D.P. Adhikari, V. Dangol, M. Ikehara, Y. Miyairi, H. Matsuzaki. 2016. Weak monsoon event at 4.2 ka recorded in sediment from Lake Rara, Himalayas. *Quaternary International* 397: 349–359. doi 10.1016/j.quaint.2015.05.053.
107	Bagasra Village	Amekawa, S., K. Kubota, Y. Miyairi, A. Seki, Y. Kawakubo, S. Sakai, P. Ajithprasad, H. Maemoku, T. Osada, Y. Yokoyama. 2016. Fossil otoliths, from the Gulf of Kutch, Western India, as a paleo-archive for the mid- to late-Holocene environment. *Quaternary International* 397: 281-288. doi 10.1016/j.quaint.2015.07.006.
108	Ashtamudi Estuary	Vishnu Mohan, S., R.B. Limaye, D. Padmalal, S.M. Ahmad, K.P.N. Kumaran. 2016 IN PRESS. Holocene climatic vicissitudes and sea level changes in the south western coast of India: Appraisal of stable isotopes and palynology. *Quaternary International*. doi 10.1016/j.quaint.2016.07.018.
109	Lonar Lake	Sarkar, S., S. Prasad, H. Wilkes, N. Riedel, M. Stebich, N. Basavaiah, D. Sachse. 2015. Monsoon source shifts during the drying mid-Holocene: Biomarker isotope based evidence from the core `monsoon zone' (CMZ) of India. *Quaternary Science Reviews* 123: 144-157. doi 10.1016/j.quascirev.2015.06.020.

SITE #	SITE/RECORD NAME	BIBLIOGRAPHIC REFERENCE
110	Sofular Cave	Göktürk, O.M., D. Fleitmann, S. Badertscher, H. Cheng, R.L. Edwards, M. Leuenberger, A. Fankhauser, O. Tüysüz, J. Kramers. 2011. Climate on the southern Black Sea coast during the Holocene: implications from the Sofular Cave record. *Quaternary Science Reviews* 30: 2433–2445. doi 10.1016/j.quascirev.2011.05.007.
111	Gol-e Zard Cave	Carolin, S., J. Morgan, E. Peckover, R. Walker, G. Henderson, P. Rowe, J. Andrews, V. Ersek, A. Sloan, M. Talebian, M. Fattahi, J. Nezamdoust. 2016. Iranian speleothems: Investigating Quaternary climate variability in semi-arid Western Asia. *Geophysical Research Abstracts* 18, EGU 2016-923.
		Carolin, S.A., Walker, R., Henderson, G.M., Maxfield, L., Ersek, V., Sloan, R.A., Talebian, M., Fattahi, M., Nezamdoust, J., 2015 Decadal-scale climate variability on the central Iranian plateau spanning the so-called 4.2 ka BP drought event, Abstract PP21D-07 presented at 2015 AGU Fall Meeting, San Francisco, CA.
112	Nar Gölü	Dean, J.R., M.D. Jones, M.J. Leng, S.R. Noble, S.E. Metcalfe, H.J. Sloane, D. Sahy, W.J. Eastwood, C.N. Roberts. 2015. Eastern Mediterranean hydroclimate over the late glacial and Holocene, reconstructed from the sediments of Nar lake, central Turkey, using stable isotopes and carbonate mineralogy. *Quaternary Science Reviews* 124: 162-174. doi 10.1016/j.quascirev.2015.07.023.
113	Gulf of Gemlik	Filikci, B., N. Çağatay, K.K. Eriş, M. Akyol, B. Yalamaz, G. Uçarkus, P. Henry. 2015. The sedimentary records of Holocene environmental changes from the Central High of the Sea of Marmara. *Geophysical Research Abstracts* 17, EGU2015-14607-2.
		Filikci, B., K.K. Eriş, M.N. Çağatay, L. Gasperini, A. Sabuncu, D. Acar, B. Yalamaz. 2016. Late Pleistocene to Holocene paleoceanographic and paleo-climatic changes in Gulf of Gemlik, Sea of Marmara, Turkey. *Geophysical Research Abstracts* 18, EGU 2016-625-1.

SITE #	SITE/RECORD NAME	BIBLIOGRAPHIC REFERENCE
114	Alepotrypa Cave	Boyd, M. 2015. "Speleothems from Warm Climates: Holocene Records from the Caribbean and Mediterranean Regions." Ph.D. Dissertation, 82pp. Stockholm University.
115	Malham Tarn Moss and Walton Moss	Morris, P.J., A.J. Baird, D.M. Young, G.T. Swindles. 2015. Untangling climate signals from autogenic changes in long-term peatland development. *Geophysical Research Letters* 42: 10788-10797. doi 10.1002/2015GL066824.
		Hughes, P.D.M., D. Mauquoy, K.E. Barber, P.G. Langdon. 2000. Mire-development pathways and paleoclimatic records from a full Holocene peat archive at Walton Moss, Cumbria, England. *The Holocene* 10: 465–479, doi 10.1191/095968300675142023.
		Daley, T.J., K.E. Barber. 2012. Multi-proxy Holocene palaeoclimate records from Walton Moss, northern England and Dosenmoor, northern Germany, assessed using three statistical approaches. *Quaternary International* 268: 111-127. doi 10.1016/j.quaint.2011.10.026.
116	Dosenmoor	Daley, T.J., K.E. Barber. 2012. Multi-proxy Holocene palaeoclimate records from Walton Moss, northern England and Dosenmoor, northern Germany, assessed using three statistical approaches. *Quaternary International* 268: 111-127. doi 10.1016/j.quaint.2011.10.026.
117	Sluggan Moss, Fallahogy Bog, and Mallachie Moss	Roland, T.P., C.J. Caseldine, D.J. Charman, C.S.M. Turney, M.J. Amesbury. 2014. Was there a '4.2 ka event' in Great Britain and Ireland? Evidence from the peatland record. *Quaternary Science Reviews* 83: 11-27. doi 10.1016/j.quascirev.2013.10.024.
		Langdon, P.G. K.E. Barber. 2005. The climate of Scotland over the last 5000 years inferred from multiproxy peatland records: inter-site correlations and regional variability. *Journal of Quaternary Science* 20: 549-566. doi 10.1002/jqs.934.

CHAPTER 4 | # 3.2 ka BP Megadrought and the Late Bronze Age Collapse

DAVID KANIEWSKI AND ELISE VAN CAMPO

The collapse of Bronze Age civilizations in the Aegean, southwest Asia, and the eastern Mediterranean 3200 years ago remains a persistent riddle in Eastern Mediterranean archaeology, as both archaeologists and historians believe the event was violent, sudden, and culturally disruptive. In the first phase of this period, many cities between Pylos and Gaza were destroyed violently and often left unoccupied thereafter. The palace economy of the Aegean Region and Anatolia that characterized the Late Bronze Age was replaced by the isolated village cultures of the Dark Ages. Earthquakes, attacks of the Sea Peoples, and socio-political unrest are among the most frequently suggested causes for this phenomenon. However, while climate change has long been considered a potential prime factor in this crisis, only recent studies have pinpointed the megadrought behind the collapse. An abrupt climate shift seems to have caused, or hastened, the fall of the Late Bronze Age world by sparking political and economic turmoil, migrations, and famines. The entirety of the megadrought's effects terminated the Late Bronze Age in the eastern Mediterranean.

Introduction

In Old World history, 1200 BC is a symbolically important date. It signifies a cultural disruption that brought down the vibrant and powerful Bronze Age civilizations of the eastern Mediterranean. Corresponding events included massive migrations, the collapse of political entities, warfare (Neumann & Parpola 1987; Weiss 1982; Bryce 2005), and invasions by a group of clans commonly known as the Sea Peoples (Singer 1999; Killebrew 2005; Gilboa 2006–2007; Killebrew & Lehmann 2013). The idea that a century-long megadrought might explain this dramatic cultural discontinuity and population shift has received relatively little attention, though it was first proposed by Rhys Carpenter (1966) and gained support from Reid Bryson (1974) and Barry Weiss (1982), who identified drought patterns consistent with ancient Mycenaean migration routes.

Six major and chronologically well-defined episodes of abrupt climate change have punctuated the post-glacial Holocene period (Alley et al. 1997; Mayewski, Meeker, & Twickler 1997; Mayewski, Rohling, & Stager 2004). They occurred over periods of a few decades to a few hundred years and abruptly enough to significantly affect human societies. The more recent of these abrupt climate change events appear closely linked to major breaks in regional cultural evolution (see, for example, Weiss et al. 1993; deMenocal 2001; Weiss 2010; Medina-Elizade & Rohling 2012). For example, the societal collapses of the Harappan Indus Valley (Staubwasser, Sirocko, & Grootes 2003), the Egyptian Old Kingdom (Stanley, Krom, & Cliff 2003), and Akkadian Mesopotamia (Weiss et al. 1993) have been shown to coincide with the well-known ca. 2200–1900 BC megadrought (Weiss 2014). Likewise, the association between the cultural disruption at 1200 BC and the 3.2 ka BP megadrought event, an abrupt climate change ca. 1500–1200 BC that heavily impacted the Near East, is now supported by several lines of paleoclimate and archaeological evidence.

The 3.2 ka BP megadrought in the Eastern Mediterranean and West Asia is coupled with larger scale climate patterns. A complex abrupt climate change at 1500–500 BC was first detected in the North Atlantic atmospheric and oceanic circulation patterns. Glaciochemical series developed from the GISP2 Greenland ice core revealed a period of winter-like circulations and storm conditions, correlated with a worldwide glacier expansion (O'Brien et al. 1996; Mayewski, Meeker, & Twickler 1997). A series of increases of drift ice, reflected in percentage variations of hematite-stained grains in marine sediments, have been observed throughout the Holocene (Bond, Kromer, & Beer 2001). These ice-rafted debris (IRD) events, caused by advection of cold surface polar waters into the warmer subpolar waters, were accompanied by northerly surface winds. A double-peaked event (Bond event 2) in the IRD curve clearly indicates high-latitude cooling and climatic instability between 1200 and 850 BC.

Here we review the spatio-temporal pattern of the Late Bronze Age mega-drought as it occurred ca. 1500–500 BC in several reconstructions of relative humidity from the Eastern Mediterranean and Western Asia (fig. 4.1). Large uncertainties remain concerning the spatial coherency of the event, its duration, and how the climatic patterns were affected. Yet new data highlights an arid shift at the coastal sites of both Gibala-Tell Tweini in Syria and Hala Sultan Tekke in Cyprus between 1200 and 850 BC (fig. 4.1; Kaniewski, Paulissen, & Van Campo 2008, 2010; Kaniewski, Van Campo, & Guiot 2013). Along with climate records from the Dead Sea (Litt, Ohlwein, & Neumann 2012) and the Sea of Galilee (Langgut, Finkelstein, & Litt 2013), these new data support the hypothesis that the transitional period between the Late Bronze Age and the Early Iron Age in the Eastern Mediterranean was related to both natural and social stress.

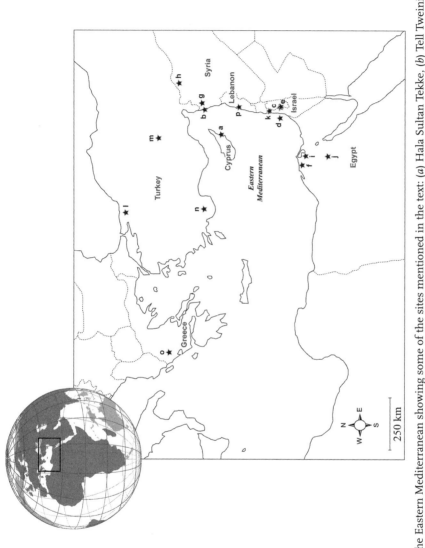

FIGURE 4.1 Map of the Eastern Mediterranean showing some of the sites mentioned in the text: (*a*) Hala Sultan Tekke, (*b*) Tell Tweini, (*c*) Soreq Cave, (*d*) Ashdod Coast, (*e*) Dead Sea, (*f*) Burullus Lagoon, (*g*) Tell Breda, (*h*) Ras El-Ain, (*i*) Manzala Lagoon, (*j*) Qarun Lake, (*k*) Nahal Qanah Cave, (*l*) Sofular Cave, (*m*) Eski Açigöl, (*n*) Gölhisar, (*o*) Ioannina, (*p*) Jeita Cave. (D. Kaniewski)

The Sea Peoples and the Crisis of the Ancient Near East: 1200 BC

The decades around 1200 BC were disastrous for nearly all eastern Mediterranean societies. The Late Bronze Age civilizations of the Aegean and Near East collapsed. In Greece, the Mycenaeans, who had conquered the Minoans and developed an effective polity and trade economy, particularly with the Hittites, were nevertheless an agriculturally vulnerable society dependent upon unpredictable weather patterns. Isolated city-states, possibly lacking unity and integrative political structures, were vulnerable to destabilizing external forces at the end of the thirteenth century (Hutchinson 1977). The collapse of the Mycenaean culture has traditionally been blamed on the invasion of the Dorians, who migrated south into the Peloponnese, but an alternative theory focuses on the Sea Peoples, an enigmatic group of maritime raiders.

In his mortuary temple at Medinet Habu (Thebes, Upper Egypt) and on the Great Harris Papyrus, the Pharaoh Ramses III commemorated a major battle from the eighth year of his reign with reliefs (fig. 4.2) and inscriptions that have been studied for over 150 years, but remain puzzling in many respects (Roberts 2008). A section of engraved text referring to the great battle proclaims: "Those who came together on the sea, the full flame was in front of them at the river-mouths, while a stockade of lances surrounded them on the shore. They were dragged in enclosed, and prostrated on the beach, killed and made into heaps from tail to head. Their ships and their goods were as if fallen into the water" (Pritchard 1969). This chaotic scene of boats and warriors entwined in battle in

FIGURE 4.2 Boats of the Sea Peoples during the battle of the Nile Delta. (Reworked by D. Kaniewski from Nelson, 1930)

the Nile delta can be placed in time according to the first regnal year of Ramses III, variously given from the historical dates of 1176 BC (Shaw 2000) and 1179 BC (Hornung, Krauss, & Warburton 2006) to a radiocarbon-based date of 1188–1177 cal BC (Bronk-Ramsey, Dee, & Rowland 2010). Inscriptions accompanying the reliefs make reference to the origins of warriors in *"islands in the midst of the sea,"* a confederation of clans referred to now as the Sea Peoples (Singer 2000; Gilboa 2006–2007; Killebrew & Lehmann 2013), a name introduced in 1881 by the French Egyptologist Gaston Maspero. The Sea Peoples were foes of varied tribal origins, including the Peleshet (Philistines), Akawasha (Achaeans), Tjekker (Teucrians), Lukka (Lycians), Shekelesh (Anatolians), Denyen (Danaans), Sherden (Sardinians) and Weshwesh (Stiebing 1980, Artzy 1987, Gilboa 2006–2007, Killebrew & Lehmann 2013).

The Nile delta battle against Ramses III's army was, however, by no means the first assault on Egypt by coalitions of seafaring and inland tribes. During the reign of Pharaoh Ramses II (1279–1213 BC), Sherden sea pirates, who had been wreaking havoc along Egypt's Mediterranean coast by attacking the cargo-laden vessels travelling the sea routes to Egypt, were defeated in the mouth of the Nile (Year 2; Shaw 2000, Grimal 1992). The pirates, who were allied with Lukka and Shekelesh tribes (Tyldesley 2000), are remembered on the Tanis Stela II: "the unruly Sherden whom no one had ever known how to combat, they came boldly sailing in their warships from the midst of the sea, none being able to withstand them" (Kitchen 1983). During the reign of Pharaoh Merneptah (1213–1203 BC), the "Nine Bows" coalition, led by the Libyans and joined by the Sea Peoples (Ekwesh, Teresh, Lukka, Sherden, Shekelesh), marched on Egypt and threatened Memphis and Heliopolis (Year 5; Shaw 2000). The battle of Merneptah's army is depicted on a pylon of the Great Karnak Temple (Luxor), and its victory is narrated on the Merneptah "Israel Stele" (Thebes): "The wretched, fallen chief of Libya, Mereye, son of Ded, has fallen upon the country of Tehenu with his bowmen–Sherden, Shekelesh, Ekwesh, Lukka, Teresh" (Manassa 2003).

The last incursion of the Sea Peoples was first documented on cuneiform tablets from the kingdom of Ugarit, at the northern coastal Syrian site of Ras Shamra (Yon 2006). The cuneiform tablet RS 20.238, sent by the king of Ugarit to the king of Alašia, describes the ravages inflicted by the enemy: "Now the ships of the enemy have come. They have been setting fire to my cities and have done harm to the land. . . . Now the seven ships of the enemy which have come, have done harm to us" (Singer 1999). The letter describes one of a series of Sea Peoples' attacks along the coastline of the kingdom of Ugarit before they turn their weapons against the southern Levant and Egypt. Other cuneiform tablets, documenting the final days of the coastal city, derive from the Rapanu archives at Ugarit. The cuneiform tablet RS 19.011, one of the last letters from Ugarit, describes the dramatic urban situation: "When your messenger arrived, the army was humiliated and the city was sacked. Our food in the threshing floors was burnt and the vineyards were also destroyed. Our city was sacked. May you know it! May you know it!"

The documentary evidence thus indicates that the Sea Peoples and their allies threatened the coastlines of the Eastern Mediterranean throughout the thirteenth century BC, with Egypt as their ultimate goal. Through combined land-sea invasions they succeeded in penetrating the western delta of the Nile during the reign of Merneptah, but they failed to settle and control Egyptian territory. The last wave of Sea People invasions, around 1200 BC, was the most decisive, as it affected not only Egypt, but also weakened the Hittite realm and may have been the decisive agent in its collapse (Singer 2000).

Evidence of Drought-induced Food Shortages and Famine

The earliest evidence of food shortages and famine derives from a rich trove of correspondence found among the late thirteenth century BC ruins of Ugarit. Letters from the pharaoh Merneptah mention consignments of grain sent from Egypt to relieve the famine in Ugarit (RS 88.2158 & RS 94.2002, Singer 1999). The first mention of grain shortage in Hittite domains appears during the reign of Ramses II, whose rule is radiocarbon dated to have begun in 1292–1281 BC (Bronk-Ramsey, Dee, & Rowland 2010). A letter (KUB 21.38) from the Hittite queen Puduhepa declares, "I have no grain in my lands." Subsequently, several shipments of famine aid including barley and wheat were sent from Egypt (Warburton 2003; Bryce 2005). The pharaoh Merneptah would export sizable quantities of the vital food: *"grain to be taken in ships, to keep alive this land of Hatti"* (KRI IV 5, 3; Singer 1999). The Hittite king Arnuwanda III described the terrible hunger suffered in Anatolia in his father's time, mentioning drought as the cause (Warburton 2003).

Across the region, including Syria and Mesopotamia, the situation was famously similar. The clay tablet RS 34.152 from Emar offers a vivid testimony to the deteriorating conditions and severe food shortage around 1190 BC along the Syrian Euphrates River. The Emar year-names bear witness to a staggering increase in grain prices in the "year of hardship/famine." Impoverished families were forced to sell their children to wealthy merchants in order to sustain themselves (Singer 2000; Cohen & Singer 2006). The clay tablet RS 18.38, dated to the late thirteenth century BC, indicates grain shipments again from Egypt to the Hittites, suggesting grain shortages in Eastern Anatolia (Bryce 2005). A particular note of urgency is struck in a letter (clay tablet RS 20.212) sent from the Hittite court to the Ugaritic king, either Niqmaddu III or Hammurabi (1215–1194/1175 BC), demanding ship and crew for the transport of 2000 *kor* of grain (about 450 tons) from the Syrian coastal district Mukish to Ura. This is a matter of life or death, the letter concludes (Nougayrol et al. 1968). Famine impacted Egypt as well during the reign of Merneptah (1213–1203 BC; Bryson et al. 1974). The drop of Nile discharges during the reign of Ramses III led to crop failures, low harvests (Butzer 1976), and riots (Faulkner 1975). In Mesopotamia, cuneiform documents from Babylon and Assyria describe crop failures, famine, and the outbreak of plague (Brinkman 1968; Neumann & Parpola 1987). These accounts of food crises in the late thirteenth to early twelfth centuries BC in the eastern Mediterranean may serve as historical anchors for paleoclimatic examinations of the deteriorating conditions at 3.2 ka BP.

A Refined Radiocarbon Chronology for the Late Bronze Age Collapse

The Sea Peoples' event is traditionally associated with the end of a long and complex spiral of eastern Mediterranean decline. Whereas the raids constituted a major turning point in eastern Mediterranean history, attested to by both epigraphic and archaeological evidence (at Ugarit, Enkomi, Kition, and Byblos, for example), our ability to date them has depended on the difficult epigraphic chronology derived from cuneiform tablets predating the invasions and the historical chronology of Ramesses III's reign. The site at Gibala-Tell Tweini (fig. 4.3) represents the first archaeological source of independent, stratified radiocarbon-based evidence to date the Sea Peoples event in the Northern Levant precisely (Kaniewski, Van Campo, & Van Lerberghe 2011). Short-lived samples (olive stones, seeds, and a young branch) found in the destruction debris from eight key loci were initially dated by accelerator mass spectrometry. We add here three new, unpublished samples (Beta-400692, Beta-400693, Beta-400695, olive stones) from our database. The samples pooled in the matrix are statistically the same at the 95 percent confidence level using a Chi-square ($\chi2$) test. The weighted average date (2962±11 ^{14}C yr BP) gives a 1 sigma (σ) calibrated age range of 1215–1190 BC with 40.5 percent relative probability, 1180–1160 BC with 28.4 percent, 1145–1130 BC with 31.1 percent, and a 2σ calibrated age range of 1220–1130 BC with 100 percent relative probability, using Calib-Rev. 7.0.2 with IntCal13 and R software. The "weighted average date" method was selected because this process is widely used (Bruins, van der Plicht, & Mazar 2003; Bruins, van der Plicht, & MacGillivray 2009; Finkelstein & Piasetzky 2009), recommended (Telford, Heegaard, & Birks 2004), and fits well with the stratified radiocarbon-based archaeology in the eastern Mediterranean (Mazar & Carmi 2001).

The local conflagration at Tell Tweini has been dated to 1215–1190 BC. This calibrated date conforms to the radiocarbon data published for the end of the Late Bronze Age at Hala Sultan Tekke (Cyprus; Fischer et al. 2012): 2971±15 ^{14}C BP (1σ calibration 1220–1190 BC and 2σ calibration 1230–1130 BC). The presence of the Sea Peoples immediately after the destruction at Tell Tweini is indicated by the material culture of the new settlements on the site—namely, the appearance of Aegean-type architecture, locally made Mycenaean IIIC Early pottery, handmade burnished pottery, and Aegean-type loom-weights. These materials, also known from Philistine settlements (Yasur-Landau 2010), are the cultural markers of Aegean-derived settlers—most probably the Sea Peoples.

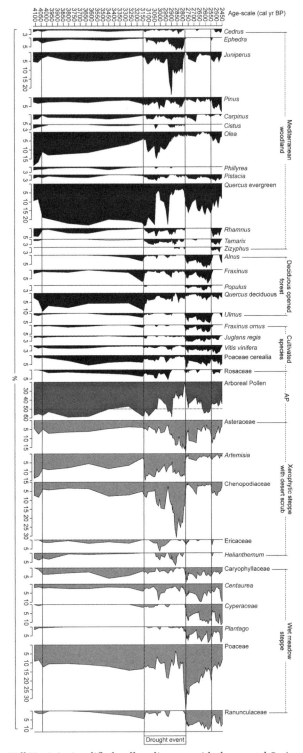

FIGURE 4.3 Tell Tweini: simplified pollen diagram with the coastal Syrian drought event at 3.2 ka BP. (D. Kaniewski)

Neutron activation analyses suggest that the thirteenth century BC Late Helladic IIIB vessels found at Gibala originate from the northern Peloponnese area (Al-Maqdissi et al. 2011). Gibala's Late Cypriot IIC ceramics were directly imported from Cyprus. The Late Helladic IIIB–IIIC (Early) and Late Cypriot IIC–IIIA transitions dated, respectively, to 1210–1175 BC in mainland Greece (Weninger & Jung 2009) and to 1220–1190 BC in Cyprus (Manning, Kromer, & Kuniholm 2001; Manning 2006–2007) are the chronological markers for the end of palace culture in the Aegean and Cyprus.

The radiocarbon-dated terminal occupations of Late Bronze Age cultures in Syria, Cyprus, and in the Aegean, as described above, are closely synchronous. The radiocarbon determinations, with anchor points in the epigraphic records, Hittite-Levantine-Egyptian regnal years, and astronomical observations, suggest an absolute age range of 1192–1190 BC for the terminal destructions and cultural collapse in the northern Levant. This proposed chronology for the invasion in the northern Levant integrates radiocarbon, written/historical, archaeological, and astronomical data. (For a full detailing and explanation of the data from diverse fields, see Kaniewski, Van Campo, & Van Lerberghe 2011.)

Overview of the 1500–1200 cal BC Climate Event

The Climatic Setting of the Eastern Mediterranean and West Asia

The interpretative framework for past climate shifts relies on the current climate of the Eastern Mediterranean and West Asia, which is characterized by cool-wet winters and hot-dry summers. Winter rainfall is controlled by the dynamics of the mid-latitude cyclones from the Atlantic during the cold season (Ziv et al. 2010). Eastward moving air masses gain moisture over the warm Mediterranean Sea and generate the Cyprus low-pressure system usually located south of Turkey, which governs the spatial and temporal rainfall variability over the Near East. Inland precipitation is modulated by orographic effects and decreases from north to south and west to east, with latitude and continentality. Hot and dry summer conditions result from the northward shift of the descending branch of the Hadley Cell from the Sahara. Rodwell and Hopkins (1996) proposed that part of the summer descent over the eastern Mediterranean and Sahara could be induced by the northward shift of monsoon heating.

The eastern Mediterranean and West Asia were not directly affected by monsoonal rainfall during the Holocene and the last glacial and interglacial (Arz, Lamy, & Pätzold 2003; Felis, Lohmann, & Kuhnert 2004). Only the hydrologic characteristics of the Nile delta region depend more upon the discharge of the river upstream—that is, upon the East African monsoon regime—than on local Mediterranean conditions. Exceptional rainfall events, commonly linked to a strong Red Sea trough, occur in the fall and spring over the southern Levant and northwest Arabia (Kahana, Baruch, & Enzel 2002). Most windstorms occur in winter following the dominant direction of the westerlies. In spring, the warm and dry *Sharav* (or *Khamsin*) winds originating over Libya and Egypt move eastward along the North African coast and occasionally carry vast quantities of desert dust and sand. Strong easterly *Sharqiya* winds can also bring dust storms from Arabia during winter and early spring (Alpert & Ziv 1989; Saaroni et al. 1998). To estimate the extent to which the spiral of decline in the eastern Mediterranean and West Asia could have been related to an abrupt arid climate change, we must examine the environmental dynamics in the Near East during the Late Bronze Age crisis (figs. 4.4 and 4.5).

Abrupt Climate Change in Coastal Syria and Cyprus ca. 1200 BC

The first clearly identified and well-dated arid climate-change event at the Late Bronze Age-Early Iron Age transition comes from the ancient site of Gibala (present Tell Tweini) in coastal Syria (Kaniewski, Paulissen, & Van Campo 2008, 2010). A pollen-based record of climatic variability close to the archaeological site (fig. 4.3) was obtained from an 800-centimeter long core retrieved from the alluvial deposits of the Rumailiah River bordering the northern edge of the site (fig. 4.4). The AMS ^{14}C ages allowed us to reconstruct the evolution of the environmental conditions from 2150 to 550 BC, covering the time interval of Gibala's Late Bronze Age collapse. Numerical analysis of the pollen data revealed an abrupt and severe drought period between 1175 and 850 BC, following a humid episode around 1250 BC, and interrupted by a short half-century (ca. 1000–950 BC) wetter pulse. The drought period suddenly ended after 850 BC. The two aridity peaks at ca. 1100–1000 BC and 900–850 BC are well correlated with a lower frequency of aquatic and cultivated pollen types, indicating drier conditions and reduced agriculture. These results show that the first destruction of Gibala (Kaniewski, Van Campo, & Van Lerberghe 2011) occurred quickly after the onset of a severe climatic crisis, or megadrought, which induced crop failures, famines, and economic stress.

The nearby island of Cyprus was also located at the heart of the ancient civilizations and trade routes of the eastern Mediterranean during the Late Bronze Age. An integrated radiocarbon-based chronology of archaeological and pollen-derived climatic proxies from the site of Hala Sultan Tekke details the environmental context along the southeastern Cypriot coast during the Late Bronze Age crisis (Kaniewski, Van Campo, & Guiot 2013). An 820-centimeter long core includes the period ranging from 1600 BC to AD 1500 (fig. 4.4). The marine embayment of the Hala Sultan Tekke harbor, which had served as the port of entry for elite goods from the beginning of the Late Cypriot period ca. 1600 BC, gradually evolved into a lagoon between 1450 and 1350 BC. This first environmental shift is mirrored in the gradual transformation of the coastal Mediterranean woodland into a dry steppe, and was also partly due to agriculture-related fire activities. A further development of the dry-saline steppe at the expense of wetter Mediterranean woodland was reached ca. 1200 BC, when no fire activity or other changes in the lagoon are attested. The rich

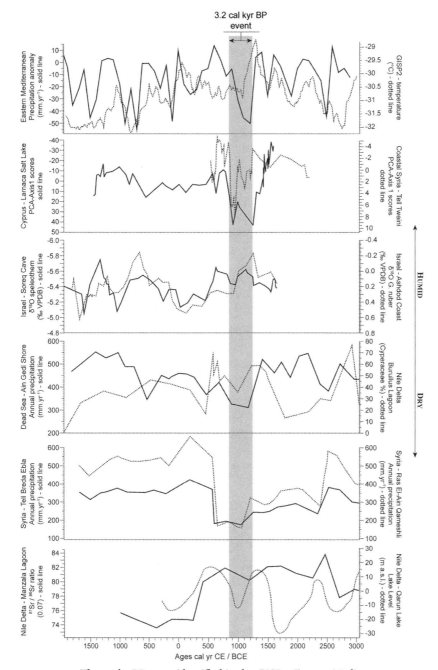

FIGURE 4.4 The 3.2 ka BP event identified in the GISP2, Eastern Mediterranean, Syria, Cyprus, Israel, and Egypt. See text for references. (D. Kaniewski)

agricultural activity around the site also declined strongly at that time. The climate proxies from Hala Sultan Tekke and Gibala-Tell Tweini firmly link the island and the mainland through comparable climatic evolutions (fig. 4.4). We must go beyond this evidence, however, before we propose a climate-change based model for cultural change throughout the Levant and Southwest Asia at the end of the Late Bronze Age. Varieties of proxy data indicate increasingly and significantly drier environments in the eastern Mediterranean and West Asia during the latter part of the Late Bronze Age (fig. 4.4).

Additional Marine Evidence

Marine sediments usually record natural changes without anthropogenic forcing. A detailed reconstruction of marine conditions in the southeastern Mediterranean Sea during the last 3600 years was based on the $\partial^{18}O$ record of the planktonic foraminifera G. *ruber* (fig. 4.4) derived from two cores retrieved in the southern Levantine Basin off the Israeli Ashdod coast (Schilman, Bar-Matthews, & Almogi-Labin 2001). Fourteen AMS ^{14}C dates of planktonic foraminifera species corrected for a reservoir effect of about 400 years robustly constrain a composite time series formed by the $\partial^{18}O$ values and total organic carbon of the two cores. The $\partial^{18}O$ values reflect both sea-surface temperature (SST) and salinity (SSS). During the past 3600 years, SST changed slightly and variations in sea-surface $\partial^{18}O$ mainly reflect changes in the SSS—as in the fresh-water budget—resulting from evaporation (E) over precipitation plus river runoff (P), mainly from the Nile. The contribution of Nile water does not significantly affect the $\partial^{18}O$ values, which mainly reflect changes in the E/P ratio. Onset of aridification is shown by a gradual increase in $\partial^{18}O$ values from ca. 1200 BC onward.

Additionally, a record of relative SST changes based on the percentage of warm versus cool planktonic foraminifera species was derived from the SE Aegean Sea radiocarbon-dated core LC21 (Rohling, Mayewski, & Abu-Zied 2002). A cool event representing winter SST reductions of 2 to 4°C is registered between ca. 1500 and 1000 BC. Holocene cool events in LC21 could have been caused by more intense and/or frequent wintertime northerly air outbreaks that reduced evaporation. Fewer and less intense storms, in turn, would have resulted in a significant decrease in precipitation over the EM region (Bartov, Goldstein, & Stein 2003).

The Dead Sea Evidence

The Dead Sea can be viewed as a terminal lake that drains one of the largest hydrological systems in the Near East. Its level depends primarily on the precipitation received in its northern headwaters, and its variations are indicative of the climate variations in the region (Enzel, Bookman, & Sharon 2003). Dead Sea levels have fluctuated within a range of 390 to 415 meters below sea level during the past 4000 years (Bookman, Enzel, & Agnon 2004). A comprehensive record of sea level changes was reconstructed for the late

Holocene period from dense radiocarbon dating of sedimentary sections and level indicators in the Ze'elim plain on the western shore of the Dead Sea. An erosional unconformity marked by pebbles and aragonite crusts suggests a major drop in the lake level after 1445 BC. Overlying lacustrine sediments indicate relatively low and fluctuating levels from 1000 to 550 BC. A pollen diagram was derived from the section used for palaeo-lake reconstructions. One pollen sample from the 110-centimeter thick beach ridge top of the unconformity shows low tree and high Chenopod values, which also point to arid conditions at the end of the Late Bronze Age (Neumann, Schölzel, & Litt 2007).

Additionally, a 2-meter long core was recovered at the Ein Gedi shore, north of Ze'elim (fig. 4.4). Twenty radiocarbon dates indicate mainly continuous sedimentation of fine-grained material for the last 10,000 years. A gypsum crust followed by silt-sand laminations marks a significant level drop ca. 1300 BC (Migowski, Stein, & Prasad 2006). Temporal variations in rainfall and temperature were reconstructed from pollen data. A winter temperature increase of about 4°C and an annual precipitation decrease of about 100 millimeters are suggested at the transition between the Late Bronze Age and the Iron Age. These shifts were relatively abrupt (within about 200 years) at the time resolution used (the time interval between each sample is 150–200 years). The interval from 1200 BC to present is characterized by a clear trend of –1.5°C (Litt, Ohlwein, & Neumann 2012).

Northern and Southern Levant

Accelerator Mass Spectrometry (AMS) techniques for the simultaneous analysis of carbon-stable isotopic values and ^{14}C dating on charred plant remains from the archeological site of Tell Mardikh/Ebla in western Syria (fig. 4.4) reveal a maximum reduction in rainfall and overall water availability ca. 1200 BC (Fiorentino et al. 2008). The estimated rainfall at Ebla confirms the past precipitation trend proposed by Bryson and Bryson (1997) at Ras el-Ain-Qamishli (fig. 4.4), near the site of Tell Leilan (northeastern Syria).

Glacial-interglacial $\partial^{18}O$ variations in the speleothem $\partial^{18}O$ record of a stalagmite from Soreq Cave, central Israel (fig. 4.4), mainly depend upon the isotopic composition of the source seawater, the fractionation processes between the vapor source and the site of precipitation, and the temperature of deposition (Kolodny, Stein, & Machlus 2005), whereas short-term changes appear to be more directly linked to wet/dry conditions. Bar-Matthews, Ayalon, and Gilmour (2003) developed a model in order to estimate paleorainfall amounts, based on the present-day relationship between the annual rainfall amount, its $\partial^{18}O$ composition, and the $\partial^{18}O$ of the cave water. They estimated temperatures from alkenone SST derived from a core in close proximity to the Israeli coast, since the annual land temperature today at the Soreq site is the same as the average winter and spring SST. A systematic decrease in rainfall is observed ca. 2500 to 500 BC.

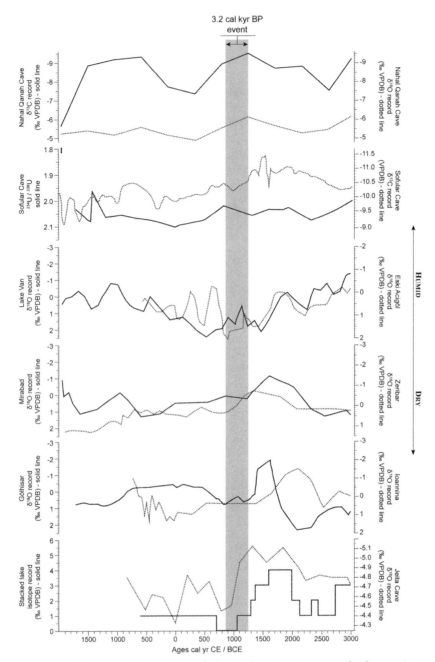

FIGURE 4.5 The 3.2 ka BP event identified in Turkey, Iran, Greece and Lebanon. See text for the references. (D. Kaniewski)

At nearby Jeita Cave, Beirut, high values of the $\partial^{18}O$ and $\partial^{13}C$ profiles of an U/Th-dated Holocene stalagmite (fig. 4.5) were related to drier conditions after 3000 cal BP (Verheyden et al. 2008). The change to dry conditions is defined by the decrease of the stalagmite diameter and the drop in growth rate. Changes in the carbon and oxygen isotopes of a 20-centimeter speleothem from the Nahal Qanah Cave in central Israel also indicate that a 1300 BC peak of wet climate (fig. 4.5) comes to an abrupt end at the Late Bronze Age transition (Frumkin et al. 1999). The stalagmite was growing when collected, giving the upper age limit. The basal age limit is provided by eight ^{14}C dates of charcoal associated with Chalcolithic artifacts deposited under the speleothem. Lastly, several sites in the Nile Delta—Burullus Lagoon (Bernhardt, Horton, & Stanley 2012), Qarun Lake (Baioumy, Kayanne, & Tada 2010), and Manzala Lagoon (Krom, Stanley, & Cliff 2002)—show a shift to more arid conditions during the 1200 BC event, though these proxy-based paleoclimate series more likely reflect changing environments in the Nile headwaters due to fluctuations in monsoon intensity.

Turkey, Iran, and Greece

Sofular Cave is situated 10 kilometers south of the Black Sea coast in northern Turkey. The age model of the Holocene speleothem record here is based on 41 ^{230}Th dates (Göktürk et al. 2011). Summer rainfall can be much higher than in the eastern Mediterranean and more negative values of the $\partial^{13}C$ record correlate with the spring-summer-fall precipitation excess. The short-term positive anomaly centered at ca. 1000 BC is interpreted as a decrease in rainfall (fig. 4.5).

Lake Van, in the eastern Anatolian highlands, is characterized by annually varved sediments. Varve counting on thin sections of the entire core was used to establish a continuous, non-floating varve chronology back to the late Glacial. A hydrologic/isotopic balance model was used to show that the $\partial^{18}O$ of Lake Van carbonates (fig. 4.5) is mainly affected by changes in the relative humidity over the lake area. Periods of high Mg/Ca values corresponding with phases of $\partial^{18}O$ enrichment are assumed to indicate very low lake levels, because only extended shelf areas provide a reactive zone for the formation of authigenic, high Mg containing carbonates (dolomite). The most prominent feature of the late Holocene period at Lake Van is a lake-level recession caused by a period of decreasing humidity from 3500 to 2000 ka BP (Lemcke & Sturm 1997). Elemental (Ti, Fe, K) and magnetic susceptibility profiles from new records, dated by radionuclide (^{210}Pb and ^{137}Cs) analysis and varve-counting, extend back to about 3600 BP. They indicate low detrital input and high carbonate contents under relatively dry conditions before 1150 BC (Barlas-Şimsek & Cağatay 2012).

In central Turkey, Eski Açigöl lake sediment cores (fig. 4.5) show increasing aridity with falling lake levels and maximum salinity levels between 1500 and 1000 BC (Roberts et al. 2001, 2011). $\partial^{18}O$ records from cores at Lake Mirabad and Lake Zeribar (fig. 4.5) in the central western Zagros Mountains of Iran also show a drying trend during this period (Stevens et al. 2006).

An Alternate View

In a recent article, Knapp and Manning (2015) questioned the climate-change hypothesis developed here, deeming it not yet fully proven, but overlooked much of the corroborative physical, ecological, and climatic data available for the 3.2 cal BP event (compiled, in part, in Kaniewski et al. 2015 and in this paper), as well as newly published data (see, for example, Boyd 2015; Langgut et al. 2015; Olsvig-Whittaker et al. 2015; Soto-Berelov et al. 2015). Additionally, Knapp and Manning use the radiocarbon dates from Kaniewski et al. (2008, 2010, 2011, 2013) to suggest a looser chronology for some events, but the radiocarbon dating of events around 1200 BC is particularly challenging (Manning 2006). As well, the Bayesian age modeling used is not incontrovertible (Parnell et al. 2011). Although more or less important standard deviations have always affected the precise chronology of this event (due to the plateau effect in the calibration curve around 1200 BC), the associated climate event is clearly attested in numerous sequences and coincident with site destructions from Greece to Israel, as Knapp and Manning agree.

Conclusions and Perspectives

By combining historical and environmental data, this review shows that the Late Bronze Age crisis was a long and complex spiral of decline that coincided with the onset of a ca. 300-year megadrought 3200 years ago. This climate shift caused crop failures and famine, which precipitated or hastened socio-economic crises and forced eventful regional human migrations in the Aegean and Eastern Mediterranean. The 3.2 ka BP megadrought underlines the agro-production sensitivity of ancient Mediterranean societies to climate change and demystifies the crisis at the Late Bronze Age-Early Iron Age transition. Drought was a primary causal trigger within the nexus of relations and events that produced the Late Bronze Age collapse in the east Mediterranean. No doubt more light will be shed on the climatic features and human impacts of the 3.2 ka BP megadrought as further paleoclimatic and archaeological evidence becomes available. Interdisciplinary investigations (as demonstrated here) can thoroughly revise our understanding of causal agents within complex historical processes.

Acknowledgments

Support was provided by the Institut Universitaire de France, CLIMSORIENT program, and by the PAI PVI/34 (Belspo) project.

References

Alley, R. B., P. A. Mayewski, T. Sowers, M. Stuiver, K. C. Taylor, and P. U. Clark. 1997. Holocene climatic instability: a prominent, widespread event 8200 yr ago. *Geology* 25: 483-486.

Al-Maqdissi, M., Bretschneider, K. Van Lerberghe, and M. Badawi. 2011. *Tell Tweini: Onze Campagnes de Fouilles Syro-Belges (1999–2010).* Damascus: Documents d'Archéologie Syrienne.

Alpert, P., and B. Ziv. 1989. The Sharav cyclone: observations and some theoretical considerations. *Journal of Geophysical Research* 94: 18495–18514.

Artzy, M. 1987. On boats and Sea Peoples. *Bulletin of American Schools of Oriental Research* 266: 75–84.

Arz, H. W., F. Lamy, and J. Pätzold. 2003. Mediterranean moisture source for an early-Holocene humid period in the northern Red Sea. *Science* 300: 118–121.

Baioumy H. M., H. Kayanne, and R. Tada. 2010. Reconstruction of lake-level and climate changes in Lake Qarun, Egypt, during the last 7000 years. *Journal of Great Lakes Research* 36: 318–327.

Barlas-Şimsek, F., and M. N. Çağatay. 2012. Late Holocene high resolution multi-proxy climate and environmental records from Lake Van, eastern Turkey. *Geophysical Research Abstracts* 14: EGU2012–6276.

Bar-Matthews, M., A. Ayalon, and M. Gilmour. 2003. Sea-land oxygen isotopic relationship from planktonic foraminifera and speleothems in the eastern Mediterranean region and their implication for paleorainfall during interglacial intervals. *Geochimica et Cosmochimica Acta* 67: 3181–3199.

Bartov, Y., S. L. Goldstein, and M. Stein. 2003. Catastrophic arid episodes in the eastern Mediterranean linked with the North Atlantic Heinrich events. *Geology* 31: 439–442.

Bernhardt, C. E., B. P. Horton, and J. D. Stanley. 2012. Nile Delta vegetation response to Holocene climate variability. *Geology* 40: 615–618.

Bond, G., B. Kromer, and J. Beer. 2001. Persistent solar influence on North Atlantic climate during the Holocene. *Science* 294: 2130–2136.

Bookman, R., Y. Enzel, and A. Agnon. 2004. Late Holocene lake levels of the Dead Sea. *Bulletin of the Geological Society of America* 116: 555–571.

Boyd, M. 2015. "Speleothems from warm climates: Holocene records from the Caribbean and Mediterranean regions." Stockholm University: PhD thesis, Department of Physical Geography.

Brinkman, J. A. 1968. *A Political History of Post-Kassite Babylonia, 1158–722 BC.* Ville, Italy: Analecta Orientalia.

Bronk-Ramsey, C., M. W. Dee, and J. M. Rowland. 2010. Radiocarbon-based chronology for dynastic Egypt. *Science* 328: 1554–1557.

Bruins, H. J., J. van der Plicht, and A. Mazar. 2003. ¹⁴C dates from Tel Rehov: Iron-Age chronology, pharaohs, and Hebrew kings. *Science* 300: 315–318.

Bruins, H. J., J. van der Plicht, and J. A. MacGillivray. 2009. The Minoan Santorini eruption and tsunami deposits in Palaikastro (Crete): dating by geology, archaeology, 14c, and Egyptian chronology. *Radiocarbon* 51: 397–411.

Bryce, T. 2005. *The Kingdom of the Hittites.* Oxford: Oxford University Press.

Bryson, R. A., H. H. Lamb, and D. Donley. 1974. Drought and the decline of the Mycenae. *Antiquity* 43: 46–50.

Bryson, R. A., and R. U. Bryson. 1997. High resolution simulations of regional Holocene climate: North Africa and the Near East. In *Third Millennium BC Climate Change and Old World Collapse*, ed. H. Nüzhet-Dalfes, G. Kukla, and H. Weiss. NATO ASI Series I, Global Environmental Change 49. Berlin: Springer, 565–594.

Butzer, K. W. 1976. *Early hydraulic civilization in Egypt.* Chicago: University of Chicago Press.

Carpenter, R. 1966. *Discontinuity in Greek civilization.* Cambridge: Cambridge University Press.

Cohen, Y., and I. Singer. 2006. Late synchronism between Ugarit and Emar. In *Essays on Ancient Israel in Its Near Eastern Context. A Tribute to Nadav Na'aman*, ed. Y. Amit, E. B. Zvi, I. Finkelstein, and O. Lipschits. Winona Lake, Indiana: Eisenbrauns, 123–139.

deMenocal, P. B. 2001. Cultural responses to climatic change during the late Holocene. *Science* 292: 667–673.

Enzel, Y., R. Bookman, and D. Sharon. 2003. Late Holocene climates of the Near East deduced from the Dead Sea level variations and modern regional winter rainfall. *Quaternary Research* 60: 263–273.

Faulkner, R. O. 1975. Egypt, from the inception of the nineteenth dynasty to the death of Ramesses III. In *Cambridge Ancient History.* 3rd ed., Vol. 2, ed. I. E. S. Edwards, C J. Gadd, and E. Sollberger. Cambridge: Cambridge University Press, 217–251.

Felis, T., G. Lohmann, and H. Kuhnert. 2004. Increased seasonality in Middle East temperatures during the last interglacial period. *Nature* 429: 164–168.

Finkelstein, I., and E. Piasetzky. 2009. Radiocarbon-dated destruction layers: a skeleton for Iron Age chronology in the Levant. *Oxford Journal of Archaeology* 28: 255–274.

Fiorentino G, V. Caracuta, and L. Calcagnile. 2008. Third millennium B.C. climate change in Syria highlighted by carbon stable isotope analysis of ^{14}C-AMS dated plant remains from Ebla. *Palaeogeography, Palaeoclimatology, Palaeoecology* 266: 51–58.

Fischer, P. M., T. Bürge, L. Franz, and R. Feldbacher. 2012. The new Swedish Cyprus Expedition 2011. Excavations at Hala Sultan Tekke. *Opuscula* 5: 89–112.

Frumkin, A., D. C. Ford, and H. P. Schwarcz. 1999. Continental oxygen isotopic record of the last 170,000 years in Jerusalem. *Quaternary Research* 51: 317–327.

Gilboa, A. 2006–2007. Fragmenting the Sea Peoples, with an emphasis on Cyprus, Syria and Egypt: a Tel Dor perspective. *Scripta Mediterranea* 27–28: 209–244.

Göktürk, O. M., D. Fleitmann, and S. Badertscher. 2011. Climate on the Black Sea coast during the Holocene: implications from the Sofula Cave record. *Quaternary Science Reviews* 30: 2433–2445.

Grimal, N. 1992. *A History of Ancient Egypt.* Oxford: Blackwell.

Hornung, E., R. Krauss, and D. A. Warburton, eds. 2006. *Ancient Egyptian Chronology.* Handbuch der Orientalistik, erste Abteilung: Der Nahe und Mittlere Osten 83. Leiden: Brill.

Hutchinson, J. S. 1977. Mycenaean kingdoms and medieval estates (an analogical approach to the history of LH III). *Historia: Zeitschrift für Alte Geschichte Band* 26: 1–23.

Kahana, R., Z. Baruch, and Y. Enzel. 2002. Synoptic climatology of major floods in the Negev desert, Israel. *International Journal of Climatology* 22: 867–882.

Kaniewski, D., E. Paulissen, and E. Van Campo. 2008. Middle East coastal ecosystem response to middle-to-late Holocene abrupt climate changes. *Proceedings of the National Academy of Sciences of the United States of America* 105: 13941–13946.

Kaniewski, D., E. Paulissen, and E. Van Campo E. 2010. Late Second-Early First Millennium BC abrupt climate changes in coastal Syria and their possible significance for the history of the eastern Mediterranean. *Quaternary Research* 74: 207–215.

Kaniewski, D., E. Van Campo, and K. Van Lerberghe. 2011. The Sea Peoples, from cuneiform tablets to carbon dating. *PLoS ONE* 6(6): e20232. doi:10.1371/journal. pone.0020232

Kaniewski, D., E. Van Campo, and J. Guiot. 2013. Environmental roots of the Late Bronze Age crisis. *PLoS ONE* 8(8): e71004. doi:10.1371/journal.pone.0071004

Killebrew, A. E. 2005. *Biblical peoples and ethnicity: an archaeological study of Egyptians, Canaanites, Philistines, and early Israel, 1300–1100 BC.* Atlanta: Society of Biblical Literature Archaeology and Biblical Studies.

Killebrew, A. E., and G. Lehmann. 2013. *The Philistines and Other Sea Peoples in Text and Archaeology.* Atlanta: Society of Biblical Literature.

Kitchen, K. 1983. *Pharaoh triumphant: the life and times of Ramesses II, King of Egypt.* London: Aris and Phillips.

Knapp, A. B., and S. W. Manning. 2015. Crisis in context: the end of the Late Bronze Age in the eastern Mediterranean. *American Journal of Archaeology* 120: 99–149.

Kolodny, Y., M. Stein, and M. Machlus. 2005. Sea-rain-lake relation in the Last Glacial East Mediterranean revealed by $\delta^{18}O$–$\delta^{13}C$ in Lake Lisan aragonites. *Geochimica et Cosmochimica Acta* 69: 4045–4060.

Krom, M. D., J. D. Stanley, and R. A. Cliff. 2002. Nile River sediment fluctuations over the past 7000 yr and their key role in sapropel development. *Geology* 30: 71–74.

Langgut, D., I. Finkelstein, and T. Litt. 2013. Climate and the Late Bronze collapse: new evidence from the southern Levant. *Tel Aviv* 40: 149–175.

Langgut, D., I. Finkelstein, T. Litt, F. H. Neumann, and M. Stein. 2015. Vegetation and climate changes during the Bronze and Iron ages (~3600–600 BCE) in the southern Levant based on palynological records. *Radiocarbon* 57: 217–235.

Lemcke, G., and M. Sturm.1997. $\partial^{18}O$ and trace element measurements as proxy for the reconstruction of climate changes at Lake Van (Turkey): preliminary results. In *Third Millennium BC Climate Change and Old World Collapse*, ed. H. Nüzhet-Dalfes, G. Kukla, and H. Weiss. NATO ASI Series I, Global Environmental Change 49. Berlin: Springer, 653–678.

Litt, T., C. Ohlwein, and F. H. Neumann. 2012. Holocene climate variability in the Levant from the Dead Sea pollen record. *Quaternary Science Reviews* 49: 95–105.

Manassa, C. 2003. *The Great Karnak Inscription of Merneptah: Grand Strategy in the 13th Century BC.* YES 5. New Haven: Yale Egyptological Seminar.

Manning, S. W., B. Kromer, and P. I. Kuniholm. 2001. Anatolian tree rings and a new chronology for the East Mediterranean Bronze-Iron ages. *Science* 294: 2532–2535.

Manning, S. W., C. Bronk-Ramsey, and W. Kutschera. 2006. Chronology for the Aegean Late Bronze Age 1700–1400 BC. *Science* 312: 565–569.

Manning, S. W. 2006–2007. Why radiocarbon dating 1200 BC is difficult: a sidelight on dating the end of the Late Bronze Age and the contrarian contribution. *Scripta Mediterranea* 27–28: 53–80.

Mayewski, P. A., L. D. Meeker, and M. S. Twickler. 1997. Major features and forcing of high-latitude northern hemisphere atmospheric circulation using a 110,000 year-long glaciochemical series. *Journal of Geophysical research* 102: 26345–26366.

Mayewski, P. A., E. E. Rohling, and J. C. Stager. 2004. Holocene climate variability. *Quaternary Research* 62: 243–255.

Mazar, A., and I. Carmi. 2001. Radiocarbon dates from Iron Age strata at Tel Beth Shean and Tel Rehov. *Radiocarbon* 43: 1333–1342.

Medina-Elizade, M., and E. J. Rohling. 2012. Collapse of Classic Maya civilization related to modest reduction in precipitation. *Science* 335: 956–959.

Migowski, C., M. Stein, and S. Prasad. 2006. Holocene climate variability and cultural evolution in the Near East from the Dead Sea sedimentary record. *Quaternary Research* 66: 421–431.

Nelson, H. H. 1930. *Medinet Habu: the earliest historical records of Ramses III, volume I.* Chicago: Oriental Institute Publications.

Neumann, F., C. Schölzel, and T. Litt. 2007. Holocene vegetation and climate history of the northern Golan Heights (Near East). *Vegetation History and Archaeobotany* 16: 329–346.

Neumann, J., and S. Parpola. 1987. Climatic change and the eleventh-tenth-century eclipse of Assyria and Babylonia. *Journal of Near Eastern Studies* 46: 161–182.

Nougayrol, J., E. Laroche, C. Virolleaud, and C. Schaeffer. 1968. *Ugaritica V, Nouveaux textes accadiens, hourrites et ugaritiques des archives et bibliothèques privées d'Ugarit.* Mission de Ras Shamra, 16. Paris: Librairie Orientaliste Paul Geuthner.

O'Brien, S. R., P. A. Mayewski, L. D. Meeker, D. A. Meese, M. S. Twickler, and S. I Whitlow. 1996. Complexity of Holocene climate as reconstructed from a Greenland ice core. *Science* 270: 1962–1964.

Olsvig-Whittaker, L., A. M. Maeir, E. Weiss, S. Frumin, O. Ackermann, and L. Kolska Horwitz. 2015. Ecology of the past – Late Bronze and Iron Age landscapes, people and climate change in Philistia (the southern coastal plain and Shephelah), Israel. *Journal of Mediterranean Ecology* 13: 57–75.

Parnell, A. C., C. E. Buck, and T. K. Doan. 2011. A review of statistical chronology models for high-resolution, proxy-based Holocene palaeoenvironmental reconstruction. *Quaternary Science Reviews* 30: 2948–2960.

Pritchard, J. B. 1969. *Ancient Near Eastern texts related to the Old Testament.* 3rd ed. with supplement. Princeton: Princeton University Press.

Roberts, R. G. 2008. Identity, choice, and the year 8 reliefs of Ramesses III at Medinet Habou. In *Forces of Transformation: The End of the Bronze Age in the Mediterranean,* ed. C. Bachhuber and R. G. Roberts. Oxford: Oxbow/BANEA, 60–68.

Roberts, N., J. M. Reed, and M. J. Leng. 2001. The tempo of Holocene climate change in the eastern Mediterranean region: new high-resolution crater-lake sediments data from central Turkey. *The Holocene* 11: 721–736.

Roberts, N., W. J. Eastwood, C. Kuzucuoğlu, G. Fiorentino, and V. Caracuta. 2011. Climatic, vegetation and cultural change in the eastern Mediterranean during the mid-Holocene environmental transition. *The Holocene* 21: 147–162.

Rodwell, M. J., and B. J. Hopkins.1996. Monsoons and the dynamics of deserts. *Quarterly Journal of the Royal Meteorological Society* 122: 1385–1404.

Rohling, E. J., P. A. Mayewski, and R. H. Abu-Zied. 2002. Holocene atmosphere ocean interactions: records from Greenland and the Aegean Sea. *Climate Dynamics* 18: 587–593.

Saaroni, H., B. Ziv, and A. Bitan.1998. Easterly wind storms over Israel. *Theoretical and Applied Climatology* 59: 61–77.

Schilman, B., M. Bar-Matthews, and A. Almogi-Labin. 2001. Global climate instability reflected by eastern Mediterranean marine records during the late Holocene. *Palaeogeography, Palaeoclimatology, Palaeoecology* 176: 157–176.

Shaw, I. 2000. *The Oxford History of Ancient Egypt*. Oxford: Oxford University Press.

Singer I. 1999. A political history of Ugarit. In *Handbook of Ugaritic Studies*, ed. W. G. E. Watson and N. Wyatt. Leiden: Handbuch der Orientalistik, Erste Abteilung, 603–733.

Singer, I. 2000. New evidence on the end of the Hittite Empire. In *The Sea Peoples and Their World: a Reassessment*, ed. E. D. Oren. Philadelphia: University of Pennsylvania, 21–33.

Soto-Berelov, M., P. L. Fall, S. E. Falconer, and E. Ridder. 2015. Modeling vegetation dynamics in the southern Levant through the Bronze Age. *Journal of Archaeological Science* 53: 94–109.

Stanley, J. D., M. D. Krom, and R. A. Cliff. 2003. Short contribution: Nile flow failure at the end of the Old Kingdom, Egypt: strontium isotopic and petrologic evidence. *Geoarchaeology* 18: 395–402.

Staubwasser, M., F. Sirocko, and P. M. Grootes. 2003. Climate change at the 4.2 ka BP termination of the Indus valley civilization and Holocene south Asian monsoon variability. *Geophysical Research Letters* 30: 1425–1428.

Stevens, L. R., E. Ito, and A. Schwalb. 2006. Timing of atmospheric precipitation in the Zagros Mountains inferred from a multi-proxy record from Lake Mirabad, Iran. *Quaternary Research* 66: 494–500.

Stiebing, W. H. 1980. The end of the Mycenean Age. *The Biblical Archaeologist* 43: 7–21.

Telford, R. J., E. Heegaard, and H. J. B. Birks. 2004. The intercept is a poor estimate of a calibrated radiocarbon age. *The Holocene* 14: 296–298.

Tyldesley, J. 2000. *Ramesses: Egypt's Greatest Pharaoh*. London: Viking/Penguin Books.

Verheyden, S., F. H. Nader, and H. J. Cheng. 2008. Paleoclimate reconstruction in the Levant region from the geochemistry of a Holocene stalagmite from the Jeita cave, Lebanon. *Quaternary Research* 70: 368–381.

Warburton, D. 2003. Love and war in the later Bronze Age: Egypt and Hatti. In *Ancient Perspectives on Egypt*, ed. R. Matthews and C. Roemer. Lond: UCL Press, 75–100.

Weiss, B. 1982. The decline of the Late Bronze Age civilization as a possible response to climate change. *Climate Change* 4: 173–198. doi: 10.1007/bf00140587

Weiss, H. 2014. Altered trajectories: the Intermediate Bronze Age in Syria and Lebanon 2200-1900 BCE. In *Oxford Handbook of the Archaeology of the Levant,* ed. A. Killebrew and M. Steiner. Oxford: Oxford University Press.

Weiss, H., M.-A. Courty, W. Wetterstrom, F. Guichard, L. Senior, R. Meadow, and A. Curnow. 1993. The genesis and collapse of third millennium north Mesopotamian civilization. *Science* 261: 995–1004.

Weninger, B., and R. Jung. 2009. Absolute chronology of the end of the Aegean Bronze Age. In *LH IIIC Chronology and Synchronisms III: LH IIIC Late and the Transition to the Early Iron Age,* ed. S. Deger-Jalkotzy and A. Bächle. Proceedings of the International Workshop Held at the Austrian Academy of Sciences . . . 2007. Vienna: VÖAW, 373–416.

Yasur-Landau, A. 2010. *The Philistines and Aegean Migration at the End of the Late Bronze Age.* Cambridge: Cambridge University Press.

Yon, M. 2006. *The city of Ugarit at Tell Ras Shamra.* Winona Lake, Indiana: Eisenbrauns.

Ziv, B., H. Saaroni, M. Romem, and A. Baharad. 2010. Analysis of conveyor belts in winter Mediterranean cyclones. *Theoretical and Applied Climatology* 99: 441–455.

CHAPTER 5 | AD 550–600 Collapse at Teotihuacan

Testing Climatic Forcing from a 2400-Year Mesoamerican Rainfall Reconstruction

MATTHEW S. LACHNIET AND

JUAN PABLO BERNAL-URUCHURTU

We analyze a 2400-year rainfall reconstruction from an ultra-high-resolution absolutely-dated stalagmite (JX-6) from southwestern Mexico. Oxygen isotope variations correlate strongly to rainfall amount in the Mexico City area since AD 1870 and for the wider southwestern Mexico region since 1948, allowing us to quantitatively reconstruct rainfall variability for the Basin of Mexico and Sierra Madre del Sur for the past 2400 years. Because oxygen isotopes integrate rainfall variations over broad geographic regions, our data suggest substantial variations in Mesoamerican monsoon strength over the past two millennia. As a result of low age uncertainties (≤ 11 yr), our stalagmite paleoclimate reconstruction allows us to place robust ages on past rainfall variations with a resolution an order of magnitude more precise than archeological dates associated with societal change. We relate our new rainfall reconstruction to the sequence of events at Teotihuacan and to other pre-Colombian civilizations in Mesoamerica. We observe a centuries long drying trend that culminated in peak drought conditions ca. AD 750 related to a weakening monsoon, which may have been a stressor on Mesoamerican societies. Teotihuacan was an ideal location to test for links between climate change and society, because it was in a semi-arid highland valley with limited permanent water sources and relied upon spring-fed irrigation to ensure a reliable maize harvest. The city of Teotihuacan was one of the largest Mesoamerican cities, apparently reaching population sizes of 80,000 to 100,000 inhabitants by AD 300. Following the "Great Fire," which dates approximately to AD 550, the population decreased and many buildings were abandoned. Because of the apparent reliance on rainwater capture and spring-fed agriculture in the Teotihuacan valley, food production and domestic water supplies would have been sensitive to rainfall variations.

Climate and Civilization

The relationship between climate and civilizations is an important topic as societies face an increasingly altered climate caused by the rise in long-lived greenhouse gas concentrations (CO_2 and CH_4) in the atmosphere (Pachauri & Reisinger 2007). Model projections for Mexico, for example, suggest an overall drying and a weakening of the monsoon (Karmalkar et al. 2011), and Mexico has recently been afflicted by pronounced droughts. However, paleoclimatic evidence for rainfall variations indicates that monsoon strength has varied dramatically on interannual to millennial time scales. In most of southwestern and central Mexico, summer rainfall associated with the Mesoamerican sector of the North American Monsoon delivers moisture sourced from the Gulf of Mexico and eastern Pacific Ocean (Lachniet et al. 2013, Schmitz & Mullen 1996). The monsoon is affected by proximity to the eastern Pacific Intertropical Convergence Zone and the strength of the ocean to land temperature gradient, and results in a pronounced dry season between December and May. During the wet season, monsoon rainfall typically begins in May or June and lasts until November or December, providing the moisture to nourish crops and recharge regional aquifers. Thus, variations in the strength of the Mesoamerican monsoon may have implications for agriculture and societal use of fresh water supplies. Further, these monsoon rainfall variations have been linked to societal changes in Mesoamerica, and most prominently to the role of aridity in the Classic Period of the Maya lowlands (Hodell, Brenner & Curtis 2005, 2001; Kennett et al. 2012; Medina-Elizalde et al. 2010). Recently, climatic variation has been placed within a complex set of interacting feedbacks and forcings on civilizations that include environmental degradation and political and socioeconomic changes and do not allow for a simple and single forcing of societal "collapse" (Turner II and Sabloff 2012).

Much paleoclimate research has focused on the Maya lowlands (Hodell, Brenner & Curtis 2005, 2007; Hodell et al. 2001; Kennett et al. 2012; Medina-Elizalde et al. 2010; Medina-Elizalde and Rohling 2012). The role of climate change on Mesoamerican societies in the semi-arid highlands has received less attention (fig. 5.1). Herein, we focus on the climate history relevant to the rise and fall of Teotihuacan, located in the Basin of Mexico and the most highly urbanized New World civilization (Cowgill 2008, 2015a; Millon 1970). In the absence of inscriptions and significant documentary evidence that might explain Teotihuacan's history, we must rely on field data and the various hypotheses associated with population demographics in the Basin in general and the Teotihuacan valley in particular. Population estimates for Teotihuacan were first estimated at its peak in the Classic Period are a minimum of 75,000, a probable 125,000, and a possible 200,000 (Millon 1970). Later estimates used here (based on sherd counts and subject to much uncertainty) suggested a peak population of 80,000 to 100,000 (Cowgill 1997, 2015a) that was relatively stable until declining to ca. 45,000 following the decline of the Teotihuacan state around AD 600. Teotihuacan was the largest and most important civilization

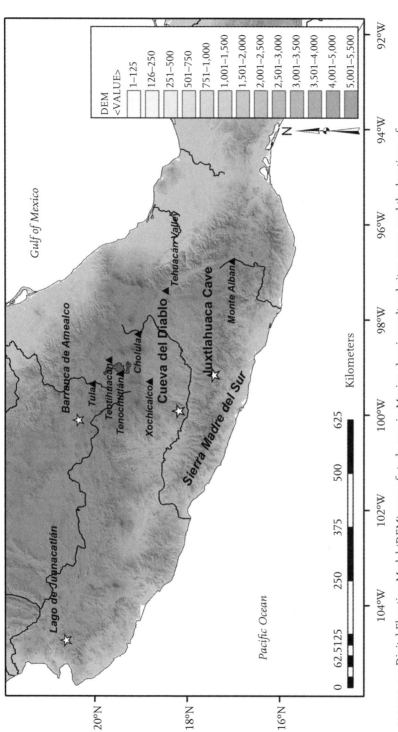

FIGURE 5.1 Digital Elevation Model (DEM) map of study area in Mexico showing cultural sites, caves, and the locations of the Lago de Juanacatlán and Barranca de Amealco tree-ring records. (M. S. Lachniet)

and monumental society in the Basin of Mexico during the Classic Period. Evidence of productive agricultural soils (Sanchez-Perez et al. 2013), spring-fed irrigation (Mooser 1968, Sanders 1977), stream channelization of the Rio San Juan, and rainfall capture and reservoirs (Linné 2003, Millon 1970, Sanders 1977) suggests that Teotihuacan was an "irrigation state." The intense use of irrigation in the Teotihuacan valley was likely the result of central state control of freshwater supplies for the cultivation of maize and other crops on canal-fed (irrigated raised fields) plots.

Previous Paleoclimate Research in Highland Mesoamerica

Although few data have been available to constrain the impact of climate change on the irrigation state of Teotihuacan, some of the earliest paleoenvironmental reconstructions were carried out in the Basin of Mexico. Measuring ratios of oak and pine pollen from soils in the Basin, the investigators suggested that increased oak pollen was a proxy for greater effective moisture, whereas increased pine pollen represented decreased effective moisture (Sears 1952). The data indicated that the early Teotihuacan period was associated with dry conditions and that the Aztec (Nahua) period was wet, an interpretation that correlates exceptionally well with our new stalagmite $\delta^{18}O$ rainfall reconstruction described below which shows dry conditions between 350 and 50 BC, and wetter conditions AD 900–1500.

More recently, a tree-ring reconstruction links climate and cultural change in central Mexico. A 1238-year tree-ring reconstruction from Montezuma baldcypress (also known as Ahuehuete trees in Mexico; *Taxodium mucronatum*) at the Barranca de Almeaco, Querétero (fig. 5.1), provides an early wet season rainfall reconstruction for central Mexico and the Mexican altiplano (Stahle et al. 2012, 2011). The tree-ring chronology is well correlated with central Mexico June Palmer Drought Severity Index (PDSI) values over the 1950–2003 instrumental record, a proxy for available soil moisture. The presence of droughts and megadroughts coincident with thin rings was inferred from this correlation. The tree-ring chronology defines droughts coincident with the Classic (ca. AD 910), Toltec (ca. AD 1160), Aztec (ca. AD 1400), and Conquest (ca. AD 1530) periods. However, because statistical processing removed variance at greater than the centennial scale, it is difficult to compare the severity of droughts separated by centuries directly. The Barranca de Amealco tree chronology does not extend far enough back in time to investigate a climatic forcing for the fall of Teotihuacan.

Another high-resolution paleoclimate rainfall reconstruction comes from Lago Juanacatlán (Metcalfe et al. 2010), a small lake in the western Mexican highlands. The authors measured titanium, potassium, calcium, iron, and magnetic susceptibility on the lake sediment at high resolution over ca. the past 2000 years. The sediment core is well dated, and the zero-aged sediment collected in a sediment trap returns a "modern" value of 105 pmc (percent modern carbon), which indicates that organic matter delivery to the lake is of recent origin. This point is important because it makes Lago Juanacatlán one of the few Mexican lakes known to have been tested rigorously for radiocarbon systematics (to measure reservoir effects, old carbon, etc.). As one of the few well-dated and high-resolution lacustrine sediment core from the Mexican highlands, the Lago Juanacatlán record is an important contribution to our understanding of climate over the past two millennia. Most notably, the period from ca. AD 1100 to 1900 was characterized by significantly higher titanium (Ti) concentrations, which was interpreted to represent wetter conditions. Prior to this time interval, and spanning the rise and fall of Teotihuacan, conditions were apparently drier. The authors interpreted the Ti record to be a proxy for stream runoff, and hence precipitation, whereby heavy rainfall events resulted in greater slope erosion into the catchment and an increase in bedrock-derived Ti relative to other sediment sources. Based on the authors' interpretations, the Late Classic was relatively dry, particularly between ca. AD 400 and 600, an interval that spans the apparent population crash of Teotihuacan following the "Big Fire" of AD 550. However, the record appears to be uncorrelated to the rainfall amount in nearby Guadalajara over the modern period, which suggests that titanium proxy data may have a complicated relationship to climate change.

Cave Stalagmites as Paleo-rain Gauges

The paucity of well-dated and high-resolution paleoclimate records for central and southwestern Mexico makes additional, climate-calibrated paleoclimate proxy records highly desirable. To that end, we have been completing paleoclimate reconstructions of the summer monsoon in the Sierra Madre del Sur (SMS)—a region within the informally termed "Mesoamerican monsoon" (Lachniet et al. 2013)—using radiometrically dated cave calcite deposits (speleothems) from Juxtlahuaca and del Diablo Caves. Our data provide a discontinuous 22,000-year reconstruction of Mesoamerican monsoon rainfall from several dated stalagmites (Lachniet et al. 2013; 2017). Here we focus on the last 2400 years of Mesoamerican monsoon history, and in particular on the sequence of climate change events associated with the rise and fall of Teotihuacan between ca. 350 BC and AD 900.

The utility of speleothems as paleoclimate proxies is anchored by the radiometric ages that can be determined from uranium-series decay chains (Richards and Dorale 2003). In contrast to radiocarbon, which has a variable production rate in the atmosphere, ^{234}U decays at a known rate into its daughter product ^{230}Th, which itself has a precisely known decay rate. The principle of ^{234}U/^{230}Th dating relies on the fact that uranium is highly soluble in water, whereas thorium is insoluble. Thus, cave drip waters containing uranium and $CaCO_3$ (calcium carbonate) will precipitate speleothems when the waters become saturated in $CaCO_3$, and the uranium is precipitated in the

CaCO$_3$ crystal lattice, at which time it begins its decay to ^{230}Th. The ratio of ^{234}U to ^{230}Th then decreases until they attain a secular equilibrium after about 500,000 years. Measurement of ^{234}U/^{230}Th thus provides a robust estimate of the age of the CaCO$_3$ after correcting for any non-radiogenic detrital ^{230}Th. In ideal systems where the uranium concentration is high and any detrital thorium contamination is low (as in our study), age precisions can be better than ± 1 percent. Under such ideal conditions, age uncertainties resulting from analytical and decay-constant uncertainties are on the order of only ± 1 to 5 years over the Common Era. On the other hand, dating derived from calibration into calendar years of radiocarbon (^{14}C), which can produce age ranges of 100 to 300 years over the Common Era, may be too broad to be useful when the pace of societal change, as inferred from such archaeological materials as ceramics, may be on the order of one or two centuries.

The rainfall proxy in speleothem analysis is the ratio of stable oxygen isotopes incorporated into the stalagmite CaCO$_3$. The ratio of ^{18}O to ^{16}O in samples relative to a standard is denoted δ^{18}O and is defined as δ^{18}O = (^{18}O/^{16}O$_{sample}$ − ^{18}O/^{16}O$_{standard}$) ÷ ^{18}O/^{16}O$_{standard}$ × 1000, where the values are in "per mil" variations, or ‰, relative to a standard of known ^{18}O/^{16}O ratios. The δ^{18}O standards are Vienna Standard Mean Ocean Water (VSMOW) for waters, and Vienna Pee Dee Belemnite (VPDB) for carbonates; the VPDB standard is approximately 30 per mil higher than VSMOW when expressed on the same standard scale due to the enhanced incorporation of the heavy isotope (^{18}O) in stalagmite carbonate during CaCO$_3$ precipitation.

The δ^{18}O values of tropical rainfall are inversely proportional to rainfall amount, such that more intense rainout of an air mass results in rainwater with lower δ^{18}O, usually during rainier months (Dansgaard 1964). We have shown that the δ^{18}O variations in rainfall for central Mexico exhibit this standard "amount effect" on seasonal and interannual time scales, with lower δ^{18}O values during rainier months, and vice versa (Lachniet et al. 2012, 2017). Low δ^{18}O values in rainfall are thus associated with air masses that have experienced large rainout amounts, such as strong convection cells and tropical storms. As a result, the δ^{18}O value of rainfall falling over the cave sites in the SMS is inversely correlated to the strength of air mass rainout. Additional controls on atmospheric δ^{18}O values in Mesoamerica are changes in moisture source and sea surface temperature anomalies in nearby source regions (Lachniet 2009b, Lachniet et al. 2007). These rain waters then infiltrate into the soils and the epikarst and unsaturated zone, which is the zone in limestone bedrock above the cave that is largely air-filled and defined by secondary porosity, and ultimately reach the cave. The soil and epikarst become charged with biogenic CO$_2$ as water passes through them, which allows for dissolution of limestone. Once within the cave, lower cave-passage CO$_2$ concentrations permit the drip waters to degas carbon dioxide, which results in the precipitation of CaCO$_3$ due to calcite or aragonite saturation to form speleothems. Because the oxygen in the speleothem CaCO$_3$ has been equilibrated with the large oxygen reservoir in the drips, the δ^{18}O values of the precipitated CaCO$_3$ will record variations in the $\delta^{18}O$ of the drip (and hence rain) waters, separated by the aforementioned ca. + 30 per mil fractionation (offset) due to the precipitation of solid $CaCO_3$ from the waters. These well-known processes (Lachniet 2009a) render tropical stalagmites robust paleorainfall climate indicators.

One of the strengths of tropical stalagmite $\delta^{18}O$ is that it is not merely a local proxy. Rather, the $\delta^{18}O$ values of atmospheric moisture record rainfall linked to climate processes on a regional (synoptic) scale spanning hundreds of kilometers. For example, at-a-site rainfall amount reflects just a portion of total air mass rainout, whereas the $\delta^{18}O$ value of precipitation reflects the cumulative rainout both at the sample site and upstream along the air mass trajectory. This characteristic of $\delta^{18}O$ makes the proxy more ideal for studying large-scale phenomena like monsoons than some other records that may only record a local climatic signal (such as titanium concentrations as a proxy for local runoff in a lake, or tree rings at a single site). However, the geographical extent of the climate/$\delta^{18}O$ relationship must be quantified with statistical analyses. Because summer rainfall associated with convective storms may be isolated geographically, the $\delta^{18}O$ proxy gives a better indicator of regional rainfall amounts in the Mesoamerican monsoon. Strictly interpreted, however, the $\delta^{18}O$ values of tropical stalagmites represent the strength of monsoon convection (that is, average air mass rainout amount) integrated over a summer rainy season. The strength of convection and total rainfall amount are likely to be correlated, but total rainfall amount over the rainy season is not necessarily linked to low $\delta^{18}O$ values, because numerous small rainfall events of high $\delta^{18}O$ values can be aggregated into a high rainfall total. As a result, stalagmite $\delta^{18}O$ paleorainfall reconstructions are best interpreted as a proxy of regional air mass rainout amount, which is likely though not necessarily strongly correlated with total at-a-site wet seasonal rainfall.

Methods

We collected stalagmite JX–6 from the La Sorpresa room of Juxtlahuaca Cave (Gay 1967). The cave is located 90 kilometers from the Pacific Coast, near the city of Chilpancingo in the heart of the Sierra Madre del Sur (17.44°N, 99.16°W, 927 meters above sea level). The stalagmite was cut in half and a working face was prepared on a central slab along the growth axis (1044 mm, or millimeters). We subsampled the stalagmite for twenty uranium-series disequilibrium dates, and at about 1-mm resolution for $\delta^{18}O$ over most of the stalagmite and at a 0.25 mm resolution on the upper 80 mm of growth, for a total of 1230 stable isotope analyses over the past 2400 years. The high-resolution stable isotope data over the upper 80 mm have an average age resolution of 0.59 years. The subsampled powders were analyzed for U-series dating at the University of New Mexico, and $\delta^{18}O$ samples were analyzed at the Las Vegas Isotope Science Laboratory at the University of Nevada Las Vegas (see Lachniet et al. 2012) for more methodological details.) We quantified the relationship between stalagmite δ¹⁸O values and rainfall amount by comparing our data to the instrumental record at the Tacubaya station, which is located in what is now the western edge of the Mexico City Metropolitan area and has a record extending back to AD 1878. To test the hypothesis that stalagmite δ¹⁸O represents rainfall amount over a wide geographic area, we also compared our data to the NCEP (National Centers for Environmental Prediction) reanalysis data (Kalnay et al. 1996) of rainfall totals for several domains in southern and southwestern Mexico. Finally, and new to this study, we provide a climate calibration between stalagmite JX–6 and rainfall in the vicinity of the Oaxaca Valley to show relevance to the site of Monte Albán. The stalagmite δ¹⁸O data were interpolated to an annual resolution, smoothed using a 5-year running average, and then correlated to the summed May through October wet season rainfall rate (kg/m²/sec) extracted from the NCEP reanalysis data for the grid cell (16.2°N, 96.9°W) nearest to Oaxaca City (17.05°N, 96.7°W). The correlation interval was 2008 to 1948 to account for the 5-year running average on the δ¹⁸O data, and the correlations determined with δ¹⁸O lagging Oaxaca region rainfall by zero to + 15 years to account for transit time through the epikarst.

Results

The dating shows that stalagmite JX–6 grew continuously over the past ca. 2400 years (fig. 5.2). The U-series dates on stalagmite JX–6 are ultra-precise, with typical uncertainties less than ca. 10 years, and show regular, continuous growth with a lack of hiatuses. The very low age uncertainties are a result of the high uranium concentrations in the aragonite stalagmite, which allowed us to obtain a high precision on the analysis of daughter ²³⁰Th. Further, because the stalagmite was very fast growing (0.44 millimeter per year), the bulk sampling over most of the stalagmite had an age resolution of one sample approximately every two years. It is important to note that this level of chronological control and age resolution is an order of magnitude more precise and detailed than most existing archaeological or paleoenvironmental studies and approaches the resolution of tree rings and historical sources. The stalagmite δ¹⁸O values vary by 1.5 per mil about a relatively stable mean and are inversely correlated with rainfall amount (see below).

Because the cave sampling location is overlain by about 160 meters of limestone bedrock, the drip waters feeding the stalagmite are mixed prior to reaching the stalactite that is dripping onto the JX–6 stalagmite. As a result, much if not most of the annual- to possibly inter-annual scale climate variability has been smoothed out or potentially lost despite the high drilling resolution. Together with the observation that rainfall δ¹⁸O values integrate air mass rainout amounts over wide geographic areas, the stalagmite δ¹⁸O record is thus best interpreted at the multi-annual scale as a proxy for the intensity of the Mesoamerican monsoon.

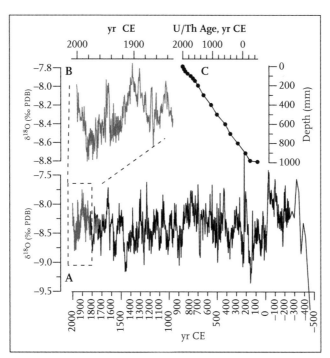

FIGURE 5.2 Stalagmite JX–6 U-series and δ¹⁸O plot. The δ¹⁸O values are inversely correlated with rainfall amount via the amount effect. Lower δ¹⁸O values represent wetter periods. (M. S. Lachniety & J. P. Bernal)

Climate Calibration

The results of the climate calibration indicate that stalagmite $\delta^{18}O$ is a robust proxy for rainfall amount at the Tacubaya station in the Basin of Mexico, yielding a correlation of $r = -0.79$ when 5-year running averages of JX–6 $\delta^{18}O$ and rainfall are compared with a JX–6 lag of 9 years to account for drip transit time through the epikarst between AD 1880 and 2008 (fig. 5.3). The correlation strengthens to $r = -0.89$ over the 1880–1988 interval, after removal of the most recent rainfall data which may be influenced by anthropogenic effects in the Basin of Mexico (Juaregui 1990/1991). Further, the rainfall amounts at Tacubaya were significantly correlated to rainfall amount in broader regions of the Sierra Madre del Sur, ensuring that the climate signal contained with stalagmite JX–6 is representative of the regional strength of the Mesoamerican monsoon and not just a local proxy. (See data in supporting information of (Lachniet et al., 2012)). We also completed a new climate calibration against the NCEP grid cell that contains the Oaxaca valley to provide a rainfall reconstruction for Monte Albán (fig. 5.3). A maximum negative correlation resulted between stalagmite $\delta^{18}O$ and Oaxaca domain rainfall of -0.62 between 5-year smoothed JX–6 $\delta^{18}O$ and rainfall at zero lag. The correlation remains stronger than -0.54 to a JX–6 lag of five years. If the 1999–2008 data are excluded for possible anthropogenic effects as above, the

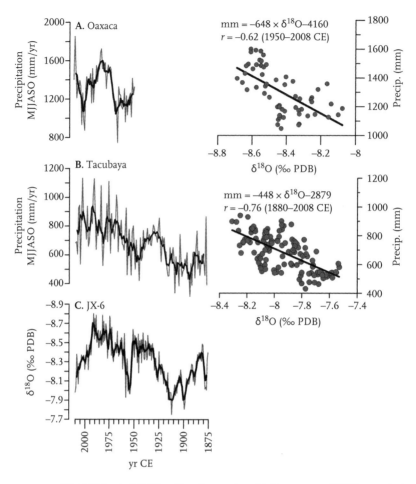

FIGURE 5.3 JX–6 $\delta^{18}O$ rainfall (C) calibration against (*a*) Oaxaca area NCEP reanalysis wet-season rainfall (May through October) and (*b*) Tacubaya instrumental data for wet season (May through October) rainfall. Plots to the right show quantitative calibrations using the five-year smoothed $\delta^{18}O$ and rainfall data (solid lines). The correlations are for lags of zero (Oaxaca) and 9 years (Tacubaya). (M. S. Lachniet & J. P. Bernal)

correlation improves to r = –0.65. The high negative correlations at 0 to 5 year lag against the Oaxaca data are similar to the lags of maximized negative correlations at Tacubaya of 5 to 11 years, indicating shared climate responses in areas ranging from the Oaxaca Valley, through the SMS, and to the Basin of Mexico, albeit accounting for some variation in the timing of inter-annual rainfall variability. The correlations suggest that the JX–6 paleoclimate reconstruction should be valid for a wide geographic range of the Sierra Madre del Sur, including Xochicalco, Oaxaca Valley, and the Mexican highlands, the Basin of Mexico, and Puebla Valley. Thus, our data have a statistical underpinning that permits the extrapolation of the Juxtlahuaca Cave (Guerrero) data to a wider geographic region and it thus reflects a regional rather than local signal (see Cowgill 2015b).

One strength of the JX–6 record is that the $\delta^{18}O$ data were calibrated to rainfall amount in the full wet season (June through October; see fig. 5.3), which is the season of peak aquifer recharge (Carrera-Hernández & Gaskin 2008) in the valleys of the Basin of Mexico (Mexico and Teotihuacan valleys) and Sierra Madre del Sur (Oaxaca, Morelos, and Tehuacán valleys). Thus, the stalagmite JX-6 proxy record should also be representative of groundwater recharge and spring discharge on multi-annual timescales. The quantitative climate/$\delta^{18}O$ calibration is only considered to be valid for the past ca. 4000 years, when modern climatic boundary conditions were established (low summer insolation, modern sea levels, active El Niño/Southern Oscillation, active ocean circulation, etc. (Lachniet et al. 2013).

The Mesoamerican Monsoon Over 22,000 Years

Our 22,000-year Mesoamerican monsoon reconstruction (fig. 5.4) provides a long-term context for the development of societies in the Basin of Mexico and elsewhere (Lachniet et al. 2013). Following a wet early Holocene at ca. 11 ka BP, rainfall amounts decreased to the present on millennial time scales, in

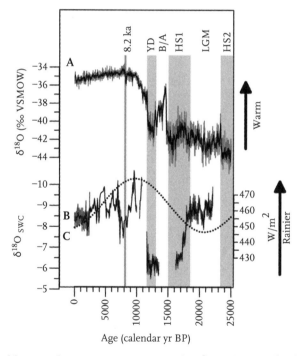

FIGURE 5.4 Mesoamerican monsoon reconstruction from Sierra Madre del Sur stalagmites for the last 22,000 years. The Greenland Ice Sheet $\delta^{18}O$ record (A) and prominent climate events over the late glacial for reference are also shown. Blue bars indicate cold periods and the yellow bar, a warm period; YD = Younger Dryas; B/A = Bølling/Allerød; HS1 and 2 = Heinrich stadials 1 and 2; LGM = Last Glacial Maximum. See Lachniet et al. 2013 for further details.

concert with decreasing northern hemisphere insolation. This decreasing rainfall trend is marked by abrupt climate transitions, including a long dry period between ca. 9.5 and 7 ka BP. The monsoon strengthened thereafter to a relative peak at 6.5 ka BP, followed by decreasing rainfall to the relatively low values of the past ca. 2000 years. Within this context, the advanced societies of the last two millennia in the Basin of Mexico and elsewhere mostly arose against the context of a weakened monsoon relative to the Holocene average. This "modern" monsoon was, however, significantly stronger than the millennial-scale monsoon collapses centered around Heinrich stadial 1 (ca. 17,000 year BP) and the Younger Dryas (ca. 13,000 year BP). Our rainfall reconstruction for the past ca. 2400 years documents abrupt rainfall variations that switch from wet to dry on scales of 10 to 30 years, superimposed on longer rainfall variations ranging from a single to a few human generations to several hundred years (fig. 5.5).

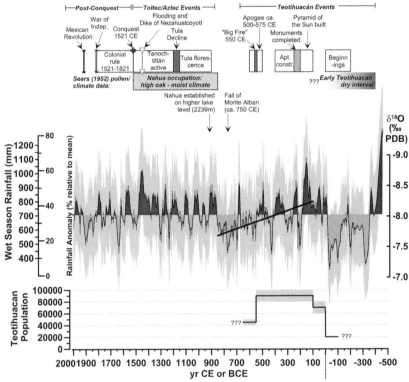

FIGURE 5.5 Paleorainfall reconstruction from stalagmite JX–6, based on the δ¹⁸O-rainfall calibration at the Tacubaya station. Represented at the top is the sequence of societal change in the Basin of Mexico, the middle panel is the rainfall reconstruction, and the bottom panel is the estimated population at Teotihuacán. A striking relationship is the inverse correlation between decreasing rainfall amount and increasing human population during the efflorescence of Teotihuacán, culminating in a 25 percent drying at the time of the AD 550 population collapse following the "Big Fire." (M. S. Lachniet & J. P. Bernal)

The 350–50 BC Dry Period: Early Teotihuacan

Our reconstruction indicates an extended dry interval between 350 and 50 BC (fig. 5.5) that coincides with the "Early Teotihuacan dry period" (Sears 1952). The timing of Teotihuacan's establishment at ca. 100 BC (Cowgill 1997, 2015a) was associated with a dry interval (–10 to –40 percent below the long-term mean of 705 millimeters per year). Some of the population increases at Teotihuacan have been suggested to have resulted from migrants arriving from the wetter southern Basin of Mexico, associated with a massive VEI–6 (on a scale of 0–8 on the volcanic explosivity index) eruption of Popocatéptl in the first century AD (Plunket & Urunuela 2006, Sears 1952, Siebe et al. 1996). A later eruption at Xitle volcanoes in the southern basin dates to ca. AD 245–315 (Siebe 2000) and may have been associated with a second migration to Teotihuacan (Manzanilla 2015). A climatic drying is unlikely to be an important factor for this migration, because the movement was to a more drought-prone area in the Teotihuacan valley relative to the better-watered southern basin. The hypothesis of a societal migration to Teotihuacan is also supported by analyses of strontium isotopes in bones from human burials, which indicate that several of the individuals were not born in the Teotihuacan valley (Manzanilla 2015; Price, Manzanilla, & Middleton 2000). In contrast to the southern Valley of Mexico, the Teotihuacan valley has few volcanic hazards.

The AD 1–350 Pluvial: Teotihuacan Growth

The Early Teotihuacan dry period was abruptly terminated in about 40 years with a transition to wetter conditions, after which rainfall was 10 to 30 percent above average for about two centuries between AD 1 and 200, with a relative rainfall maximum (+ 40 percent relative to the long-term mean) at AD 150. This abrupt transition to wetter conditions represents an approximately 60 percent increase in rainfall (relative to the preceding dry period), or from about 500 to 800 mm/yr. The first and second century AD pluvial was the largest and most sustained of the past 2000 years. Mostly wet conditions continued through about AD 350, spanning the estimated time interval ca. AD 1–300 during which most of the monumental and apartment construction at Teotihuacan was completed (Cowgill 1997, 2008). Expansion of early monumental architecture at Cholula in the Puebla Valley east of the Sierra Nevada also appears to have taken place in the first and second centuries AD (Plunket & Urunuela 2006). Perhaps not coincidentally, the Great Pyramid at Cholula was initially sited next to a lake fed by springs (Dumond & Muller 1972), which would be consistent with the relatively wet conditions that we document at that time. Also, copious rainfall in the Teotihuacan valley during this pluvial period would have resulted in enhanced groundwater recharge and spring discharge, favoring the sustenance of irrigation agriculture. The estimated population of the valley rose significantly from the early Teotihuacan dry period to the end of the AD 350 pluvial reaching ca. 80,000 to 100,000 (Cowgill 1997, 2015a) or higher (Millon 1967).

The AD 350–800 Desiccation and Teotihuacan Population Collapse

Following the AD 1–350 pluvial, our rainfall reconstruction indicates a four-century long rainfall decrease spanning the interval of the population collapse at Teotihuacan around the time of the "Big Fire," most recently estimated to be around AD 550 from a Bayesian analysis combining stratigraphy and multiple radiocarbon ages (Beramendi-Orosco et al. 2009) and associated archaeomagnetic age estimates of burned lime plasters (Soler-Arechalde et al. 2006). The long-term rainfall trend is defined by a ca. 30 percent decrease, about which abrupt rainfall minima and maxima varied by ± 10–20 percent. The result of the centennial-scale variability would have been accentuated dry periods over the declining trend followed by brief returns to wetter conditions, of which each subsequent "wet" interlude appears to have been drier than preceding events. The result of the centuries-long drying would have been less aquifer recharge in the Teotihuacan valley, manifested by decreased spring discharge and hence less water delivery to the agricultural center southwest of the ancient city. The rainfall amount in the century concluding with the AD 550 "Big Fire" is similar to that in the early twentieth century, when spring discharge was also insufficient to meet all irrigation needs (Millon 1962). The trends to decreasing rainfall may have resulted in a climatically forced stressor on Teotihuacan agriculture, with consequent effects on society. Although there is substantial evidence that Teotihuacan was maintained as a population center following the "Big Fire," its regional influence appears to have been significantly diminished.

The AD 550–850 Epiclassic Megadrought

Between AD 550 and 850, abnormally dry conditions prevailed with rainfall reaching a minimum (−30 percent) during a century-long drought centered on AD 750. Because the timing of this dry period coincides approximately with the Epiclassic (ca. AD 600–900), we term it the Epiclassic megadrought. If viewed in terms of long-term mean and abrupt centennial-scale variability, respectively, the rainfall reductions varied from −30 to −50 percent at Teotihuacan. Perhaps it is not a coincidence that numerous smashed artifacts of the storm god (similar to the later Aztec rain god Tlaloc) associated with rain, fertility, and devastating storms are found destroyed at the city's fall ca. AD 550 (Manzanilla 2003, Beramendi-Orosco et al. 2009, Soler-Arechalde et al. 2006), though this could have been as late as AD 650 (Cowgill 2015a). Considering that most of the agriculture was concentrated around Teotihuacan and not in the southern Basin of Mexico (Cowgill 1997), the food supply of the valley's population may have been highly vulnerable to climate variability. That the peak dry conditions in our reconstruction (ca. AD 700–850) follow the apparent collapse at Teotihuacan ca. 550 AD suggests that the remaining Teotihuacan population would have been faced with more extreme climatic conditions than their predecessors.

The timing of the Epiclassic megadrought coincides with the apparent rise of the fortified hilltop city of Xochicalco in the Morelos valley at AD 650 to 900 (Hirth 2000). The rise of this city on a topographic high without local water sources—though nearby natural lakes south of Xochicalco possibly provided drinking water—may have been a strategic response to the fall of Teotihuacan's regional power, as well as to decreasing rainfall. Xochicalco was abandoned at ca. AD 900 (Hirth 1984), coincident with the onset of increased rainfall.

Our data showing peak dry conditions during the Epiclassic megadrought at ca. AD 750 also appear to coincide with the decline of cities in Oaxaca Valley, including the abandonment of public buildings and occupational terraces at Monte Albán (Marcus & Flannery 2000). Further, our new climate calibration between stalagmite JX–6 and rainfall near Oaxaca supports a climate-change stressor hypothesis on the fate of Zapotec society during the late Classic and Epiclassic. Given the shallow-well irrigation practiced in the Oaxaca valley combined with a reliance on low summer rainfall amounts and access to groundwater resources for hand-irrigation and irrigation control structures (Marcus & Flannery 1996, 2000), decreasing summer rainfall would likely have made local agriculture more difficult.

Finally, the timing of droughts in the highlands is distinct from those in the Yucatan lowlands. Peak late Classic dry conditions in southwestern Mexico happened at ca. AD 650–850, whereas dry conditions lagged by a century or two in the Yucatan lowlands (Hodell, Brenner, & Curtis 2005). We suggest that such differences in timing may relate to a variable monsoon response to similar boundary conditions, potentially through forcing of strong tropical storms via the eastern Pacific Ocean and Caribbean Sea/Gulf of Mexico. However, the specific controls on this timing difference warrant further evaluation.

AD 900–1450: The Toltec/Aztec Wet Period

Although it is beyond the scope of this discussion, briefly, the Toltec/Aztec period coincided with the onset of wet conditions at ca. AD 900, when rainfall amounts were consistently higher than in the preceding Classic Period (fig. 5.5). This wet period, following rapidly from the Epiclassic megadrought and peaking around AD 1450, coincides with the rise of first the Toltec and later Aztec (Nahua) civilizations, when *chinampa* agriculture became prevalent in the southern and western Basin of Mexico. Increased rainfall beginning at ca. AD 900 is also consistent with population shifts to higher altitudes in the Basin of Mexico at ca. AD 950–1150 (Sanders, Parsons, & Santley 1979). A large population increase in the southern Basin Lakes between AD 1150 and 1400 (Sanders, Parsons, & Santley 1979) was associated with wetter conditions, which would have decreased salinity in the lake waters, thus permitting more *chinampa* agriculture. It was noted by Drewitt (1967: 111) that the size of the canal system in the Teotihuacan valley appears much larger than the amount of water available to be carried in it in the late 1950s. This observation is consistent with a hypothesis that much of the irrigation system in the Teotihuacan valley was associated with a formerly wetter climate during the Aztec pluvial or later periods (Sanders 1977; Sanders, Parsons, & Santley 1979). The presence of large *Ahuehuete* trees along canals in the Teotihuacan valley suggest they may be as old as 700 to 800 years (Drewitt

1967: 127), supporting a formation during Aztec times or earlier. The transition to a wetter climate in southwestern Mexico also coincided with the expansion of agricultural sites in Oaxaca Valley to the higher piedmont zones and lower slopes of the mountains, where farmers would have relied on rainfall rather than shallow-well irrigation (Flannery et al. 1967, Marcus & Flannery 2000).

There was a small dry period (approximately −20 percent) at ca. AD 1120–1160 that coincides with the apparent fall of Tula (Healan 2012). The fall of the Toltec state has been linked to a similar drought (ca. 1150–1170) inferred from the Barranca de Amealco dendrochonology data (Stahle et al. 2011). Rainfall variations were certain to have affected discharge in the Tula River, which would also have influenced water-table height in the surrounding alluvial plain. However, at least two similar rainfall minima evident in our reconstruction occurred over previous centuries (at ca. AD 1040–1080 and 920–940) with no apparent ill effects on Tula. Each of these three relative rainfall minima is less severe than the Epiclassic megadrought and dry periods of the preceding millennia. A possible reason for the apparent discrepancy between the droughts inferred from Juxtlahuaca Cave and Barranca de Amealco is that the stalagmite record retains centennial-scale and greater climate variability, whereas the tree-ring record does not because of statistical processing. Without a record of centennial-scale climate variability, the magnitude of the Toltec drought in the tree-ring data may be overestimated. The discrepancy may also stem from the ways climate processes in the cave (calibration is for full wet season rainfall) and tree rings (calibration is for June drought severity) are recorded (Lachniet et al. 2012).

Discussion

The effect of precipitation variability on water capture from rainfall and run-off in the Teotihuacan valley would have been immediate, though the spring discharge used for irrigation would have been buffered by residence time in the surficial aquifer, somewhat ameliorating abrupt climate changes. Viewed in the longer context of the Holocene (Bernal et al. 2011, Lachniet et al. 2013), the last two millennia have been characterized by a relatively weak monsoon (fig. 5.4). Thus, some of the groundwater resources available to the founding Teotihaucanos may have been stored recharge from prior wet periods. But how rapidly would groundwater levels respond to rapid (less than 100 year) rainfall variations? Observations in the Valley of Mexico suggest that spring levels may respond to individual rainfall events in hours to days (Durazo and Farvolden 1989). Yet the Valley of Mexico is the lowest spot in the Basin and, therefore, possibly more susceptible to minor variations in recharge. A similar but likely more muted rapid response was possible in the springs of the Teotihuacan valley (Carrera-Hernandez & Gaskin 2007).

However, observation of Teotihuacan spring discharge reveals that steady annual flow is apparently not influenced by increased rainfall during the wet season (Lorenzo 1968). Although little data has been published on the response times of groundwater levels in the Teotihuacan valley to modern rainfall variability, flow in the now-dry Rio San Juan persisted until at least the late 1960s (Mooser 1968), and a conversation with a resort operator near the Teotihuacan ruins indicated to us that water levels continue to drop due to well pumping in the valley, similar to draw downs observed in the Valley of Mexico (Durazo & Farvolden 1989). Indeed, today the groundwater table at Teotihuacan is significantly below the ground surface, and the main spring discharge was higher in the 1920s than in the 1970s (Sanders 1977). According to observations in the early 1960s, the water table was within a few meters of the ground surface in the vicinity of the modern city of San Juan Teotihuacan (Millon 1962), where there are (were) more than eighty springs (Drewitt 1967). Thus, we posit that the Teotihuacan springs appear to have been buffered from interannual climate change, but spring discharge would likely have changed on decadal or multi-decadal timescales. There may also have been a positive feedback in anthropogenic influence in the Teotihuacan valley, with deforestation for construction materials and fuel leading to flashier stream responses and diminished groundwater infiltration (Mooser 1968). These factors in concert with decreasing rainfall would have enhanced the effect of climate change on groundwater resources in the Teotihuacan valley.

Was Rainfall a Driving Force in the Rise and Fall of Teotihuacan?

Some have suggested that the collapse of Teotihuacan may have been related to climatic drying, such environmental changes as soil erosion and deforestation (Lorenzo 1968, Mooser 1968), whereas others have suggested more complex societal changes (Cowgill 2015a, García 1974, Manzanilla 2003, 2015). That rainfall and climate were important consideration to the residents of Teotihuacan is suggested by their reference to images of the storm god (Cowgill 1997), the nature of their irrigation-intensive existence in this semi-arid valley (Drewitt 1967), and smashed storm-god artefacts (Manzanilla 2003). These driving factors may have been either "push" or "pull" forces that resulted in in- or out-migration. Also suggested is that socioeconomic, environmental, and climatic changes may have had complex feedbacks (Turner II and Sabloff 2012), and "collapse" (in the broadest sense of population decrease and loss of regional power) would unlikely have stemmed from a single factor. However, our data provide strong evidence that monument construction and population decline at Teotihuacan were associated with wet and dry conditions, respectively. Following the pioneering work of Sanders, Parsons, and Santley (1979) who correctly realized that agricultural output in the Basin of Mexico and Teotihuacan valley was a function of climate, soils, irrigation, and land-use, our data suggest that a climatic stressor on Teotihuacan society via potential agricultural output in this semi-arid location should not be neglected.

Although a correlation between climate and societal change need not imply cause and effect, our rainfall reconstruction provides compelling evidence that societal change coincided with a variable and capricious climate (fig. 5.5) that reached driest conditions in the Epiclassic megadrought. One interpretation of our data would be that climate was associated with societal change at Teotihuacan, either directly via water resource availability, or indirectly through a stressor effect on other natural and human systems. The diminution of flow in the Teotihuacan springs would have reduced available irrigation water for the crops that fed the Teotihuacan population. However, although the consistency of our rainfall reconstruction showing a centuries-long desiccation trend prior to the fall of Teotihuacan is suggestive of a climate-society link, it does not by itself preclude other hypotheses (socioeconomic, disease, etc.). Indeed, climatic change may have accelerated a societal change already in process, but such contentions are difficult to test in the absence of well-dated reconstructions of other non-climatic forcings, which lack the precise chronology of our climate-calibrated data.

A discussion of Mesoamerican archaeological chronology is merited here. Part of the reason for the poor chronological control of archaeological and many other paleoenvironmental records is inherent to the physics of radiocarbon, which has been the main source of archaeological dating. First, the production rate of radio-carbon in the atmosphere has varied over time, such that the calibration of radio-carbon dates into calendar years may introduce considerable uncertainty (on the order of several hundred years) because of calibration curve wiggles and plateaus (Beramendi-Orosco et al. 2009). Second, it is likely that old wood was used in the construction of younger buildings (Manzanilla 2003). *Ahuehuete* trees, for example, may live for hundreds of years (Kovar 1966). Large *Ahuehuete* trees line the Teotihuacan valley irrigation canals today, suggesting the canals are at least several hundreds of years old (Drewitt 1967), and if this tree species was used in architecture, its ^{14}C ages would pre-date the age of construction considerably. Old wood may also have been re-used by later builders—a distinct possibility in semi-arid highland Mexico. The effects of "old" carbon in younger deposits, particularly in Mesoamerican lake systems, have rarely been studied sufficiently (with the exception of the Juanacatlán sediment core) and may also exert a signif-icant centennial-scale bias in the estimated times of paleoenvironmental change. Similarly, radiocarbon age constraints on soil development are also broad, because soils form over timescales of a century or greater and integrate climate variations over similarly long time intervals, thus providing only a loose idea of environ-mental change on human time scales. Finally, establishing accurate radiocarbon chronologies of archeological material is hindered by the lack, in many cases, of deposition of contemporaneous carbon at the time of artefact burial. Combined, these uncertainties result in a wide range of possible ages for cultural periods of at best 50 to 100 years (Cowgill 2015a). In contrast, the age uncertainties on our speleothem age model are typically better than a decade.

How we attribute and measure the forcings of climatic variability in Mesoamerica must be investigated in more detail. Previous work has invoked solar forcing (Hodell et al. 2001), El Niño/Southern Oscillation variability (Lachniet et al. 2012, 2015, 2017; Stahle et al. 2012), and North Atlantic Ocean circulation anomalies (Bernal et al. 2011) on Mesoamerican monsoon rainfall. However, most of the high-resolution paleoclimate records from lakes in the region appear quite dissimilar in detail to one another, suggesting that the importance of the different forcings at each location differ, or that there are problems with the paleoclimate reconstructions. More detailed modern climate calibrations, such as the excellent example by (Stahle et al. 2012), should be attempted. Finally, additional high-resolution and precisely dated paleoclimate records have been developed, including replicated cave speleothem $\delta^{18}O$ time series (Lachniet 2015; Lachniet et al. 2017).

Acknowledgments

This research was supported by a collaborative National Science Foundation grant ATM-1003558 to UNLV and the University of New Mexico (Drs. Asmerom and Polyak) and National Geographic Society grant 88-2810. We thank Prof. Andrés Ortega in Colotlipa, Gro., for permission to complete work in Juxtlahuaca Cave.

References

Beramendi-Orosco, L. E., G. Gonzalez-Hernandez, J. Urrutia-Fucugauchi, L. R. Manzanilla, A. M. Soler-Arechalde, A. Goguitchaichvili, N. A. Jarboe. 2009. High-resolution chronology for the Mesoamerican urban center of Teotihuacan derived from Bayesian statistics of radiocarbon and archaeological data. *Quaternary Research* 71(2): 99–107.

Bernal, J. P., M. S. Lachniet, M. McCullogh, G. Mortimer, P. Morales, and E. Cienfuegos. 2011. A speleothem record of Holocene climate variability from southwestern Mexico *Quaternary Research* 75: 104–113.

Carrera-Hernández, J. J., and S. J. Gaskin. 2007. The Basin of Mexico aquifer system; regional groundwater level dynamics and database development. *Hydrogeology Journal* 15(8): 1577–1590.

Carrera-Hernández, J. J., and S. J. Gaskin. 2008. Spatio-temporal analysis of potential aquifer recharge: application to the Basin of Mexico. *Journal of Hydrology* 353: 228–246.

Cowgill, G. L. 1997. State and society at Teotihuacan, Mexico. *Annual Review of Anthropology* 26: 129–161.

Cowgill, G. L. 2008. An update on Teotihuacan. *Antiquity* 82(318): 962–975.

Cowgill, G. L. 2015a. Ancient Teotihuacan: early urbanism in central Mexico. Cambridge, UK: Cambridge University Press.

Cowgill, G. L. 2015b. The Debated Role of Migration in the Fall of Ancient Teotihuacan in Central Mexico. In *Migration and Disruptions: Unifying Themes in Studies of Ancient and Contemporary Migrations*, ed.T. Tsuda and B. Baker.

Dansgaard, W. 1964. Stable isotopes in precipitation. *Tellus* 16: 438–468.

Drewitt, R. B. 1967. "Irrigation and agriculture of the Valley of Teotihuacan." PhD diss. Berkeley: University of California.

Dumond, D. E., and F. Muller. 1972. Classic to postclassic in highland Central Mexico *Science* 175: 1208–1215.

Durazo, J., and R. N. Farvolden. 1989. The groundwater regime of the Valley of Mexico from historic evidence and field observations. *Journal of Hydrology* 112: 171–190.

Flannery, K. V., A. V. T. Kirkby, M. J. Kirkby, and A. W. Williams, Jr. 1967. Farming systems and political growth in ancient Oaxaca. *Science* 158: 445–454.

García, E. 1974. Situaciones climáticas durante el auge y la caída de la cultura Teotihuacana. *Boletin del Instituto de Geografia* 5: 35–69.

Gay, C. T. E. 1967. Oldest paintings of the New World. *Natural History* 4: 28–35.

Healan, D. M. 2012. The archaeology of Tula, Hidalgo, Mexico. *Journal of Archaeological Research* 20(1): 53–115.

Hirth, K. 1984. Xochicalco: urban growth and state formation in Central Mexico. *Science* 225: 579–586.

Hirth, K. 2000. *Archaeological Research at Xochicalco.* Salt Lake City: University of Utah Press.

Hodell, D. A., M. Brenner, and J. H. Curtis. 2005. Terminal classic drought in the northern Maya lowlands inferred from multiple sediment cores in Lake Chichancanab (Mexico). *Quaternary Science Reviews* 24: 1413–1427.

Hodell, D. A., M. Brenner, and J. H. Curtis. 2007. Climate and cultural history of the northeastern Yucatan Peninsula, Quintana Roo, Mexico. *Climatic Change* 83: 215–240.

Hodell, D. A., M. Brenner, J. H. Curtis, and T. Guilderson. 2001. Solar forcing of drought frequency in the Maya lowlands. *Science* 292: 1367–1370.

Juaregui, E. 1990/1991. Influence of a large urban park on temperature and convective precipitation in a tropical city. *Energy and Buildings* 15–16: 457–463.

Kalnay, E., M. Kanamitsu, R. Kistler, W. Collins, D. Deaven, L. Gandin, M. Iredell, S. Saha, G. White, J. Woollen, Y. Zhu, M. Chelliah, W. Ebisuzaki, W. Higgins, J. Janowiak, K. C. Mo, C. Ropelewski, J. Wang, A. Leetmaa, R. Reynolds, R. Jenne, and D. Joseph. 1996. The NCEP/NCAR 40-year reanalysis project. *Bulletin of the American Meteorological Society* 77(3): 437–471.

Karmalkar, A., R. S. Bradley, and H. F. Diaz. 2011. Climate change in Central America and Mexico: regional climate model validation and climate change projections. *Climate Dynamics* 37: 605–629.

Kennett, D. J., S. F. M. Breitenbach, V. V. Aquino, Y. Asmerom, J. Awe, J. U. L. Baldini, P. Bartlein, B. J. Culleton, C. Ebert, C. Jazwa, M. J. Macri, N. Marwan, V. Polyak, K. M. Prufer, H. E, Ridley, H. Sodemann, B. Winterhalder, and G. H. Haug. 2012. Development and disintegration of Maya political systems in response to climate change. *Science* 338: 788–791.

Kovar, A. J. 1966. Problems in radiocarbon dating at Teotihuacan. *American Antiquity* 31(3): 427–430.

Lachniet, M. S. 2009a. Climatic and environmental controls on speleothem oxygen isotope values. *Quaternary Science Reviews* 28: 412–432.

Lachniet, M. S. 2009b. Sea surface temperature control on the stable isotopic composition of rainfall in Panama. *Geophysical Research Letters* 36: L03701. doi:10.1029/2008GL036625.

Lachniet, M. 2015. Are aragonite stalagmites reliable paleoclimate proxies? Tests for oxygen isotopic time series replication and equilibrium: Geological Society of America Bulletin.

Lachniet, M. S., Y. Asmeron, V. Polyak, and J. P. Bernal. 2017. Two millennia of Mesoamerican monsoon variability driven by Pacific and Atlantic synergistic forcing. *Quaternary Science Reviews* 155: 100–113.

Lachniet, M. S., Y. Asmerom, J. P. Bernal, V. Polyak, and L. Vazquéz-Selem. 2013. Orbital pacing and ocean circulation-induced collapses of the North American Monsoon over the past 22,000 y. *Proceedings of the National Academy of Sciences* 110: 9255–9260.

Lachniet, M. S., J. P. Bernal, Y. Asmerom, V. Polyak, and D. Piperno. 2012. A 2400-yr Mesoamerican rainfall history links climate and cultural change in Mexico. *Geology* 40 (3): 259–262.

Lachniet, M. S., W. P. Patterson, S. J. Burns, Y. Asmerom, and V. J. Polyak. 2007. Caribbean and Pacific moisture sources on the Isthmus of Panama revealed from stalagmite and surface water $\delta^{18}O$ gradients. *Geophysical Research Letters* 34: L01708. doi:10.1029/2006GL028469

Linné, S. 2003. Mexican highland cultures: archaeological researches at Teotihuacan, Calpulalpan, and Chalchicomula in 1934–35. Tuscaloosa, AL: The University of Alabama Press.

Lorenzo, J. L.1968. Clima y agricultura en Teotihuacan. In *Materiales para la arqueo-logia de Teotihuacan. Serie Invesitagciones 17*, ed. J. L. Lorenzo. Mexico: Instituto Nacional de Antropologia e Historia, 51–72.

Manzanilla, L. 2003. The abandonment of Teotihuacan. In *The archaeology of settle-ment abandonment in Middle America*, ed.T. Inomata and R. W. Webb. Salt Lake City: University of Utah Press, 91–101.

Manzanilla, L. 2015. Cooperation and tensions in multiethnic corporate societies, using Teotihuacan, central Mexico, as a case study. *Proceedings of the National Academy of Sciences* Early Edition: www.pnas.org/cgi/doi/10.1073/pnas.1419881112

Marcus, J., and K. V. Flannery. 1996. *Zapotec Civilization.* London: Thames and Hudson.

Marcus, J., and K. V. Flannery. 2000. Cultural evolution in Oaxaca: the origins of the Zapotec and Mixtec civilizations. In *The Cambridge History of the Native Peoples of the Americas, Volume II: Mesoamerica*, ed. R. E. W. Adams and M. J. MacLeod. Cambridge: Cambridge University Press, 358–406.

Medina-Elizalde, M., S. J. Burns, D. W. Lea, Y. Asmerom, L. von Gunten, V. Polyak, M. Vuille, and A. Karmalkar. 2010. High resolution stalagmite climate record from the Yucatán Peninsula spanning the Maya terminal classic period. *Earth and Planetary Science Letters* 298(1–2): 255–262.

Medina-Elizalde, M., and E. J. Rohling. 2012. Collapse of Classic Maya civilization related to modest reduction in precipitation. *Science* 335: 956–959.

Metcalfe, S. E., M. D. Jones, S. J. Davies, A. Noren, and A. MacKenzie. 2010. Climate variability over the last two millennia in the North American Monsoon region, recorded in laminated lake sediments from Laguna de Juanacatlán, Mexico. *The Holocene* 20(8): 1195–1206.

Millon, R. 1962. Conflict in the modern Teotihuacan irrigation system. *Comparative Studies in Society and History* 4(4): 494–524.

Millon, R. 1967. Teotihuacan. *Scientific American* 216: 38–48.

Millon, R. 1970, Teotihuacan: completion of map of giant ancient city in the Valley of Mexico. *Science* 170: 1077–1082.

Mooser, F. 1968. Geologia, naturaleza y desarrollo del Valle de Teotihuacan. In *Materiales para la arqueología de Teotihuacan,* ed. J. L. Lorenzo. Mexico: INAH, 29–37.

Pachauri, R. K., and A. Reisinger, eds. 2007. *Climate Change 2007: Synthesis Report: Contribution of Working Groups I, II, and III to the Fourth Assessment Report of the Intergovernmental Panel on Climate Change.* Geneva, Switzerland: IPCC.

Plunket, P., and G. Urunuela. 2006. Social and cultural consequences of a late Holocene eruption of Popocatepetl in central Mexico. *Quaternary International* 151: 19–28.

Price, T. D., L. Manzanilla, and W. D. Middleton. 2000. Immigration and the ancient city of Teotihuacan in Mexico: a study using strontium isotope ratios in human bone and teeth. *Journal of Archaeological Science* 27: 903–913.

Richards, D., and J. Dorale. 2003. Uranium-series chronology and environmental applications of speleothems. *Reviews in Mineralogy* 52: 407–460.

Sánchez-Pérez, S., E. Solleiro-Rebolledo, S. Sedov, E. M. Tapia, A. Golyeva, B. Prado, and E. Ibarra-Morales. 2013. The Black San Pablo Paleosol of the Teotihuacan Valley, Mexico: Pedogenesis, Fertility, and Use in Ancient Agricultural and Urban Systems. *Geoarchaeology* 28: 249–267.

Sanders, W. T. 1977. Resource utilization and political evolution in the Teotihuacan Valley. In *Explanation of Prehistoric Change,* ed. J. N. Hill. Albuquerque: University of New Mexico Press, 231–257.

Sanders, W. T., J. R. Parsons, and R. S. Santley. 1979. *The Basin of Mexico: Ecological Processes in the Evolution of a Civilization.* New York: Academic Press.

Schmitz, J. T., and Mullen, S. L., 1996, Water vapor transport associated with the summertime North American Monsoon as depicted by ECMWF analyses. *Journal of Climate* 9: 1621–1634.

Sears, P. B. 1952. Palynology in southern North America. Part 1: Archaeological horizons in the basin of México. *Geological Society of America Bulletin* 63(3): 241–254.

Siebe, C. 2000. Age and archaeological implications of Xitle volcano, southwestern Basin of Mexico-City. *Journal of Volcanology and Geothermal Research* 104(1–4): 45–64.

Siebe, C., M. Abrams, J. L. Macias, and J. Obenholzner. 1996. Repeated volcanic disasters in prehispanic time at Popocatepetl, central Mexico; past key to the future? *Geology* 24(5): 399–402.

Soler-Arechalde, A. M., F. Sanchez, M. Rodriguez, C. Caballero-Miranda, A. Goguitchaichvili, J. Urrutia-Fucugauchi, L. Manzanilla, and D. H. Tarling. 2006. Archaeomagnetic investigation of oriented pre-Columbian lime-plasters from Teotihuacan, Mesoamerica. *Earth, Planets and Space* 58: 1433–1439.

Stahle, D. W., D. J. Burnette, J. Villanueva Díaz, R. R. Heim, Jr., D. J. Fye, J. Cerano-Paredes, R. Acuna-Soto, and M. K. Cleaveland. 2012. Pacific and Atlantic influences on Mesoamerican climate over the past millennium. *Climate Dynamics* 39: 1431–1446.

Stahle, D. W., J. Villanueva-Diaz, D. J. Burnette, J. Cerano-Paredes, R. R. Heim, Jr., F. K. Fye, R. Acuna-Soto, M. D. Therrell, and M. K. Cleaveland. 2011, Major Mesoamerican droughts of the past millennium. *Geophysical Research Letters.* DOI: 10.1029/2010GL046472.

Turner, II, B. L., and J. A. Sabloff. 2012, Classic Period collapse of the Central Maya Lowlands: insights about human-environment relationships for sustainability. *PNAS* 109 (35):13908–13914.

| AD 750–1100 Climate Change and
Critical Transitions in Classic Maya
Sociopolitical Networks

DOUGLAS J. KENNETT AND DAVID A. HODELL

Multiple palaeoclimatic reconstructions point to a succession of major droughts in the Maya Lowlands between AD 750 and 1100 superimposed on a regional drying trend that itself was marked by considerable spatial and temporal variability. The longest and most severe regional droughts occurred between AD 800 and 900 and again between AD 1000 and 1100. Well-dated historical records carved on stone monuments from forty Classic Period civic-ceremonial centers reflect a dynamic sociopolitical landscape between AD 250 and 800 marked by a complex of antagonistic, diplomatic, lineage-based, and subordinate networks. Warfare between Maya polities increased between AD 600 and 800 within the context of population expansion and long-term environmental degradation exacerbated by increasing drought. Nevertheless, in spite of the clear effects of drought on network collapse during the Classic Period, one lingering question is why polities in the northern lowlands persisted and even flourished between AD 800 and 1000 (Puuc Maya and Chichén Itzá) before they too fragmented during an extended and severe regional drought between AD 1000 and 1100. Here we review available regional climate records during this critical transition and consider the different sociopolitical trajectories in the South/Central versus Northern Maya lowlands.

Introduction

The societal effects of climate change on the ancient Maya are of great interest amid current debates regarding the resilience of global social, economic, and political networks to abrupt climate change. Because of the indelible archaeological record of this complex civilization preserved in the tropical forests of Mexico, Guatemala, Belize, Honduras, and El Salvador, coupled with its well-known collapse, particular emphasis is placed on the Classic Maya (AD 250–1000; Goodman-Martínez-Thompson [GMT] correlation; Kennett et al.

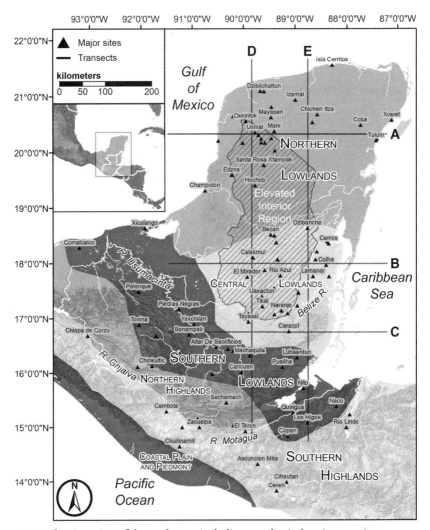

FIGURE 6.1 Overview of the study area including geophysical regions, major Maya centers, and the spatial extent of profile transects A through E (illustrated in Figure 5.2).

2013). The largest and most influential Classic Period polities developed in the central and southern lowlands (fig. 6.1) and were connected via trade with communities in the northern lowlands and the highland regions to the south and west. By the Late Classic Period (ca. AD 650–800) the urbanized centers of the central and southern lowlands boasted large temple complexes (some 70 meters high), expansive public plazas, elaborate palaces, and administrative buildings (Sharer & Traxler 2006). Elevated causeways connected temple complexes within centers, as in Tikal (Harrison 1999), or to outlying communities, as with Caracol (Chase et al. 2011). Reservoirs and other water control systems were used to store water for dry-season months (Lucero 2002; Scarborough et al. 2012) or, as in Palenque, to control excess flow throughout the year (French et al. 2012). Major investments in agricultural infrastructure and landesque cultivation systems (particularly terraces and raised fields) were developed in a complex agrarian mosaic that varied in accordance with topographic, geological, and hydrological differences in the region (Fedick 1996; Beach et al. 2002, 2009, 2015; Dunning et al. 2012; Kennett & Beach 2013; Luzzadder-Beach et al. 2012; Chase et al. 2011).

A divine king ruled the largest polities and the dynastic lineages for the most influential political centers are known from historical texts carved on stone monuments (Martin & Grube 2000). Kings were focal political actors whose reigns and achievements were glorified on the monuments and anchored by the Long Count calendar system, one of the defining features of Classic Maya civilization. These records provide an exceptional chronicle of rituals, wars, and hierarchical relationships among individuals in these polities. Network connectivity was greatest among polities in the central and southern lowlands, but also extended into the northern lowlands and parts of the highlands. The calendar and writing systems persisted in the northern lowlands during the Terminal Classic (ca. AD 800–1000) with the Puuc Maya florescence and the development of Chichén Itzá as a regional power. This occurred as urban centers to the south collapsed. Archaeologists have identified a 100-year long "dark age" in the wake of Chichén Itzá's collapse ca. AD 1000 (Andrews et al. 2003; Hoggarth et al. 2016) and the development of Mayapán, the largest and final Late Postclassic Period Maya polity in the northern lowlands (Kennett et al. 2016; Peraza Lope et al. 2006; Masson and Peraza Lope 2014). Political cycling of this kind is characteristically Maya (see Marcus 1993; 2012) and is evident in the region starting as early as the Late Preclassic (ca. 1000 BC–AD 250; for example, the rise and fall of El Mirador; Hansen et al. 2002).

More than 100 monument bearing urban centers developed at different times during the Classic Period and the relative importance of these centers within the sociopolitical fabric of the lowlands waxed and waned. Forty-four of these centers featured emblem glyphs that are now used by epigraphers to track the interactions among the largest polities (Martin & Grube 2000; Munson & Macri 2009). Many other small centers existed and emblem glyphs are recorded in texts that have not been connected to specific polities identified archaeologically. Weakened political vigor of key centers produced conspicuous political hiatuses following major defeats in war that on occasion coincided with short-term droughts (for example, in AD 536; Dahlin & Chase 2014). The number of urban centers peaked at the height of the Classic Period between AD 700 and 800 when populations were highest across the lowlands. These centers were characterized by dispersed or low-density urbanism, with smaller settlements and household compounds continuously distributed across the landscape. In many areas, population densities reached 600–1200 people per square kilometer (Drennan 1988; Witschey & Brown 2013). Population estimates for monument bearing urban centers range from 10,000 to 60,000 people (Culbert & Rice 1990; Kennett & Beach 2013), with total populations in the lowlands between 3 and 4 million (See Kennett & Beach 2013 for a recent overview).

The political "collapse" was a staggered decline in the number of monument bearing urban centers across the Maya lowlands. It played out over several hundred years between ca. AD 750 and 1000 and started as a contraction in the number of centers in several key regions (Neiman 1997; Ebert et al. 2014). This process appears to start in the Usumacinta-Pasión region within the context of heightened internecine warfare in the western periphery of the southern Maya lowlands (Inomata 1997; O'Mansky 2014). Collapse of these urban centers was most abrupt between AD 800 and 900 (50 percent loss) in the central/southern lowlands where Classic Period polities were the largest and most politically networked. The last recorded long count calendar date associated with one of these polities in the south/central lowlands is ca. AD 900 (Tonina, AD 909). Urban-centered polities persisted in the northern lowlands and along the coast in some areas (Turner & Sabloff 2012; Hoggarth et al. 2016). There is also a cultural fluorescence in the northern lowlands in the Puuc region (AD 800–900) and Chichén Itzá (AD 900–1000). Here we are interested in the differing regional responses to drought between AD 750 and 1000 and the resilience of certain centers in different cultural and environmental contexts (Iannone 2014).

In the following sections we also summarize the available paleoclimatic data for the Maya region. Climate change is often treated as an external stress impacting past human societies at specific times, with emphasis placed on determining the magnitude and duration of a given climatic event (for example, drought) and how it might relate to societal change (for example, political collapse or the abandonment of settlements). In fact, climate change is a continuous process with considerable spacio-temporal variability and thus needs to be considered as one of multiple variables (such as population size and technological availability) that constrain or provide opportunities for humans under changing environmental and societal conditions. Therefore, we consider the available evidence for the severity and timing of droughts between AD 800 and 1000 in relation to cultural context and other environmental changes through the Classic Period. Within this context we compare structural differences in sociopolitical networks between the south/central region and the northern lowlands, because this may help explain why these regions responded differently during this critical transition.

Environmental Context and Climate Variability

Environmental variability over the Maya region is shaped by differences in elevation, geological structure, vegetation, and gradients in temperature and rainfall (Dunning et al. 2012), in addition to alteration of these landscapes by humans due to the expansion of agriculture beginning in the early Holocene (ca. 7000 BC; see Brenner et al. 2002 and Kennett & Beach 2013 for reviews). Regional differences in agricultural potential largely resulted from variability in precipitation that, in turn, generally correlated with topography and ultimately geological substrate and soil productivity. Lowland karstic landscapes of the Peten and Yucatan regions give way to more rugged terrain along the western and southern periphery of the region (Central and Southern highlands). The arc of mountains (northern and southern Maya Highlands) along the southern periphery of the region form between the North American and Caribbean plates, with Guatemala's Motagua Valley/Fault forming the most visible geomorphic manifestation of this tectonic boundary. Topographic variability even exists, although more subtly, in the lowlands (for example, Maya mountains of Belize, see fig. 6.2 and the slightly elevated interior portions off the lowlands (40–300 m; Elevated Interior Areas; Dunning et al. 2012; see fig. 6.1).

The Motagua, Usumacinta, and Grijalva rivers are the largest in the region flowing out of the northern and southern highlands. The Motagua River (Length: 486 km; Flow Rate: 208.7m³/s) flows out of the southern highlands and into the Caribbean Sea, while the Usumacinta River (Length: 1000 km; Flow Rate: 900 m³/s) flows out of the northern highlands through the southern lowlands to meet the Grijalva River (Length: 600 km) in the Gulf Coast lowlands. Perennial rivers and lakes are more prevalent in the southern and central lowlands compared to the northern lowlands. These water sources were augmented at some locations—for example, in Tikal—during the Classic Period with rainwater from natural depressions (*bajos*), artificially constructed storage pits (*chultunob*), and larger water control systems and reservoirs at some locations (at Tikal: Scarborough & Gallopin 1991; Scarborough et al. 2012; in Palenque: French & Duffy 2010; French, Duffy, & Bhatt 2012). Surface water is scarce in the drier northern lowlands and some of the larger lakes, such as Lake Chichancanab, are relatively saline (Hodell, Brenner, & Curtis 1995). Rivers and streams are smaller and more likely to flow underground, but the water table is generally closer to the surface and exposed in karst windows (*cenotes*) or wells (Luzzadder-Beach 2000). The quality of water in *cenotes* varies spatially and becomes more saline during drier climatic intervals. Because of the scarcity of surface water, sophisticated water control systems were developed early in the northern lowlands (for example, see Matheny 1976), possibly in association with elaborate systems to tap underground water supplies in caves (Peterson & Haug 2005).

The availability of water for human use in the Maya region is also mediated by access to groundwater (Yaeger & Hodell 2008). Rainfall in the northern lowlands rapidly percolates through the karst, but the water table is shallow (about 30 m) and accessible in some areas through *cenotes*. The water table is deeper at higher elevations in the Puuc Hills and hence more complex water capture and storage systems were required for this region (Dunning 1992; McAnany 1990; Matheny 1976; Peterson & Haug 2005). Rivers and lakes are more abundant in the south/central lowlands, but the water table is hundreds of meters below the surface and hence was inaccessible to prehistoric Maya populations. Water capture during the wet season supplemented other sources and these systems were particularly well developed in larger urban centers (Scarborough et al. 2012; French & Duffy 2014). The higher water table in the north, accessible through *cenotes*, contributed to the persistence and fluorescence of cities in the northern lowlands between AD 800 and 1000 along with early technological adaptions to water storage in this more arid zone (Dahlin 2002; Peterson & Haug 2005).

Instrumental records for the last 30 years indicate large variations in average annual rainfall across the Maya lowlands (fig. 6.3). Annual average rainfall is generally lower in the north (ca. 800 millimeters) and much higher in the south (ca. 4600 millimeters). This parallels a modest temperature gradient from an average of 26°C in the north to 24°C in the south. Northern Honduras and the Copan Valley tend to be slightly drier and warmer than other parts of the southern Maya lowlands. The prevailing easterly trade winds blowing across the region also produce an east-to-west gradient in precipitation associated with storms tracking the Caribbean low-level jet (Mestas-Nuñez et al. 2002; Giddings & Soto 2003).

Precipitation across the region is highly monsoonal, with the most rain falling during summer months between May and October, and a peak usually in September. Annual temperatures range between 20° and 28°C and the warmest conditions coincide with the boreal summer monsoon. Rainfall variability is related to the intensity of the seasonal cycle caused by the north-south migration of the Intertropical Convergence Zone (ITCZ), which in turn is linked to meridional shifts in the Azores-Bermuda high (Hastenrath 1984, 1991). During boreal summer months tropical storms and hurricanes originating in the Atlantic track more frequently across the Maya lowlands, causing increased precipitation (Wilson 1980; Gray 1993). Little rainfall occurs during the dry-season (November-April) as the ITCZ shifts south of the equator and the Azores-Bermuda high dominates the region. The wet season is sometimes interrupted by a brief dry interval during July and August known in the region as the *canícula* (Magaña, Amado, & Medina 1999).

Climate variability in the Maya region is governed by the annual cycle and the relative position of the ITCZ (Schneider, Bischoff, & Haug 2014). Titanium (Ti) concentrations in marine sediments in the Cariaco Basin off the coast of Venezuela provide a record of ITCZ position during the last 14,000 years (Haug et al. 2001). Low Ti concentrations are associated with the dry Younger Dryas when the ITCZ was positioned to the south. Sediments are enriched in Ti during the early Holocene Thermal Maximum when the ITCZ was positioned farther north and associated with increased precipitation and Ti-rich discharge from the Orinoco River. Similarities between the Cariaco Basin Ti record and climate records in the Maya region spanning the last 2000 years indicate that rainfall variability in the region is partly modulated by the relative position of the ITCZ through time (Kennett et al. 2012; fig. 6.4).

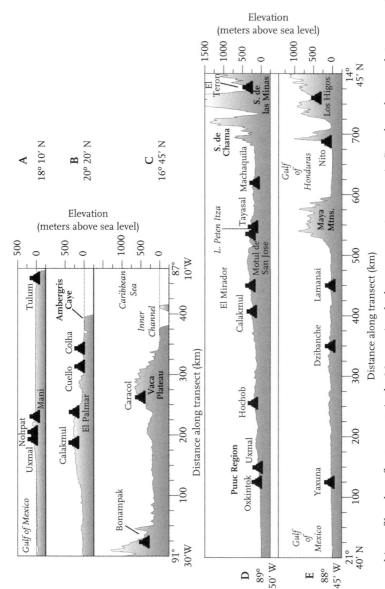

FIGURE 6.2 Topographic profiles along five transects in the Maya region, both west-to-east (transects A–C) and north-to-south (transects D and E).

Evidence for Terminal Classic Maya Droughts

Multiple scholars speculated in the 1980s and early 1990s that climate change, more specifically drought, had played a critical role in the collapse of Classic Period Maya civilization (Gunn & Adams 1981; Folan et al. 1983; Folan & Hyde 1985; Dahlin 1990; Messenger 1990). Sediments and stable oxygen isotope records from Lake Chichancanab in the northern lowlands provided some of the first physical evidence that a series of droughts had occurred at the end of the Classic Period (ca. AD 800–1000, Hodell, Curtis, & Brenner. 1995; see fig. 6.3 for locations of climate records). Because its sulphate rich waters are saturated with gypsum ($CaSO_4$), this moderately saline lake is highly sensitive to the balance of evaporation and precipitation. During drought periods, when evaporative loss from the lake exceeds gain from rainfall, gypsum precipitates from lake waters and accumulates in bottom sediments. The occurrence of gypsum layers in the sedimentary sequence is a clear sign for drier climate conditions (Hodell, Brenner, & Curtis 2005). Because gypsum is denser than other sedimentary components (e.g., organic matter and calcium carbonate), peaks in sediment density (g cm²) provide a qualitative proxy of past drought conditions (fig. 6.5). Sediments recovered from the lake indicate wetter conditions in the northern lowlands between 5000 and 1000 BC, consistent with the Cariaco Ti record (Haug et al. 2001), followed by a drying trend in the last 3000 years punctuated by several extended droughts (Hodell, Curtis, & Brenner 1995). Increased oxygen isotope values (3–3.5 per mil) in ostracod shells deposited in these gypsum-rich sediments are also consistent with increased lake salinity and aridity in the northern lowlands ca. AD 800–1100. The driest climate period occurred in the Terminal Classic and early Postclassic periods, which included an early (ca. AD 770–870) and late (ca. AD 920–1100) phase (Hodell, Brenner, & Curtis 2005). Climate during the intervening 50-year interval (ca. AD 870–920) was wet relative to these dry phases. The margin of error in radiocarbon dating of the sediment profiles is about +/-100 years and hence there is some uncertainty in the ages of these droughts. Nevertheless, the pattern of change is robust, indicating a grouping of events consisting of drier and wetter intervals during the Terminal Classic and Postclassic Periods with a recurrence interval of about 50 years (Hodell, Brenner, & Curtis 2005).

Numerous paleoclimatic records now support the hypothesis that drier than average conditions predominated during the Terminal Classic and into the Early Postclassic Periods (AD 750 to 1100; fig. 6.3). Exceptionally arid conditions in the northern lowlands were identified in the Lake Punta Laguna $\delta^{18}O$ ostracod record between ca. AD 860 and 1050 (Curtis, Hodell, & Brenner 1996; Hodell, Curtis, & Brenner, 2007). A seasonally resolved Ti record from the Cariaco Basin spanning the interval AD 750–950 indicates that multi-year droughts were centered on AD 760, 810, 860, and 910 (Haug et al. 2003). A $\delta^{18}O$ stalagmite (Tecoh) record from the northern Maya lowlands suggests that eight major droughts (3 to 18 years in duration) occurred between AD 800

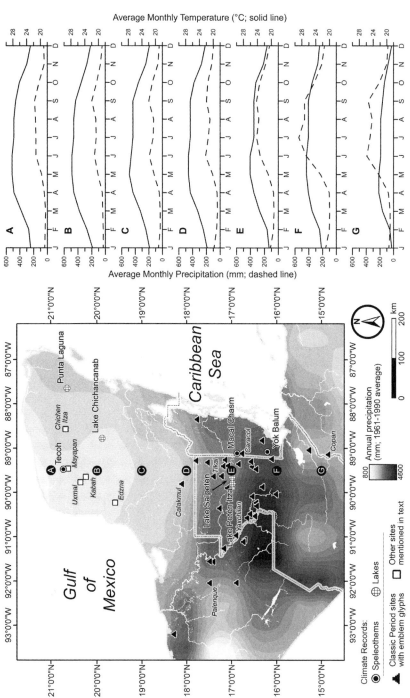

FIGURE 6.3 Map of the major sites and climate records discussed in text, with annual precipitation data derived from the Climate Research Unit Global Climate Dataset (version 2.1; Mitchell et al. 2004; Mitchell and Jones 2005). For visual effect, the cartographic surface has been interpolated to a 2-minute resolution using ordinary kriging with a spherical semivariogram model and smoothed with cubic convolution resampling. Average monthly temperature and precipitation values (non-interpolated) for the 30-year period 1961–1990 are plotted at the right, in seven one-degree increments of latitude.

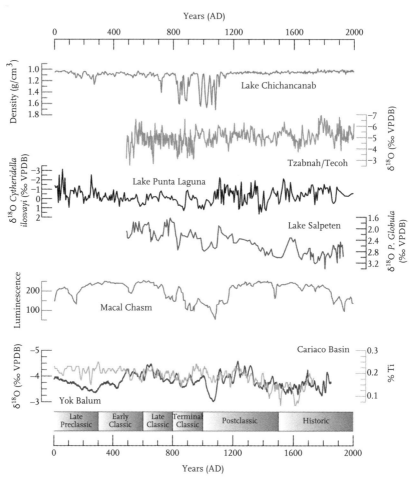

FIGURE 6.4 Palaeoclimatic records from the Maya region for the last 2000 years.

and 950 (Medina et al. 2010). This coincides with a general drying trend evident in a stalagmite luminescence record from the central lowlands (Macal Chasm, Belize; Webster et al. 2007) that has recently been confirmed with a high-resolution δ¹⁸O showing significant droughts at AD 850 and AD 1100 (Akers et al. 2016). A megadrought registers between AD 897 and 922 in ancient Montezuma bald cypress trees (*Taxodium mucronatum*) from Barranca de Amealco in the highlands of Mexico (Querétaro; Stahl et al. 2011), indicating that drought conditions reached far beyond the Maya region. Shorter droughts were also recorded in this tree-ring record at AD 810 and 860. Climate simulations indicate that the deforestation documented in the Maya lowlands towards the end of the Classic Period (see Leyden et al. 2002, for review) may also have served as an amplifier of drought (10–20 percent reduction in precipitation; Cook et al. 2012). The hydrogen and carbon isotope compositions of leaf waxes in lake sediments between AD 800 and 950 suggest that drought conditions

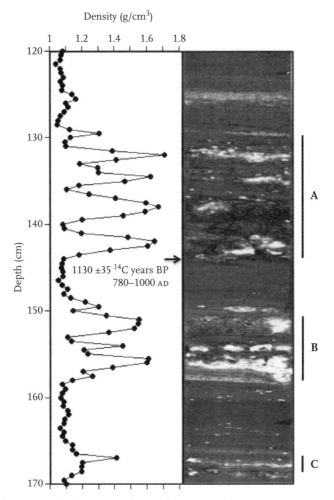

Density (g/cm³)

FIGURE 6.5 Lake Chichancanab sediment density record.

were more pronounced in the southern compared to northern Maya lowlands (Douglas et al. 2015).

Drought proxies continue to improve in the Maya region, but each of the available records also poses interpretive challenges (Hodell 2011). Some records, such as Barranca de Amealco and Cariaco Basin, have been called into question because they are located far away from the Maya region, while the chronological accuracy of others, such as Lake Chicancanab or Macal Chasm, is limited by large analytical error margins on the radiocarbon and uranium-series dates that anchor their age models. A stalagmite (YOK-I) from Yok Balum cave in the southern Maya lowlands currently provides the highest resolved and most accurately dated climate record in the region (Kennett et al. 2012). This 56-cm long stalagmite grew relatively quickly and continuously over the last 2000 years. Forty uranium-series dates anchor the subannually resolved $\delta^{18}O$ and $\delta^{13}C$ rainfall records from the stalagmite. The uranium series dates are highly accurate, averaging about +/-5 to 10 years, because this aragonitic stalagmite contains only small quantities of detrital thorium (compared with stalagmites from Macal Chasm and Tecoh). Overall, this record is highly consistent with the available paleoclimate data for the region. Both the $\delta^{18}O$ and $\delta^{13}C$ records indicate that anomalously high rainfall occurred during the Early Classic Period between AD 440 and 660 and that this was followed by a drying trend between AD 660 and 1000 that culminated in the most extended and severe drought in the 2000-year record between ca. AD 1000 and 1100. Superimposed on this drying trend are two other severe droughts: one between AD 820 and 870, and the other a brief dry interval centered on AD 930. Stalagmite growth slowed considerably between AD 820 and 900, and the severity of the droughts during this interval is underestimated in the stable isotope records. The Yok Balum climate record corresponds well with other records indicating Terminal Classic and Early Postclassic drying, with the most severe droughts occurring between AD 820 and 900 and again between ca. AD 1000 and 1100. In the Central/Southern Maya lowlands, lake hydrology can be affected by human deforestation of watersheds. For this reason, paleoclimate interpretations based on sediment sequences from Lake Salpeten were made with caution (Rosenmeier et al. 2002a, b). We note, however, a similar pattern of $\delta^{18}O$ change in both the ostracod calcite record from Lake Salpeten and the speleothem record at Yok Balum, which inspires confidence that the two regions responded to similar climatic change (fig. 6.4).

Drought, Agricultural Productivity, and Critical Societal Transitions

Multiple workers have argued that Terminal Classic Period droughts undermined agricultural productivity and destabilized economic and political institutions (Hodell, Curtis, & Brenner 1995; Webster et al. 2007; Kennett et al. 2012). However, the linkage between drought and agricultural productivity in the Maya lowlands is poorly understood. Ethnographic work with traditional Maya farmers throughout the region suggests that low yields are associated with periods of drought (Wilk 1991; Kramer 2005). The timing of the onset of the wet season is also critical, and catastrophic crop failure can occur if the predicted onset of rain does not occur (Culleton 2012). These anecdotal observations suggest that annual drought is related to decreased agricultural yields and that people use a variety of strategies to minimize the risk of periodic short-term drought (Wilk 1991). However, drought intervals during the Terminal Classic Period played out over decades or centuries rather than single growing seasons or years and present a fundamentally different type of adaptive problem.

Modern droughts are categorized as meteorological, hydrological, agricultural, or socioeconomic based on their duration, severity, and socioeconomic effects (Wilhite & Glantz 1985). Meteorological and hydrological droughts are linked and defined based on the duration and severity of dry intervals. Infrequent or lower than average rainfall in a single year may register as a meteorological drought, but may not effect stream flow, lake levels, or subsurface water supply. Agricultural drought is linked to precipitation shortages and the effects of evapotranspiration on soil moisture and the biological requirements of key cultigens. During extended dry intervals, agricultural droughts often occur well before the impact of long-term drought on hydrological systems. Hydrological drought is evident in some parts of the Maya region in the Late Classic, and water capture and control systems became more sophisticated at several centers within this context to mitigate drought (Luzzadder-Beach, Beach, & Dunning 2012; Scarborough et al. 2012), while deforestation amplified it in certain regions (Cook et al. 2012; see Van Loon et al. 2016 for more general discussion of human induced droughts). Socioeconomic droughts are population dependent and occur when the demand for ecosystem and agricultural resources, mediated by meteorological, hydrological, and agricultural droughts, exceeds supply. The multidecadal droughts evident in the Maya region between AD 820 and 900, and again between AD 1020 and 1100, were certainly socioeconomic droughts, but the details of human response to these droughts are obscured in the archaeological record. Classic Period dynastic histories are also silent on this topic (Martin & Grube 2000).

Historical documentation during the Colonial Period provides some insight into the agricultural, demographic, and socioeconomic impacts of drought (García Acosta et al. 2003; Hoggarth et al. 2017). Meteorological and agricultural drought was commonly recorded in the Yucatan starting in the early 16th century. Single year droughts impacted crop yields, but short-term stores and social networks largely buffered Colonial populations. Multi-year droughts were the most problematic for these populations. Three extended dry intervals are recorded historically between AD 1525 and 1560, AD 1610 and 1670, and AD 1720 and 1800. During each interval there are multiple accounts of declining agricultural yield, food shortages, increased food prices, starvation, and the outbreak of disease. In all three intervals there is also evidence for population decline and the dispersal of Maya populations from towns into the forest. All three of these extended droughts are recorded in the Yok Balum climate record from southern Belize, but pale in comparison to the droughts evident in the Terminal Classic period (Kennett et al. 2012; Hoggarth et al. 2017).

A variety of food production strategies provided the basis of socioeconomic and political systems during the Classic Period, and the polities involved were much larger than those encountered by the Spanish in the sixteenth century. The largest polities were also structurally more rigid and less resilient to endogamous or exogenous change (Kennett & Beach 2013). Surplus food supplies were required to support divine kings and queens and their courts, along with other members of the elite class and their families. There were also highly specialized members of Classic Period society who required support (for example, scribes, artisans, and architects). Local conditions (including topography, geology, and water availability) governed the types of agricultural systems employed to grow maize and other crops (beans, squash, chili peppers) in different environmental zones. Multiple lines of evidence indicate that maize was a staple crop (pollen, phytoliths, stable isotopes; see Kennett & Beach 2013, Beach et al. 2015 for reviews), a conclusion that is reinforced by the art and iconography of the Classic Maya (Taube 1985). However, it is also important to remember that cultivation of this important staple was embedded in a complex mosaic of food production that included tree crops, wild plant foods, and game animals (deer, peccary) from the neotropical forest, with variations depending upon different rainfall, soil types, and water tables (Fedick & Ford 1990; Puleston 1978). Salt was a vital preservative for fish and meat (McKillop 2008), but the long-term storage of maize and other plant foods was not an option because of the warm and moist conditions that prevail in the Maya lowlands. In addition, the labor demands on subsistence farmers only allowed for small amounts of food surplus (Webster 1985). Subsistence farmers were busy producing enough food for their families and saving enough seed for the following growing season. The food and labor provided to the elite sector of society placed additional demands on them (Webster 1985; Kennett & Beach 2013). This would have set up an essential tension between elite and non-elite sectors of society that was inherently unstable and vulnerable to extended drought. With the agricultural foundations of socioeconomic systems undermined locally by drought between AD 820 and 900, and again between AD 1020 and 1100, there was great potential for destabilizing Classic Period political networks.

Classic Maya Sociopolitical Networks and Critical Transitions

Classic Period civic-ceremonial centers were nodes within overlapping and interacting sociopolitical networks. Maya kings at the largest and most politically influential centers dedicated stone monuments to descriptions of their participation in a variety of antagonistic, hierarchical, and cooperative relationships with other Classic Period polities (Munson & Macri 2009; fig. 6.6). Yet, the monuments recorded only a small portion of the network activity among the largest political centers. Rulers at smaller centers rarely erected carved stone monuments with long count dates and historical information (for example, Belize Valley; Helmke & Awe 2013:65; LeCount & Yaeger 2010: 28). Smaller centers certainly participated in regional sociopolitical networks, but the details of their interactions are obscured in the archaeological record. The recovery of obsidian from virtually every Classic Period archaeological context in the Maya lowlands (see, for example, Golitko et al. 2012), sometimes associated with other non-local materials (jade), indicates broad scale economic connectivity between households. Stone monuments, therefore, only serve as an approximation for the changing vitality of sociopolitical networks between AD 800 and 1000 among the largest Classic Maya centers.

The Classic Maya polities that developed in the central and southern lowlands shared a large number of cultural traditions, as evidenced in their archaeological records, including divine kingship, monumental buildings, stone monuments, the long-count calendar, and writing. The political structure of each polity was similar (peer polities; Schele & Freidel 1990; Carmean & Sabloff 1996; Webster 1997) and they were largely autonomous. A divine king or queen governed each of the largest polities and was the primary political actor evident in historical texts. The emblem glyphs for forty-four of the most prominent polities occur in historical texts, and many other small- and medium-sized centers are known archaeologically (Witschey & Brown 2013). Hierarchical relationships were periodically recorded starting in the early Classic Period and into the Late Classic (Martin & Grube 2000). Historical texts carved on stone monuments point to the strongest connectivity between polities in the southern and central lowlands (Munson & Macri 2009). Polities in the northern lowlands and the highlands were more peripheral to Classic Period sociopolitical networks, as were smaller communities living on the coasts of the Yucatan Peninsula (Turner & Sabloff 2012). These more peripheral communities were connected via trade networks, but political network connectivity was incomplete and they were structurally more heterogeneous and exhibited greater adaptive variability to local environmental conditions and cultural contexts.

Sociopolitical networks first began to fracture in the southern lowlands (Petexbatun) between AD 760 and 800 as inter-polity warfare was peaking throughout the region (Inomata 1997, 1998; Demarest 2004; Kennett et al. 2012). A 50 percent decrease in the number of centers with dated stone monuments occurred in the southern and central lowlands between AD 800 and 850. The number of connections recorded between the largest centers also diminished as nodes in the Classic Period network weakened (see fig. 6.6; AD 800–849). Building campaigns at the largest centers in the central and southern lowlands also went into decline during this interval, and there was an asynchronous contraction of monument carving during the Terminal, Classic between AD 800 and 900 in seven core areas (Ebert et al. 2014). The decline led to the balkanization of populations and the decentralization of political networks to a more multi-nodal configuration. The process of network decentralization was largely complete in the southern and central lowlands by AD 925. Agricultural populations persisted in some regions (Laporte 1995), but there is also some evidence for regional abandonment soon after political collapse (Prufer et al. 2011; Culleton et al. 2012). Overall, this transition was marked more by major societal reorganization than collapse (McAnany & Negron, 2010), because Maya populations and some of their cultural traditions, such as writing and the derivative calendar, persisted through the Terminal Classic and into the Postclassic Period (AD 900–1520). Nevertheless, the societal and demographic transformations represent a critical transition in the way economic and political institutions were organized, with decentralization/dispersal of populations in the southern and central Maya lowlands.

Several polities persisted in the northern Maya lowlands as networks were critically reorganized to the south (fig. 6.7). The best known of these are the

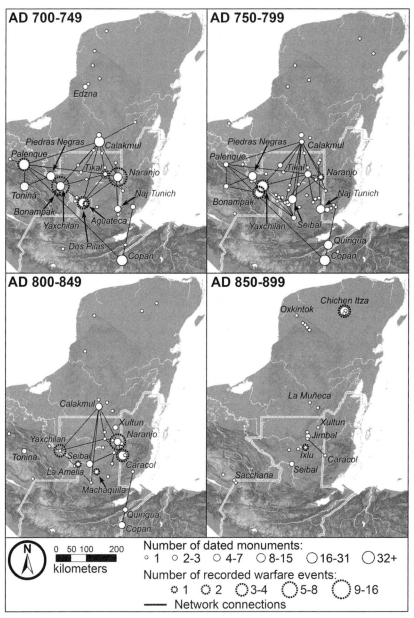

FIGURE 6.6 Space-time progression of dated monuments, recorded warfare events, and active network connections within the Maya region (after Munson and Macri 2009).

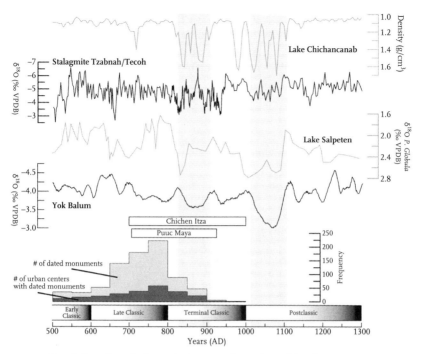

FIGURE 6.7 Palaeoclimatic data relative to total number of dated stone monuments through time and number of urban centers with dated stone monuments (data from Kennett et al. 2012).

Puuc Maya centers of Uxmal, Sayil, Xcoch, Huntchimul, Xkipché and Chac II (Hoggarth et al. 2016). Many of these centers also went into decline in the ninth century, but the process was delayed and more protracted than in the southern and central lowlands. Some Puuc Maya sites persisted until AD 1000 (for example, Xkipché). Chichén Itzá had a Puuc Maya florescence during the seventh and eighth centuries that also went into decline in the ninth century as it was eclipsed by new traditions and reinvigorated political activity, peaking between AD 900 and 1000 (Andrews et al. 2003; Hoggarth et al., 2016). Chichén Itzá was the dominant regional polity in the northern lowlands during this interval. The available evidence now suggests that it and other contemporary political centers in the interior, such as Xkipché, failed during the longest and most severe drought recorded in the region between AD 1000 and 1100. Populations shifted to the coast and were integrated into smaller polities. Archaeologists have referred to this interval as a cultural "dark age" (Andrews et al. 2003) prior to the development of Mayapán as an important regional center when conditions ameliorated after AD 1100 (Kennett et al. 2016; Peraza Lope et al. 2006; Masson & Peraza Lope 2014). Dahlin (2002) suggested that the drier conditions in the northern Yucatan necessitated early development of water storage technology and that these adaptations help explain the persistence of populations and the fluorescence of polities in the face of regional drought during the Terminal

Classic. Others have pointed to greater access to potable water in caves and *cenotes* in the region, as was the case for Chichén Itzá (Peterson & Haug 2005, Hodell et al. 2007). There is also evidence that the extended droughts were not as severe in the northern Maya lowlands (Douglas et al. 2015).

In addition to drought specific adaptations in the northern lowlands and greater access to ground water via *cenotes* and caves, we suggest that the structure of Classic Period sociopolitical networks made polities in the southern and central lowlands more vulnerable to failure. Polities in the northern lowlands were less connected and therefore less vulnerable. The structure of network connectivity makes complex systems more or less susceptible. Networks that have identical nodes and are homogenous are less likely to change gradually and are more vulnerable to abrupt transitions (fig. 6.8; Scheffer et al. 2012). This is particularly the case if the nodes are strongly interconnected. Networks that are less connected or incomplete tend to have more heterogeneous nodes that adapt to change locally and are less susceptible to catastrophic failure or abrupt transitions as external conditions change or perturbations occur. We

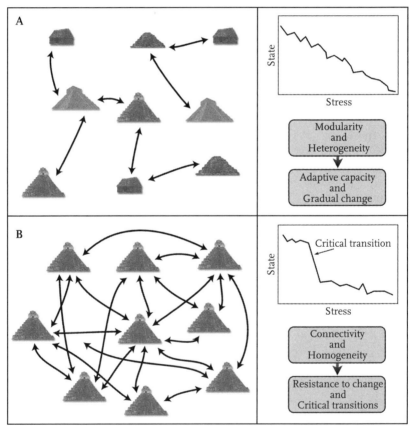

FIGURE 6.8 Schematic overview of the structure and properties of heterogeneous (*a*) versus homogeneous (*b*) networks (after Scheffer et al. 2012).

suggest that this phenomenon contributed to the gradual transition evident in the northern lowlands and on the coasts between AD 800 and 1000 compared to the more abrupt transition evident in the central and southern lowlands under severe drought conditions. The drought between ca. AD 1000 and 1100 was more consequential in the northern lowlands and may have had far reaching demographic effects in the central and southern lowlands (Kennett et al. 2012; Hoggarth et al. 2016).

Multidecadal droughts were severe in the Maya region during the transition from the Classic to Postclassic Periods, but the heterogeneity of the spatial and temporal patterns is less well known. Multiple droughts occurred between AD 800 and 1000 within a broader drying trend that started after AD 660. This was preceded by an anomalous interval of high rainfall (AD 440–660) that favored population expansion, aggregation, and the florescence of Classic Maya civilization in the southern and central lowlands (elaborate architecture, fine art, calendar, etc.). Sociopolitical network activity between the largest civic-ceremonial centers reached a peak in this region between AD 700 and 800 and then went into decline between AD 800 and 900. The failure of political centers as nodes within this extensive network was asynchronous, and the most abrupt loss of the largest centers occurred in the early ninth century within the context of severe drought between AD 800 and 900. Political networks became decentralized and multi-nodal, with the balkanization of populations throughout the region. The process was largely complete in the southern and central lowlands by AD 925.

Polities persisted in the northern lowlands and along the coast as the Puuc Maya florescence gave way to the emergence of Chichén Itzá as an important regional capital between AD 800 and 1000 (Andrews et al. 2003). We argue here that polities in the north were less vulnerable to abrupt transition because they were better adapted to dry climatic conditions and less integrated into the sociopolitical networks that failed more catastrophically to the south. Drought may have also been less pronounced in the northern lowlands (Douglas et al. 2015). Puuc Maya centers flourished and were ultimately dominated by Chichén Itzá as drought conditions ameliorated between AD 870 and 1000 (see also Hodell et al. 2007). These polities ultimately failed in the face of the second and most severe regional drought between AD 1000 and 1100. Population decline is evident throughout the Maya region at this time, and archaeologists have identified a cultural "dark age" in the northern lowlands prior to the development of Mayapán during a wet interval between AD 1200 and 1400.

The episodic formation, consolidation, and breakdown of preindustrial states occurred in multiple contexts worldwide during the last 5,000 years. Preindustrial states were inherently unstable and their decline, often after a century or two, resulted from interacting endogenous economic, demographic and political mechanisms (Gavrilets et al. 2010; Kennett and Marwan 2015). Sociopolitical cycling is evident in the Maya region starting in the Preclassic Period, and drought periodically played a contributing role in the decline of these polities (Medina et al. 2016; Hoggarth et al. 2016, Kennett et al. 2016).

We argue that the collapse of Classic Maya polities in the southern and central Maya lowlands was a product of the homogeneous sociopolitical networks that developed during the seventh and eighth centuries, combined with the severity and persistence of droughts in the ninth century.

Acknowledgements

We thank Harvey Weiss for inviting us to participate in this book project. Thanks also to Thomas Harper for graphical assistance. Funding for this work was provided by the National Science Foundation (BCS-0940744; HSD-0827205 [Kennett]) and Penn State. Jason Yaeger and Julie Hoggarth provided assistance and useful comments on the manuscript. We dedicate this chapter to James P. Kennett, Father and Doctor Father, and thank him for his comments on this manuscript.

References

Akers, P. D., G. A. Brook, L. B. Railsback, F. Liang, G. Iannone, J. W. Webster, P. P. Reeder, H. Cheng, and R. L. Edwards. 2016. An extended and higher-resolution record of climate and land use from stalagmite MC01 from Macal Chasm, Belize, revealing connections between major dry events, overall climate variability, and Maya sociopolitical changes. *Palaeogeography, Palaeoclimatology, Palaeoecology* 459: 268–288.

Andrews, A. P., E. W. Andrews, and F. Robles Castellanos. 2003. The northern Maya collapse and its aftermath. *Ancient Mesoamerica* 14: 151–156.

Beach, T., S. Luzzadder-Beach, N. Dunning, J. Hageman, and J. Lohse. 2002. Upland agriculture in the Maya Lowlands: ancient Maya soil conservation in northwestern Belize. *Geographical Review* 92: 372–397.

Beach, T., S. Luzzadder-Beach, N. Dunning, J. Jones, J. Lohse, T. Guderjan, S. Bozarth, S. Millspaugh, and T. Bhattacharya. 2009. A review of human and natural changes in Maya Lowland wetlands over the Holocene. *Quaternary Science Reviews* 28: 1710–1724.

Beach, T., S. Luzzadder-Beach, D. Cook, N. Dunning, D.J. Kennett, S. Krause, R. Terry, D. Trein, and F. Valdez. 2015. Ancient Maya impacts on the Earth's surface: An Early Anthropocene analog? *Quaternary Science Reviews* 124: 1–30.

Brenner, M., M. F. Rosenmeier, D. A. Hodel, and J. H. Curtis. 2002. Paleolimnology of the Maya lowlands. *Ancient Mesoamerica* 13: 141–157.

Carmean, K., and J. A. Sabloff. 1996. Political decentralization in the Puuc region, Yucatán, Mexico. *Journal of Anthropological Research* 52: 317–330.

Chase A. F., D. Z Chase, J. F. Weishampel, J. B. Drake, R. L. Shrestha, K. C. Slatton, J. J. Awe, and W.E. Carter. 2011. Airborne LiDAR, archaeology, and the ancient Maya landscape at Caracol, Belize. *Journal of Archaeological Science* 38: 387–398.

Cook, B. I., K. J. Anchukaitis, J. O. Kaplan, M. J. Puma, M. Kelley, and D. Gueyffier. 2012. Pre-Columbian deforestation as an amplifier of drought. *Geophysical Research Letters* 39: L16706.

Culbert, T. P., and D. S. Rice, eds. 1990. *Precolumbian Population History in the Maya Lowlands*. Albuquerque, New Mexico: University of New Mexico Press.

Culleton, B. J. 2012. "Human Ecology, Agricultural Intensification and Landscape Transformation at the Ancient Maya Polity of Uxbenká, Southern Belize." PhD diss., University of Oregon, Eugene.

Curtis, J. H., D. A. Hodell, and M. Brenner. 1996. Climate variability on the Yucatan Peninsula (Mexico) during the past 3500 years, and implications for Maya cultural evolution. *Quaternary Research* 46: 37–47.

Dahlin, B. H. 1990. Climate and prehistory on the Yucatan peninsula. *Climatic change* 5: 245–263.

Dahlin, B. H. 2002. Climate change and the end of the Classic period in Yucatán: Resolving a paradox. *Ancient Mesoamerica* 13: 327–340.

Dahlin, B. H., and A. F. Chase. 2014. A tale of three cities: effects of the AD 536 event in the lowland Maya heartland. In *The Great Maya Droughts in Cultural Context: Case Studies in Resilience and Vulnerability*, ed. G. Iannone. Boulder: University Press of Colorado, 127–156.

Demarest, A. 2004. *Ancient Maya: The Rise and Fall of a Rainforest Civilization*. Cambridge: Cambridge University Press.

Douglas, P. M. J., M. Pagani, M. A. Canuto, M. Brenner, D. A. Hodell, T. I. Eglinton, and J. H. Curtis. 2015. Drought, agricultural adaptation, and sociopolitical collapse in the Maya lowlands. *Proceedings of the National Academy of Sciences* 112: 5607–5612.

Drennan, R. D. 1988. Household location and compact versus dispersed settlement in prehispanic Mesoamerica. In *Household and Community in the Mesoamerican Past*, ed. R. R. Wilk. and W. Ashmore Albuquerque: University of New Mexico Press, 273–293.

Dunning, N. P. 1992 *Lords of the Hills: Ancient Maya Settlement in the Puuc Region, Yucatán, Mexico*. Madison, Wisconsin: Prehistory Press.

Dunning, N. P., T. P. Beach, and S. Luzzadder-Beach, 2012. Kax and kol: Collapse and resilience in lowland Maya civilization. *Proceedings of the National Academy of Sciences* 109: 3652–3657.

Ebert, C. E., K. M. Prufer, M. J. Macri, B. Winterhalder, and D. J. Kennett. 2014 Terminal Long Count Dates and the Disintegration of Classic Period Maya Polities. *Ancient Mesoamerica*, 25: 337–356.

Fedick, S. L., ed. 1996. *The Managed Mosaic: Ancient Maya Agriculture and Resource Use*. Salt Lake City: University of Utah Press.

Fedick, S. L, and A. Ford. 1990. The prehistoric agricultural landscape of the central Maya lowlands: an examination of local variability in a regional context. *World Archaeology* 22: 18–33.

Folan, W. J., E. R. Kintz, and L. A. Fletcher. 1983. *Coba: A Classic Maya Metropolis*. New York: Academic Press.

Folan, W. J., and B. H. Hyde. 1985. Climatic forecasting and recording among the ancient and historic Maya: An ethnohistoric approach to epistemological and paleoclimatological patterning. In *Contributions to the Archaeology and Ethnohistory of Greater Mesoamerica*, ed. W.J. Folan. Carbondale, IL: Southern Illinois University Press, 15–48.

French, K. D., and C. J. Duffy. 2010. Prehispanic water pressure: A new world first. *Journal of Archaeological Science* 37: 1027–1032.

French, K. D., C. J. Duffy, and G. Bhatt. 2012. The hydroarchaeological method: A case study at the Maya sites of Palenque. *Latin American Antiquity* 23: 29–50.

French, K. D., and C. J. Duffy. 2014. Understanding ancient Maya water resources and the implications for a more sustainable future. *WIREs Water* 1: 305–313.

García-Acosta, V., J. M. Pérez Zevallos, and A. M. Molina del Villar. 2003 *Desastres agrícolas en México, Catálogo histórico, Tomo I: Épocas prehispanica y colonial (958–1822)*. Mexico: Centro de Investigaciones y Estudios Superiores en Antropología Social Fondo de Cultura Económica.

Gavrilets, S., D. G. Anderson, and P. Turchin 2010. Cycling in the Complexity of Early Societies. *Cliodynamics* 1: 58–80.

Giddings, L., and M. Soto. 2003. Rhythms of precipitation in the Yucatan Peninsula. In *The Lowland Maya: Three Millennia at the Human-Wildland Interface*, ed. A. Gomez-Pompa, M. F. Allen, S. L. Fedick and J. J. Jimenez-Osornio. Binghamton, NY: Haworth Press, 77–89.

Golitko, M., J. Meierhoff, G. M. Feinman, and P. R. Williams. 2012. Complexities of collapse: the evidence of Maya obsidian as revealed by social network analysis. *Antiquity* 86: 507–523.

Gray, C. R. 1993. Regional meteorology and hurricanes. In *Climate Change in the Intra-American Sea*, ed. G.A. Maul. United Nations Environmental Program. London: Edward Arnold, 87–99.

Gunn, J., and R. E. W. Adams. 1981. Climatic change, culture, and civilization in North America. *World Archaeology* 13: 87–100.

Hansen, R. D., S. Bozarth, J. Jacob, D. Wahl, and T. Schreiner. 2002. Climate and environmental variability in the rise of Maya civilization: a preliminary perspective from the northern Peten. *Ancient Mesoamerica* 13: 273–295.

Harrison, P. D. 1999. *The Lords of Tikal: Rulers of an Ancient Maya City*. London: Thames and Hudson.

Hastenrath, S. 1984. Interannual variability and the annual cycle: mechanisms of circulation and climate in the tropical Atlantic sector. *Monthly Weather Review* 112: 1097–1107.

Hastenrath, S. 1991. *Climate Dynamics of the Tropics*. Boston: Kluwer Academic Publishers.

Haug, G. H., K. A. Hughen, D. M. Sigman, L. C. Peterson, and U. Röhl. 2001. Southward migration of the intertropical convergence zone through the Holocene. *Science* 293: 1304–1308.

Haug, G. H., D. Günther, L.C. Peterson, D. M. Sigman, K.A. Hughen, and B. Aeschlimann. 2003. Climate and the collapse of Maya civilization. *Science* 299: 1731–1735.

Helmke, C. and J. Awe. 2013. Ancient Maya Territorial Organisation of Central Belize: Confluence of Archaeological and Epigraphic Data. *Contributions in New World Archaeology* 4: 59–90.

Hodell, D. A. 2011. Maya megadrought? *Nature* (News & Views) 479: 45.

Hodell, D. A., J.H. Curtis, and M. Brenner. 1995. Possible role of climate in the collapse of Classic Maya civilization. *Nature* 375: 391–394.

Hodell, D. A., M. Brenner, J. H. Curtis, and T. Guilderson. 2001. Solar forcing of drought frequency in the Maya lowlands. *Science* 292: 1367–1370.

Hodell, D. A., M. Brenner, and J. H. Curtis. 2005. Terminal Classic drought in the northern Maya lowlands inferred from multiple sediment cores in Lake Chichancanab (Mexico). *Quaternary Science Review* 24: 1413–1427.

Hodell, D. A., M. Brenner, and J. H. Curtis. 2007. Climate and Cultural History of the northeastern Yucatan Peninsula, Quintana Roo, Mexico. *Climatic Change*. DOI: 10.1007/s10584-006-9177-4

Hoggarth, J. A., M. Restall, J. W. Wood, and D. J. Kennett. 2017. Extended Drought and its Demographic Effects in the Maya Lowlands. *Current Anthropology* 58: 82–113.

Hoggarth, J. A., S. F. M. Breitenbach, B. J. Culleton, C. E. Ebert, M. A. Masson, and D. J. Kennett. 2016. The political collapse of Chichén Itzá in Climatic and Cultural Context. *Global and Planetary Change* 138: 25–42.

Iannone, G. 2014. *The Great Maya Droughts in Cultural Context: Case Studies in Resilience and Vulnerability.* University Press of Colorado, Boulder.

Inomata, T. 1997. The last day of a fortified Classic Maya center: Archaeological investigations at Aguateca, Guatemala. *Ancient Mesoamerica* 8: 337–351.

Inomata, T., and L. R. Stiver. 1998. Floor assemblages from burned structures at Aguateca, Guatemala: A study of Classic Maya households. *Journal of Field Archaeology* 25: 431–452.

Kennett, D. J., and T. Beach. 2013. Archaeological and environmental lessons for the Anthropocene from the Classic Maya collapse. *Anthropocene* 4: 88–100.

Kennett, D. J., I. Hajdas, B. J. Culleton, S. Belmecheri, S. Martin, H. Neff, J. Awe, H.V. Graham, K. H. Freeman, L. Newsom, D. L. Lentz, F. S. Anselmetti, M. Robinson, N. Marwan, J. Southon, D. A. Hodell, and G. H. Haug. 2013. Correlating the Ancient Maya and Modern European Calendars with High-Precision AMS [14]C Dating. *Nature Scientific Reports* 3: 1597. DOI: 10.1038/srep01597

Kennett, D. J., S. F. M. Breitenbach, V. V. Aquino, Y. Asmersom, J. Awe, J. U. L. Baldini, P. Bartlein, B.J. Culleton, C. Ebert, C. Jazwa, et al. 2012. Development and disintegration of Maya political systems in response to climate change. *Science* 338: 788–791.

Kennett, D. J., M. Masson, S. Serafin, B. J. Culleton, and C. Peraza Lope. 2016. War and Food Production at the Postclassic Maya City of Mayapán. *The Archaeology of Food and War.* Springer. New York, 161–192. DOI 10.1007/978-3-319-18506-4-9.

Kennett, D. J. and N. Marwan. 2015. Climatic volatility, agricultural uncertainty, and the formation, consolidation and breakdown of preindustrial agrarian states. *Philosophical Transactions of the Royal Society* 374: 20140458.

Kramer, K. L. 2005. *Maya Children: Helpers at the Farm.* Cambridge: Harvard University Press.

Laporte, J. P. 1995. ¿Despoblamiento o problema analítico?: El Clásico Temprano en el sureste de Petén. In *VIII Simposio de Investigaciones Arqueológicas en Guatemala,* ed. J. P. Laporte, A. C. Suasnávar, and B. Arroyo. Guatemala: Museum of National Archaeology and Ethnology, 729–762.

LeCount, L. and J. Yaeger. 2010. Provincial Politics and Current Models of the Maya State. In *Classic Maya Provincial Politics: Xunantunich and its Hinterlands,* ed. L. LeCount and J. Yaeger. Tucson: University of Arizona Press, 20–45.

Leyden, B. W. 2002. Pollen evidence for climatic variability and cultural disturbance in the Maya Lowlands. *Ancient Mesopotamia* 35: 85–101.

Lucero, L. J. 2002. The Collapse of the Ancient Maya: A Case for the Role of Water Control. *American Anthropologist* 104:814–826.

Luzzadder-Beach, S. 2000. Water resources of the Chunchucmil Maya. *Geographical Review* 90: 493–510.

Luzzadder-Beach, S., T. P. Beach, and N. P. Dunning. 2012. Wetland fields as mirrors of drought and the Maya abandonment. *Proceedings of the National Academy of Sciences* 109: 3646–3651.

Magaña, V., J. A. Amado and S. Medina.1999. The midsummer drought over Mexico and Central America. *Journal of Climate* 12: 1577–1588.

Marcus, J. 1993. Ancient Maya political organization. In *Lowland Maya Civilization in the Eighth Century A.D*, ed. J. A. Sabloff and J. S. Henderson. Washington, DC: Dumbarton Oaks.

Marcus, J. 2012. Maya political cycling and the story of the *Kaan* polity. In *The Ancient Maya of Mexico: Reinterpreting the Past of the Northern Lowlands,* ed. G. E. Braswell. London: Equinox Press, 88–116.

Martin, S., and N. Grube. 2000. *Chronicle of the Maya Kings and Queens: Deciphering the Dynasties of the Ancient Maya*. New York: Thames and Hudson.

Masson, M. A., and C. Peraza Lope. 2014. *Kulkulcan's Realm: Urban Life at Ancient Mayapán*. Boulder: University Press of Colorado.

Matheny, R. T. 1976. Maya lowland hydraulic systems. *Science* 193: 639–646.

McAnany, P. A. 1990. Water storage in the Puuc region of the northern Maya lowlands: A key to population estimates and architectural variability. In *Precolumbian Population History in the Maya Lowlands,* ed. T. P. Culbert and D. S. Rice. Albuquerque: University of New Mexico Press, 263–284.

McAnany, P. A., and T. G. Negrón. 2010. Bellicose rulers and climatological peril? Retrofitting twenty-first century woes on eighth-century Maya society. In *Questioning collapse: Human resilience, ecological vulnerability, and the aftermath of empire,* ed. P. A. McAnany and N. Yoffee. Cambridge: Cambridge University Press, 142–175.

McKillop, H. 2008. *Salt: White Gold of the Ancient Maya*. Gainesville: University of Florida Press.

Medina-Elizalde, M., S. J. Burns, D. W. Lea, Y. Asmerom, L. von Gunten, V. Polyak, M. Vuille, and A. Karmalkar. 2010. High resolution stalagmite climate record from the Yucatan Peninsula spanning the Maya terminal Classic Period. *Earth and Planetary Letters* 298: 255–262.

Medina-Elizalde, M., S. J. Burns, J. M. Polanco-Martínez, T. Beach, F. Lases-Hernández, C.-C. Shen, H.-C. Wang. 2016. High-resolution speleothem record of precipitation from the Yucatan Peninsula spanning the Maya Preclassic Period. *Global and Planetary Change* 138: 93–102.

Messenger, L. C. 1990. Ancient winds of change: Climatic settings and prehistoric social complexity in Mesoamerica. *Ancient Mesoamerica* 1: 21–40.

Mestas-Nunez, A. M., C. Zhang, B. A. Albrecht, and D. B. Enfield. 2002. Warm season water vapor fluxes in the Intra-Americas sea. Paper presented at the 27th Annual Climate Diagnostics and Prediction Workshop, Fairfax, VA, 21–25 October.

Munson, J. L., and M. J. Macri. 2009. Sociopolitical network interactions: A case study of the Classic Maya. *Journal of Anthropological Archaeology* 28: 424–438.

Neiman, F. D. 1997. Conspicuous consumption as wasteful advertising: A Darwinian perspective on spatial patterns in Classic Maya terminal monument dates. In *Rediscovering Darwin: Evolutionary Theory and Archaeological Explanation*, ed. C. M. Barton and G. A. Clark. Archaeological Papers of the American Anthropological Association, No. 7. Arlington, VA: American Anthropological Association, 267–290.

O'Mansky, M. 2014. Collapse without Drought: Warfare, Settlement, Ecology, and Site Abandonment in the Middle Pasion Region. In *The Great Maya Droughts in Cultural Context: Case Studies in Resilience and Vulnerability*, ed. G. Iannone. Boulder: University Press of Colorado, 157–176.

Peraza Lope, C., M. A. Masson, T. S. Hare, and P. Candelario Delgado Ku. 2006. The chronology of Mayapan: new radiocarbon evidence. *Ancient Mesoamerica* 17: 153–175.

Peterson, L. C., and G. H. Haug. 2005. Climate and the collapse of Maya civilization. *Scientific American* 93: 322–329.

Prufer, K. M., H. Moyes, B. J. Culleton, A. Kindon, & D. J. Kennett. 2011. Formation of a complex polity on the eastern periphery of the Maya Lowlands. *Latin American Antiquity* 22: 199–223.

Puleston, D. E. 1978. Terracing, raised fields, and tree cropping in the Maya lowlands: a new perspective on the geography of power. In *Pre-Hispanic Maya Agriculture*, ed. P. D. Harrison and B. I. Turner. Albuquerque: University of New Mexico Press, 225–245.

Rosenmeier, M. F., D. A. Hodell, M. Brenner, J. H. Curtis, and J. B. Martin. 2002a. A 3500 year record of climate change and human disturbance from the southern Maya lowlands of Peten, Guatemala. *Quaternary Research* 57: 183–190.

Rosenmeier, M. F., D. A. Hodell, M. Brenner, J. H. Curtis, J. B. Martin, F. Anselmetti, D. Ariztegui, and T. P. Guilderson. 2002b. Influence of vegetation change on watershed hydrology: implications for paleoclimatic interpretation of lacustrine $\delta^{18}O$ records. *Journal of Paleolimnology* 27: 117–131.

Scarborough, V. L., N. P. Dunning, K. B. Tankersley, C. Carr, E. Weaver, L. Grazioso, B. Lane, J. G. Jones, P. Buttles, F. Valdez, and D. L. Lentz. 2012. Water and sustainable land use at the ancient tropical city of Tikal, Guatemala. *Proceedings of the National Academy of Sciences* 109: 12408–12413.

Scarborough, V. L., and G. G. Gallopin. 1991. A water storage adaptation in the Maya lowlands. *Science* 251: 658–662.

Scheffer, M., S. R. Carpenter, T. M. Lenton, J. Bascompte, W. Brock, V. Dakos, J. van de Koppel, I. A. van de Leemput, S. A. Levin, E. van Nes, M. Pascual, and J. Vandermeer. 2012. Anticipating critical transitions. *Science* 338: 344–348.

Schele, L., and D. Freidel. 1990. *A Forest of Kings*. New York: William Morrow.

Schneider, T., T. Bischoff, and G. H. Haug. 2014. Migrations and dynamics of the intertropical convergence zone. *Nature* 513: 45–53.

Sharer, R. J., and L. P. Traxler. 2006. *The Ancient Maya*. Stanford: Stanford University Press.

Stahl, D. W., J. Villanueva Diaz, D. J. Burnette, J. Cerano Paredes, R. R. Heim, Jr., F. K. Fye, R. Acuna Soto, M. D. Therrell, M. K. Cleaveland, and D. K. Stahle.

2011. Major Mesoamerican droughts of the past millennium. *Geophysical Research Letters* 38: L05703.

Taube, K. 1985. The Classic Maya Maize God: A Reappraisal. In *Fifth Palenque Round Table,* ed. M. G. Robertson and V. M. Fields. San Francisco: Pre-Columbian Art Research Institute, 171–181.

Turner II, B. L., and J. A. Sabloff. 2012. Classic Period collapse of the central Maya lowlands: insights about human-environment relationships for sustainability. *Proceedings of the National Academy of Sciences* 109: 13908–13914.

Van Loon, A. F., T. Gleeson, J. Clark, A. I .J. M. Van Dijk, K. Stahl, J. Hannadord, G. Di Baldassarre, A. J. Teuling, L. M Tallaksen, R. Uijlenhoet, et al. 2016. Drought in the Anthropocene. *Nature Geoscience* 9: 89–91.

Webster, D. 1985. Surplus, labor, and stress in late Classic Maya society. *Journal of Anthropological Research* 41: 375–399.

Webster, D. L. 1997. City States of the Maya. In *The Archaeology of City States: Cross-Cultural Approaches,* ed. D. Nichols and T. Charleton. Washington, DC: Smithsonian Institution Press, 135–154.

Webster, J. W., G. A. Brook, L. B. Railsback, H. Cheng, R. L. Edwards, C. Alexander, and P. P. Reeder. 2007. Stalagmite evidence from Belize indicating significant droughts at the time of Preclassic abandonment, the Maya hiatus, and the Classic Maya collapse. *Palaeogeography, Palaeoclimatology, Palaeoecology* 250: 1–17.

Wilhite, D., and M. H. Glantz. 1985. Understanding the drought phenomenon: the role of definitions. *Water International* 10: 111–120.

Wilk, R. R. 1991. *Household Ecology: Economic Change and Domestic Life Among the Kekchi Maya in Belize.* Tucson: University of Arizona Press.

Wilson, E. M. 1980. Physical geography of the Yucatan Peninsula. In *Yucatan: A World Apart,* ed. E. Moseley and E. D. Terry. Tuscaloosa: University of Alabama Press.

Witschey, C. T., and W. R. T. Brown. 2013. The Electronic Atlas of Ancient Maya Sites: A Geographic Information System (GIS). http://mayagis.smv.org.

Yaeger, J., and D. Hodell. 2008. The Collapse of the Maya civilization: assessing the interaction of culture, climate, and environment. In *El Nino, Catastrophism and Culture Change in Ancient America,* ed. D. H. Sandweiss and J. Quilter. Washington DC: Dumbarton Oaks, 197–251.

CHAPTER 7 | Twelfth Century AD

Climate, Environment, and the Tiwanaku State

LONNIE G. THOMPSON AND ALAN L. KOLATA

Climate is a fundamental and independent variable of human existence. Given that 50 percent of the Earth's surface and much of its population exist between 30°N and 30°S, paleoenvironmental research in the Earth's tropical regions is vital to our understanding of the world's current and past climate change. Most of the solar energy that drives the climate system is absorbed in these regions. Paleoclimate records reveal that tropical processes, such as variations in the El Niño-Southern Oscillation (ENSO), have affected the climate over much of the planet. Climatic variations, particularly in precipitation and temperature, play a critical role in the adaptations of agrarian cultures located in zones of environmental sensitivity, such as those of the coastal deserts, highlands, and altiplano of the Andean region. Paleoclimate records from the Quelccaya ice cap (5670 masl) in highland Peru that extend back ~1800 years show good correlation between precipitation and the rise and fall of pre-Hispanic civilizations in western Peru and Bolivia. Sediment cores extracted from Lake Titicaca provide independent evidence of this correspondence with particular reference to the history of the pre-Hispanic Tiwanaku state centered in the Andean altiplano. Here we explore, in particular, the impacts of climate change on the development and ultimate dissolution of this altiplano state.

Introduction

The historical record of human activities in the Andean region prior to arrival of the Spanish in AD 1532 is exceptionally rich. The role of climate variability in the rise and fall of pre-Hispanic cultures in the region is significant given that both coastal and highland populations pursued predominantly agrarian livelihoods and were located in environmentally sensitive zones. The densely populated, coastal desert cultures of Peru depended upon a reliable water supply for irrigated agriculture whereas montane Andean and *altiplano* cultures, located near the upper limits of viable agriculture, were sensitive to precipitation and temperature.

In addition to the archaeological evidence of complex cultural histories, tropical South America contains several distinct archives of past climate and environmental change, including ice core and lake sediment records. Some climate records can be extracted from ice fields in the Andean region, which contains over 70 percent of the world's tropical glaciers. Ice cores from high-altitude tropical glaciers offer long-term perspectives on the variability of precipitation, temperature, aridity, and atmospheric and sea surface conditions. Yet there are few high mountain glaciers and even fewer that preserve detailed climate histories. Moreover the unique records that do exist are rapidly disappearing because most tropical glaciers are shrinking. Potential contemporary impacts on water resources have social and economic consequences that underscore the imperative to understand the drivers and human responses to past and present tropical climate variability.

One of the largest tropical ice fields on Earth, the Quelccaya ice cap (QIC) (13°56'S; 70°50'W; 5670 masl) in the Cordillera Vilcanota, is located close to Lake Titicaca on the easternmost rise of the Andes less than 100 kilometers from the sharp descent into the Amazon Basin (fig. 7.1). In 1983 two ice cores were drilled to bedrock. Minimal post-depositional reworking of the snow surface, even during the wet season, resulted in the distinct annual layers used to reconstruct a ca. 1500 year climate history (Thompson et al. 1985). Records of oxygen isotopic ratios ($\delta^{18}O$), mineral dust concentrations, liquid conductivity (a proxy for soluble aerosol concentrations), and net mass accumulation (a proxy for precipitation) were reconstructed from these cores. The Little Ice Age (LIA) from ca. AD 1500 to 1880 was the most prominent feature in the 1983 QIC record (Thompson et al. 1986). However, the records of climate and environmental variations were limited by logistical considerations. Since these cores could not be returned frozen, they were cut into samples that were melted and bottled in the field. As a result, the non-mineral aerosols were compromised by dissolution during transport and storage, and dust concentrations were affected when aliquots were taken both in the field and the laboratory.

In 2003, two new cores were recovered to bedrock on the QIC (Thompson et al. 2013). The Summit Dome (5670 masl) core (QSD, 168.68 m) and the North Dome (5600 masl) core (QND, 128.57 m) were returned frozen and currently are stored at −30°C at Ohio State University's Byrd Polar and Climate Research Center (OSU-BPCRC). In addition to the previously measured parameters, such as dust concentrations, $\delta^{18}O$, and net accumulation, the new ice core records include hydrogen isotopic ratios (δD) and concentrations of major ions that could not be measured in the earlier cores. The spatial coherence of the climate record across the ice cap is confirmed by the highly significant correlations between the $\delta^{18}O$ profiles from the QSD and QND cores (R = 0.898, p < .0001, for decadal averages), separated by 1.92 kilometers. The reproducibility of the $\delta^{18}O$ records between the 1983 Core 1 and the 2003 QSD core from AD 1000 to 1982 is also excellent (R = 0.856, p < 0.0001, decadal averages), confirming the validity of the earlier $\delta^{18}O$ record. The 2003 QSD core provides records of climatic and environmental variations extending back

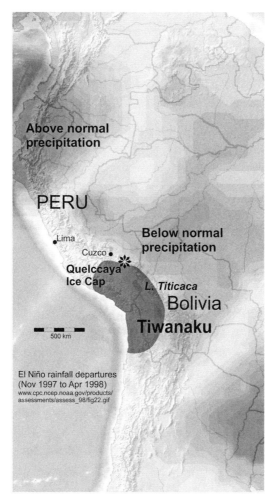

FIGURE 7.1 Map of west central South America showing location of the Quelccaya ice cap, Lake Titicaca, and Cuzco. Dark shading indicates the extent of the Tiwanaku civilization during the Middle Horizon.

to ca. AD 230 and with annual resolution back to AD 683. These new cores thus extend the Quelccaya climate history back 1800 years (fig. 7.2).

Most meteorological and climatic disturbances affecting Earth's surface and lower atmosphere originate in, or are amplified by, ocean/atmosphere interactions in tropical latitudes. Acting as the Earth's "heat engine," the warmest atmospheric and sea surface temperatures (SSTs) occur here. This energy drives intense convective precipitation and is crucial for the evolution of such phenomena as El Niño-Southern Oscillation (ENSO), the monsoon systems of Asia and Africa, and, on intra-annual time scales, hurricanes and other tropical disturbances that distribute equatorial heat energy toward the poles. ENSO dominates tropical climate variability. Linked to the position of the Intertropical Convergence Zone (ITCZ), ENSO-associated teleconnections affect the strength and direction of air masses and storm tracks, variations in convective activity that control flooding and drought, and modulation of tropical storm intensities. This relationship is especially the case for tropical South America. In Peru and Ecuador, ENSO generally brings heavy rains to the coastal deserts and drought to the *altiplano* of southern Peru and Bolivia (illustrated in fig. 7.1 for the 1997–1998 ENSO event). La Niña generally brings wetter conditions to the Peruvian highlands (Thompson, Mosley-Thompson, and Arnao 1984; Vuille 1999).

The Quelccaya record is frequently cited in the literature as primary evidence for documenting climatic and environmental changes in the region. The ice core archive has been used in the archaeological literature to evaluate the potential for climate-linked causes of significant sociocultural transformations in the Andes, including processes of civilizational collapse (Ortloff and Kolata 1993; Kolata 1993; Binford et al. 1997; Branch et al. 2007; Kolata et al. 2000; Sterken et al. 2006; Chepstow-Lusty et al. 2009; Owen 2005; Janusek 2004; Magilligan et al. 2001; Kemp et al. 2006; Arkush 2008; Manzanilla 1997). Others, however, have called into question the use of such proxy records for evaluating climate's role in social change (Stahl 1984; Calaway 2005). Thus, it is important to re-examine this relationship in light of the new higher resolution and longer ice core records drilled in 2003.

Here we discuss, in particular, the potential linkages between climate, environment, and the history of the Tiwanaku civilization that was centered ca. 300 kilometers to the south of the QIC and where water from the southern margins of the ice cap flows into Lake Titicaca (fig. 7.1). We present evidence for both decadal and century length "El Niño-like" oscillations concentrating on the role of climatic variability in Andean pre-Hispanic history. Since spatial precipitation patterns influenced by ENSO variations are pronounced, centennial-scale variability in ENSO events and ENSO-like events preserved in the QIC records can be examined with respect to their potential role in the rise and fall of pre-Hispanic cultures, as previously argued in Thompson, Davis, & Mosley-Thompson (1994) and Thompson & Davis (2014). We reference prior work to set the general context for understanding the relationship between climate and culture in the Andes, but concentrate here on examining this relationship in one pre-Hispanic society in particular, that of the Tiwanaku state, for which we have detailed archaeological and paleolimnological records that complement those of the Quelccaya ice cap.

Climate and Culture in the Andes

Paulsen (1976) examined the pre-Hispanic social and environmental relationships on the Santa Elena Peninsula in southwestern coastal Ecuador and the southern coast of Peru from 500 BC to AD 1532 (fig. 7.2). Both El Niño and the moderate advances and withdrawal of the Peru Current impact these areas

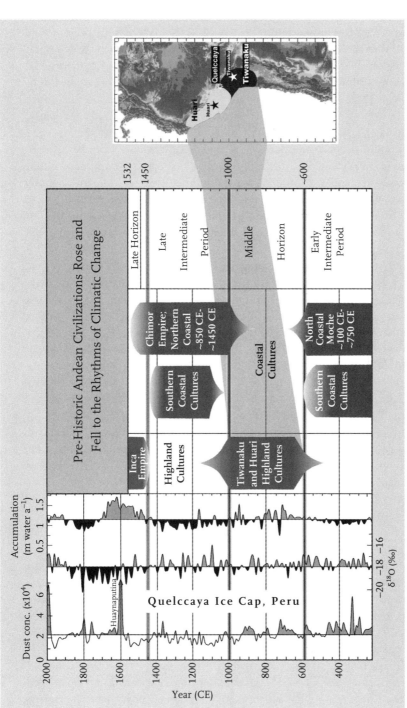

FIGURE 7.2 Decadal averages of records of mineral dust concentrations, δ⁸O (proxy for temperature), and accumulation (proxy for precipitation) from the 2003 QSD ice core. The chronology of the rise and fall of coastal and highland cultures over the four major pre-Colonial epochs shows broad patterns relative to the ice core-derived proxy records for temperature and precipitation. Also shown are the locations of major highland-culture population centers during the Middle Horizon. (Paulsen 1976; Thompson et al. 2013; Thompson & Davis 2014).

on a seasonal basis. The Santa Elena Peninsula has undergone longer duration changes with the gradual northward advance of the Peru Current from its original Pleistocene position off the north coast of Peru to its present position north of the Peninsula. Paulsen argued that the pluvial episodes on the Peninsula are best understood as exaggerated versions of the yearly cycle of advances and retreats of the Peru Current. The contraction and expansion of cultures on the Santa Elena Peninsula were largely synchronous with those along the coast of Peru between 500 BC and AD 1532 (fig. 7.2). During the Early Intermediate Period (ca. 500 BC to ca. AD 600), coastal cultures along the west coast of South America flourished. Archaeological evidence from the Middle Horizon (ca. AD 600 to ca. AD 1000) indicates that expansive highland-based polities (that is, the Wari and Tiwanaku cultures) heavily influenced many Peruvian and Chilean coastal societies, perhaps dominating some, while the Santa Elena Peninsula in Ecuador was largely deserted, especially from AD 800 to 1000. During the Late Intermediate Period (ca. AD 1000 to 1470), several complex centers of regional culture emerged on the Peruvian coast; on the Santa Elena Peninsula, the population reoccupied old sites and built new ones. During the Late Horizon from AD 1470 to 1532, as coastal climate deteriorated, the highland based Inca Empire conquered the desert littoral of Peru and adjacent coastal sections of Chile and Ecuador. Climatic deterioration was contemporaneous with the abandonment of the Santa Elena Peninsula. In an alternative reading of the evidence, Stahl (1984) argued that this apparent abandonment of the Santa Elena Peninsula may be due to a lack of records during this period and, like Calaway (2005), questioned the role of climate variability as a dominant, independent variable in prehistoric Andean culture changes. In contrast, Paulsen speculated that the Andean interpluvial that began ca. AD 600 was a direct factor in the emergence of the Wari and Tiwanaku states of the Middle Horizon and that another such Andean interpluvial beginning ca. AD 1400 was of similar importance for the establishment of the Inca Empire of the Late Horizon.

Figure 7.2 illustrates the climate record from the Quelccaya ice cap, particularly dust concentrations, $\delta^{18}O$ (a proxy for regional and Pacific Basin temperatures; see Thompson et al. 2013), and annual snow accumulation converted to water equivalent. The streams originating along the southern margins of the Quelccaya ice cap flow into Lake Titicaca, which lies only 90 kilometers to the southeast of Quelccaya. Figure 7.3 shows that, with the exception of the 1940s and the late 1970s, the accumulation record from QSD corresponds to the Lake Titicaca rise 3800 meters above sea level from 1915 to 1980. Thus the ice core record might be used as a broad proxy for past lake levels (Table 7.1) and *altiplano* precipitation. The chronologies of the highland and coastal cultures over four major pre-Hispanic periods (Early Intermediate Period, Middle Horizon, Late Intermediate Period and Late Horizon) are placed alongside this climate record. The Early Intermediate and Late Intermediate Periods are synchronous with average or higher than average $\delta^{18}O$ and lower accumulation on Quelccaya (and therefore the Peruvian *altiplano*), thus implying higher accumulation on the coast under the scenario of long-term

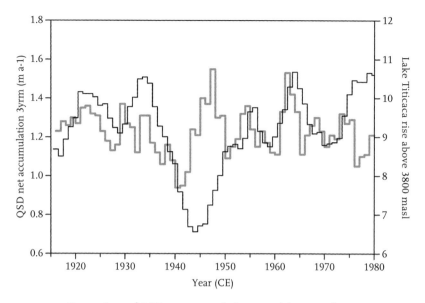

FIGURE 7.3 Comparison of QSD net accumulation record (meters of water equivalent, 3-year running means) with the Lake Titicaca rise above 3800 meters asl from AD 1915 to 1980.

ENSO type precipitation patterns. The Middle and Late Horizons are marked in the ice core record by average or lower than average δ¹⁸O and higher *altiplano* accumulation. Table 7.1 quantifies these climatic parameters during each epoch. Although dust concentrations seem to decrease towards the present and δ¹⁸O vary around the ice core mean, accumulation shows the most distinctive changes corresponding to the rise and decline of these Andean cultures.

Climate and the Tiwanaku State (ca. AD 600 to 1000)

The ice core record from Quelccaya located ca. 300 kilometers northeast of the state's heartland, along with sediment cores from Lake Titicaca that provide an independent paleoenvironmental archive (fig. 7.4), present a unique opportunity to examine climate and environmental conditions in detail during Tiwanaku's rise and ultimate dissolution. The ice accumulation record confirms that the interpluvials are out of phase with the pluvials in the highlands of southern Peru (fig. 7.2; Thompson, Davis, and Mosley-Thompson 1994). During the Early Intermediate Period on the *altiplano*, climatic conditions were relatively warm (indicated by less negative δ¹⁸O) and dry (lower accumulation) until ca. AD 500. This was followed by a period of overall, though highly variable, warm and relatively wet conditions from ca. AD 500 to 1150. According to the Quelccaya climate record, the precipitation and temperature from ca. AD 500 to 1000 underwent several short-term variations, but were relatively stable over the long term. Precipitation peaked around AD 950 and began, a steady

TABLE 7.1 Average values of dust concentrations, δ¹⁸O, and accumulation in the QSD core during Andean cultural periods. The average Lake Titicaca water rise is derived from the QSD accumulation by multiplying QSD in each period by the ratio between their modern values from AD 1915 to 1980 (9.20 m lake rise / 1.23 m ice-core water equivalent, or 7.48). Quelccaya data are archived at the National Oceanic and Atmospheric Administration, US Dept. of Commerce. Lake Titicaca data is supplied by the *Servicio Nacional de Meteorología e Hidrología del Peru.*

DATES	HORIZON	AVERAGE DUST CONCENTRATION (PER ML).	AVERAGE δ18O (‰)	AVERAGE ACCUMULATION (M WATER PER YR).	AVERAGE LAKE TITICACA WATER RISE ABOVE 3800 MASL)
AD 1451 to 1532	Late Horizon	16146	–17.93	1.23	9.20
AD 1001 to 1450	Late Intermediate Period	19596	–17.79	1.04	7.78
AD 601 to 1000	Middle Horizon	24070	–17.84	1.18	8.83
AD 226 to 600	Early Intermediate Period	28201	–17.64	1.10	8.23
Ice core mean		25724	–17.92	1.15	

decline until ca. 1200, after which it remained below average until the middle of the fifteenth century. Dust concentrations also decreased over this span.

Altiplano landscape and environment afforded adaptive opportunities for human populations, but also placed significant constraints on the primary productive capacity of pre-Hispanic civilizations. Archaeological research in the Lake Titicaca basin has shown that human populations in this unique, high altitude environment pursued several adaptive pathways including, most prominently, camelid pastoralism and the development of flooded, intensive raised-field agricultural systems (Kolata 1993, 1996, 2003; Janusek & Kolata 2004). In the circum-Lake Titicaca core area of the Tiwanaku state, these systems consisted of canals alternating with elevated planting beds that ranged from 1.5 to 10 meters in width. Natural springs or river networks fed canals in some areas, but in low-lying floodplains with high water tables, or *pampas* zones, many were fed by percolating groundwater. The elevated beds provide well-drained topsoil and optimal edaphic conditions for plant growth.

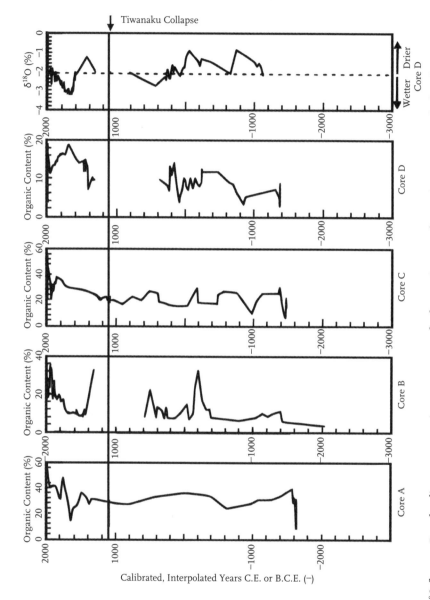

FIGURE 7.4 $\delta^{18}O$ for core D and sedimentary organic matter content for four major cores from Lake Titicaca (Binford et al. 1997).

Large-scale expansion of raised field agriculture in the *altiplano* during the seventh to early twelfth century AD florescence of the Tiwanaku state (designated the Tiwanaku IV and Tiwanaku V phases in the cultural chronology of the Bolivian *altiplano*) may be reflected in two periods of relatively high aerosol concentrations from ca. AD 650 to 850 and from ca. AD 870 to 940 in the 1983 Quelccaya ice core record (Thompson et al. 1988; Thompson, Davis, & Mosley-Thompson 1994) when wind-blown sediments derived from the construction of spatially extensive agricultural fields were deposited on the ice cap. The development of raised field agriculture into integrated, regional-scale systems occurred during this five-century long period of relatively warm and wet conditions (which corresponds with the Early Intermediate Period of the standard Peru-based cultural chronology). During this period, Tiwanaku political, cultural, and ideological norms and innovations were widely distributed throughout the *altiplano* and southern Andean region. In sum, a five-century long period of climatic conditions that favored water-intensive raised field agriculture facilitated the emergence and consolidation of Tiwanaku's economic, social, and political power.

The Decline of the Tiwanaku State and its Social Aftermath

Significant drought on the *altiplano* from ca. AD 1000 to 1100, can be inferred from the Quelccaya ice-core data, as well as sediment records from Lake Titicaca (fig. 7.4; Thompson et al. 1988; Abbott et al. 1997; Binford et al. 1997). The lake and ice cores independently demonstrate a major reduction in precipitation and a greater than 12 meter drop in lake levels after AD 1100 (fig. 7.5; Abbott et al. 1997; Binford et al. 1997; Kolata and Ortloff 1996), which would have caused the shores of Lake Titicaca to recede for kilometers and local water tables to drop far below the surface. Decreasing lake levels would effectively have stranded large tracts of raised fields and rendered their hydraulic sustaining systems inoperable, particularly in inland areas. Decreasing lake levels would have opened the *pampas* to occupation on a major scale, which is precisely what happened after AD 1150. The *pampas* zone changed from a marshy landscape of intensive agricultural production into drier lowlands devoted both to habitation and to less intensive, more diversified strategies of production. The Quelccaya accumulation record indicates that the maximum sustained drought that began at the end of the Middle Horizon persisted into the thirteenth and fourteenth centuries (figs. 7.2, 7.5).

Protracted drought conditions destroyed the viability of the large-scale, integrated agricultural systems managed predominantly by and for state rulers, thereby inducing sociopolitical fragmentation by aggravating pre-existing and emerging tensions in the region. Janusek and Kolata (2004: 418) note that "significant changes occurred in the several generations between AD 1000 and 1150, or Late Tiwanaku V. Research at Tiwanaku indicates that an elite

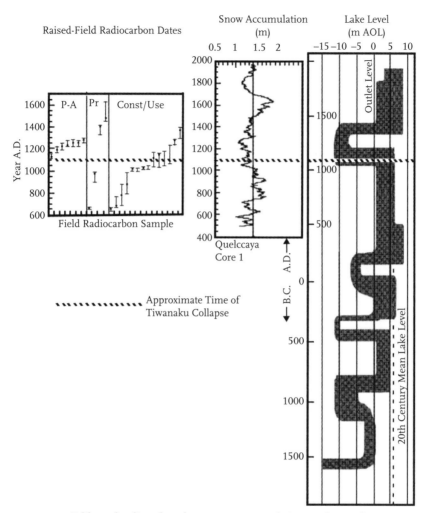

FIGURE 7.5 Calibrated radiocarbon dates, snow accumulation on the Quelccaya ice cap (redrawn from Thompson et al. 1985 and smoothed with a 10-yr moving mean), and proposed lake-level curve (redrawn from Abbott et al. 1997). P-A, Pr, and Const/ Use denote material taken from post-abandonment, problematic, and construction/ use contexts, respectively. The cross-hatched line at AD 1100 denotes the approximate beginning of reduced precipitation at the Quelccaya ice cap and the beginning of the collapse of the Tiwanaku civilization. The lake-level line is drawn thickly to illustrate century-scale variability. Because the twentieth century variation has been about 6 m, the line as illustrated is 6 m thick (Abbott et al. 1997).

residence was razed to the ground in the late eleventh century (Couture and Sampeck 2003), and not long after large-scale construction in monumental structures apparently ceased (Kolata 2003). Nevertheless, residential sectors in Tiwanaku and many sites in the Katari Valley [an extensive agricultural hinterland of the Tiwanaku capital]. . . continued to be occupied into the early

12th century. . . . Inhabitants of these sites continued to use Tiwanaku-style goods and engage in characteristic Tiwanaku residential and ritual practices. Tiwanaku collapse was a protracted and, most likely, chaotic process of state disintegration."

This was the first of many droughts over the Late Intermediate Period, the longest of which occurred throughout the thirteenth and fourteenth centuries AD as documented in the Quelccaya accumulation proxy record. The deterioration of Middle Horizon cultures was contemporaneous with this increasing aridity, although according to the $\delta^{18}O$ record this was a period of overall, though variable, warming (the "medieval climate anomaly," or MCA), with the warmest episode occurring ca. AD 1150. This period witnessed the political dissolution of the Wari and Tiwanaku highland cultures and the emergence of several complex coastal cultures, the most prominent of which were the Kingdom of Chimor and the Sicán polities centered on Peru's north coast (Moseley 1990; Shimada 2000). In competition with Sicán, the Kingdom of Chimor ruled the north coast of Peru beginning ca. AD 850 and ending ca. 1470. After ca. AD 1370, Chimor was the largest kingdom of the Late Intermediate Period, encompassing about 1000 kilometers of coastline.

During the decline of highland cultures that began ca. AD 1000, regional settlement patterns shifted locally from valley floor and lower valley slopes to higher altitude locations (Kolata 1983; Arkush 2008; Covey 2008; Bauer & Kellett 2010). Populations dispersed through the rural landscape and colonized new environmental niches unoccupied during the Tiwanaku florescence. As noted by Binford et al. (1997: 246): "The drier climate and much smaller lake of this period would have created a harsher environment similar to that of the *altiplano* south of the Tiwanaku heartland, where mineral extraction and pastoralism rather than agriculture are the principal economic activities. In addition to dispersal, some measure of population decline on the *altiplano* was an element of human adaptation to changed environmental conditions." Settlement patterns changed significantly following Tiwanaku state collapse. During the Late Intermediate Period in the Katari Valley, Janusek and Kolata indicate that: "while the number of sites increased by a factor of three, average settlement size decreased by more than four times that of Tiwanaku Period sites . . . . Settlement hierarchy fragmented and populations in the region dispersed as people left the nucleated centers once focused on Tiwanaku political activity, monumental constructions, rituals practices, and ceremonial feasting" (2004: 418).

In an early publication synthesizing the archaeological record of the south Andes, Kolata (1983) noted that the period after the dissolution of the Tiwanaku state ca. AD 1100 entailed a retrenchment and dispersal of populations from the agriculturally rich *pampas* regions around Lake Titicaca into defensible locations in higher altitude ecological zones that were the source of remnant mountain springs that supported irrigation and domestic consumption. More recently Arkush (2008; 2011) identified this same settlement pattern during the Late Intermediate Period farther to the north on the Peruvian *altiplano* arguing that populations moved away from (formerly) rich agricultural lands to settle in defensive sites high on hills and ridges. She notes, as many other scholars have before her, that the construction of hilltop forts (*pukaras*) was most extensive during the Late Intermediate Period, particularly after ca. AD 1300. Arkush and Tung (2013) interpret this settlement pattern, along with associated skeletal evidence of widespread cranial trauma, as indicative of social violence and a state of chronic warfare. Resource constraints induced by chronic drought conditions after the collapse of the Tiwanaku state clearly contributed to this state of social, political, and economic instability. It can be argued that decreasing precipitation drove the local population to higher terrain that was both closer to glacier water sources and easier to defend. The capacity to produce substantial surplus on flooded agricultural fields was undercut by increasing aridity that led to a significant drop in lake level (>12 m) and, most certainly, groundwater decline. More than five centuries of climatic conditions favorable for intensive, flooded raised field agriculture had generated significant population expansion and increasing socio-political complexity (Kolata 1993, 1996, 2003). With the onset of chronic drought, dense populations living in nucleated urban centers could no longer be sustained given the critical input of intensive agricultural production to the political economy of the Tiwanaku state. As a result of economic dislocation and social instability, the *altiplano* and adjacent highlands regions experienced a period of de-urbanization that persisted until the middle of the fifteenth century AD.

Just north of the *altiplano*, a limited agrarian economy existed in the valley near Cuzco (Chepstow-Lusty et al. 2009), but, as in the *altiplano*, intense aridity during the Late Intermediate Period may have forced populations still living in the highlands to move to higher elevations to obtain water from localized spring systems. Such local trends reflect more widespread movements recorded across the Andean highlands at this time. Evidence of the Inca in the circum-Cuzco region from ca. AD 1200 (although perhaps as early as AD 1000) is provided by the presence of Killke ceramics long associated with the emergence of the Inca polity (Rowe 1944; Bauer 2004). Chepstow-Lusty et al. (2009) have speculated that the early Inca were able to take advantage of the warmer climate to exploit higher altitude agro-ecological zones and adapt their irrigation technologies to harness the year-round melt water resources. The scale of anthropogenic manipulation and transformation of the landscape in the south-central Andes may have increased after ca. AD 1100 in response to the warm and essentially stable climate. This is inferred from an apparent increase in agroforestry as indicated by the spread of fast-growing *Thisa* tree species, which can flourish in degraded soils because of its nitrogen-fixing properties (Chepstow-Lusty & Winfield 2000). Landscape manipulation entailed the construction of high altitude terraces to extract maximum benefit from glacial melt and limited rainfall. However, intensive agriculture based on flooded raised fields permanently disappeared from the Andean archaeological record.

Conclusions

The location of the Quelccaya ice cap, close to Lake Titicaca and to the center of the Tiwanaku civilization, allows for unique, high-resolution insights into climate variability during the rise and dissolution of this Andean state. The ice-core accumulation record, as well as the independently derived archive of lake sediment cores, is consistent with the histories of pre-Hispanic cultures. In the case of the Tiwanaku state, the population and polity thrived on the *altiplano* during a period of higher than average precipitation that permitted extensive raised field agriculture. The demise of highland cultures during the early twelfth century was contemporaneous with the onset of long-term drought, which persisted for several hundred years until the emergence of the Inca Empire. Although climate is not the exclusive driver of the fortunes and misfortunes of civilizations, given its pivotal effect on land and resources, it surely plays an important role.

References

Abbott, M. B., M. W. Binford, M. Brenner, and K. R. Kelts. 1997. A 3500 ¹⁴C yr high-resolution sediment record of lake level changes in Lake Titicaca, Bolivia/Peru. *Quaternary Research* 47: 169–180.

Arkush, E. 2008. War, causality, and chronology in the Titicaca Basin. *Latin American Antiquity* 19(4): 339–373.

Arkush, E. 2011. *Hillforts of the Ancient Andes: Colla Warfare, Society, and Landscape.* Gainesville: University Press of Florida.

Arkush, E., and T. Tung. 2013. Patterns of war in the Andes from the Archaic to the Late Horizon: insights from settlement patterns and cranial trauma. *Journal of Archaeological Research* 21(4): 307–369.

Bauer, B. S. 2004. *Ancient Cuzco: Heartland of the Inca.* Austin: University of Texas Press.

Bauer, B. S., and L. C. Kellett. 2010. Cultural transformations of the Chanka Heartland (Andahuaylas, Peru) during the late intermediate period (AD 1000–1400). *Latin American Antiquity* 21(1): 87–111.

Binford, M. W., A. L. Kolata, M. Brenner, J. W., M. T. Seddon, M. Abbott, and J. H. Curtis. 1997. Climate variation and the rise and fall of an Andean civilization. *Quaternary Research* 47: 235–248.

Branch, N. P., R. A. Kemp, B. Silva, F. M. Meddens, A. Williams, A. Kendall, and C. V. Pomacanchari. 2007. Testing the sustainability and sensitivity to climatic change of terrace agricultural systems in the Peruvian Andes: a pilot study. *Journal of Archeological Science* 34: 1–9.

Calaway, M. J. 2005. Ice-cores, sediments and civilization collapse: a cautionary tale from Lake Titicaca. *Antiquity* 79: 778–790.

Chepstow-Lusty, A. J., and M. Winfield. 2000. Agroforestry by the Inca: lessons from the past. *Ambio* 29: 322–328.

Chepstow-Lusty, A. J., M. R. Frogley, B. S. Bauer, M. J. Leng, K. P. Boessenkool, C. Carcaillet, A. A. Ali, and A. Gioda. 2009. Putting the rise of the Inca Empire within a climatic and land management context. *Climate of the Past* 5: 375–388.

Couture, N.C., and K. Sampeck. 2003. Putuni: a history of palace architecture in Tiwanaku. In *Tiwanaku and its Hinterland: Archaeology and Paleoecology of an Andean Civilization, Volume II, ed.* A. L. Kolata. Washington, D.C.: Smithsonian Institution, 226–263.

Covey, R. A. 2008. Multiregional perspectives on the archaeology of the Andes during the Late Intermediate Period (c. A.D. 1000–1400). *Journal of Archaeological Research* 16: 287–338.

Janusek, J. W. 2004. Tiwanaku and its precursors: recent research and emerging perspectives. *Journal of Archaeological Research* 12: 121–183.

Janusek, J. W., and A. L. Kolata. 2004. Top–down or bottom-up: rural settlement and raised field agriculture in the Lake Titicaca Basin, Bolivia. *Journal of Anthropological Archaeology* 23: 404–430.

Kemp, R. A., N. P. Branch, B. Silva, F. Meddens, A. Williams, A. Kendall, and C. Vivanco. 2006. Pedosedimentary, cultural and environmental significance of paleosols within pre-Hispanic agricultural terraces in the southern Peruvian Andes. *Quaternary International* 158: 13–22.

Kolata, A. L. 1983. The South Andes. In *Ancient South Americans, ed.* J. Jennings. San Francisco: W. H. Freeman, 241–285.

Kolata, A. L. 1993. *The Tiwanaku: Portrait of an Andean Civilization.* Cambridge, MA: Blackwell.

Kolata, A. L. 1996. *Tiwanaku and Its Hinterland: Archaeology and Paleoecology of an Andean Civilization, Volume 1: Agroecology.* A. L. Kolata (principal author and ed.), R. McC. Adams and B. Smith (series eds.), Smithsonian Series in Archaeological Inquiry. Washington, D.C.: Smithsonian Institution.

Kolata, A. L. and C. Ortloff 1996. Agroecological perspectives on the decline of the Tiwanaku state. In *Tiwanaku and its hinterland: Archaeology and paleoecology of an Andean civilization, ed.* A. L. Kolata. Washington D.C.: Smithsonian Institution, 181–202.

Kolata, A. L. 2003. *Tiwanaku and Its Hinterland: Archaeology and Paleoecology of an Andean Civilization, Volume 2: Urban and Rural Archaeology.* A. L. Kolata (principal author and ed.), R. McC. Adams and B. Smith (series eds.), Smithsonian Series in Archaeological Inquiry. Washington, D.C.: Smithsonian Institution.

Kolata, A. L., M. W. Binford, M. Brenner, J. W. Janusek, and C. Ortloff. 2000. Environmental thresholds and the empirical reality of state collapse: a response to Erickson (1999). *Antiquity* 74: 424–426.

Magilligan, F.J., and P. S. Goldstein. 2001. El Niño floods and culture change: a late Holocene flood history for the Rio Moquegua, southern Peru. *Geology* 29(5): 431–434.

Manzanilla, L. 1997. The impact of climatic change on past civilizations: a revisionist agenda for further investigation. *Quaternary International* 43/44: 153–159.

Moseley, M. E. 1990. Structure and history in the dynastic lore of Chimor. In *The Northern Dynasties Kingship and Statecraft in Chimor, ed.* M. Rostworowski and. M. E. Moseley. Washington, DC: Dumbarton Oaks, 584.

Ortloff, C. R., and A. L. Kolata. 1993. Climate and collapse: agro-ecological perspectives on the decline of the Tiwanaku state. *Journal of Archaeological Science* 20: 195–221.

Owen, B.D. 2005. Distant colonies and explosive collapse: the two stages of the Tiwanaku diaspora in the Osmore drainage. *Latin American Antiquity* 16(1): 45–80.

Paulsen, A. 1976. Environment and empire: climatic factors in prehistoric Andean culture change. *World Archaeology* 8(2): 121–132.

Rowe, J. H. 1944. An introduction to the archaeology of Cuzco. *Papers of the Peabody Museum of American Anthropology and Ethnology* 27(2).

Shimada, I. 2000. The late prehispanic coastal states. In *Pre-Inka States and the Inka World*. Norman: University of Oklahoma Press, 49–110.

Stahl, P. 1984. On climate and occupation of the Santa Elena Peninsula: implications of documents for Andean prehistory. *Current Anthropology* 25(3): 351–355.

Sterken, M., K. Sabbe, A. Chepstow-Lusty, M. Frogley, K. Vanhoutte, E. Verleyen, A, Cundy, and W. Vyverman. 2006. Hydrological and land-use changes in the Cuzco region (Cordillera Oriental, South East Peru) during the last 1200 years: a diatom-based reconstruction. *Archiv für Hydrobiologie* 165(3): 289–312.

Thompson, L. G., E. Mosley-Thompson, and B.M. Arnao. 1984. El Niño-Southern Oscillation events recorded in the stratigraphy of the tropical Quelccaya Ice Cap, Peru. *Science* 226: 50–52.

Thompson, L. G., E. Mosley-Thompson, J. F. Bolzan, and B. R. Koci. 1985. A 1500-year record of climate variability recorded in ice cores from the tropical Quelccaya Ice Cap, Peru. *Science* 229: 971–973.

Thompson, L. G., E. Mosley-Thompson, W. Dansgaard, and P. M. Grootes. 1986. The Little Ice Age as recorded in the stratigraphy of the tropical Quelccaya ice cap. *Science* 234: 361–364.

Thompson, L. G., M. E. Davis, E. Mosley-Thompson, and K.-b. Liu. 1988. Pre-Incan agricultural activity recorded in dust layers in two tropical ice cores. *Nature* 336: 763–765.

Thompson, L. G., M. E. Davis, and E. Mosley-Thompson. 1994. Glacial records of global climate: A 1500-year tropical ice core record of climate. *Human Ecology* 22(1): 83–95.

Thompson, L.G., E. Mosley-Thompson, M. E. Davis, V. S. Zagorodnov, I. M. Howat, V. N. Mikhalenko, and P.-N. Lin. 2013. Annually resolved ice core records of tropical climate variability over the past ~1800 years. *Science* 229: 945–950.

Thompson, L.G. and M. E. Davis. 2014. An 1800-year ice core history of climate and environment in the Andes of southern Peru and its relationship with highland/lowland cultural oscillations. In *Inca Sacred Space: Landscape, Site and Symbol in the Andes*, ed. F. Meddens, C. McEwan, K. Willis, and N. Branch. London: Archetype Publications, 261–268.

Vuille, M. 1999. Atmospheric circulation over the Bolivian altiplano during dry and wet periods and extreme phases of the Southern Oscillation. *International Journal of Climatology* 19: 1579–1600.

CHAPTER 8

Thirteenth Century AD

Implications of Seasonal and Annual Moisture
Reconstructions for Mesa Verde, Colorado

DAVID W. STAHLE, DORIAN J. BURNETTE, DANIEL GRIFFIN,
AND EDWARD R. COOK

*The hypothesis that a prolonged drought across southwestern North America
in the late thirteenth century contributed to the abandonment of the region
by Ancestral Pueblo populations, ultimately including the depopulation of
the Mesa Verde region, continues to be a focus of archaeological research in
the Pueblo region. We address the hypothesis through the remeasurement of
tree-ring specimens from living trees and archaeological wood at Mesa Verde,
Colorado, to derive chronologies of earlywood, latewood, and total ring width.
The three chronology types all date from AD 480 to 2008 and were used to
separately reconstruct cool and early warm season effective moisture and total
water-year precipitation for Chapin Mesa near many of the major prehistoric
archaeological sites. The new reconstructions indicate three simultaneous
cool and early growing season droughts during the twelfth and thirteenth
centuries that may have contributed to the environmental and social factors
behind Ancestral Pueblo migrations over this sector of the Colorado Plateau.
These sustained interseasonal droughts included the "Great Drought" of
the late thirteenth century, which is estimated to have been one of the most
severe regimes of cool and early summer drought in the last 1,500 years and
coincided with the end of Puebloan occupations at Mesa Verde. The elevation
of the 30 cm isohyet of water-year precipitation reconstructed for southwestern
Colorado from the new ring-width data is mapped from AD 1276–1280 and
identifies areas where dry-land cultivation of maize may not have been
practical during the driest years of the Great Drought. There is no doubt
about the exact dating of the tree-ring chronologies, but the low sample size of
dated specimens from Mesa Verde during the late thirteenth and fourteenth
centuries contributes uncertainty to these environmental reconstructions at
the time of abandonment.*

Introduction

Modern Pueblo societies in New Mexico and Arizona have deep roots in the Four Corners region of the Colorado Plateau. Ancestral Pueblo society emerged by the early first millennium BC in southeast Utah, southwest Colorado, northwest New Mexico, and northeast Arizona. Population levels appear to have reached a maximum in southwestern Colorado during the early thirteenth century AD, and by the end of the century the Ancestral Pueblo had depopulated the region (Schwindt et al. 2016). Its subsistence base—dry-land cultivation of maize, beans, and squash, supplemented with wild foods and game—rendered Ancestral Pueblo society vulnerable to low crop yields during drought, and there is evidence linking decadal pluvials to village growth and decadal drought to village decline (Burns 1983, Barry & Benson 2010). Exact tree-ring dating of occupation sites and the tree-ring reconstruction of moisture conditions concurrent with the occupations provide, in part, the tight chronological control needed to draw close correspondences between long-ago climate and social dynamics.

There is little doubt that agricultural adaptations buffered Ancestral Pueblo society against drought, because these societies survived several intense decadal droughts during their long occupation of the Colorado Plateau. And yet the Four Corners region was fully abandoned by the Ancestral Pueblo by the close of the thirteenth century AD, when they relocated to north-central New Mexico and northern Arizona (Schwindt et al. 2016). Thus the role of the Great Drought (Douglass 1935) in Ancestral Pueblo abandonment of the Four Corners region remains a compelling focus of research, along with much that remains to be learned about past climate and these societies' social interactions (Kohler et al. 2010).

Mesa Verde, Colorado, in particular, has been an important area for dendrochronological research since the development of the tree-ring dating method in the early twentieth century (fig. 8.1). It is considered to be a type-site for tree-ring dating (Schulman 1946), including applications in dendroclimatology and dendroarchaeology. The first tree-ring dating of archaeological wood and charcoal from Mesa Verde was reported by A. E. Douglass during his development of Southwestern dendrochronology and the "Central Pueblo Chronology." Using tree-ring analysis, he dated prehistoric construction activity, identified the eventual cessation of construction, and advanced the hypothesis that prolonged drought in the late thirteenth century—the "Great Drought"—contributed to the abandonment of the region by Ancestral Pueblo populations (Douglass 1929, 1935, 1936). These ancient migrations, including the ultimate depopulation of the Mesa Verde region, continue to provide an important framework for archaeological research in the Pueblo region (see, for example, Kohler et al. 2010).

Many subsequent studies have substantiated the occurrence of severe, sustained drought over the southwestern United States during the late thirteenth

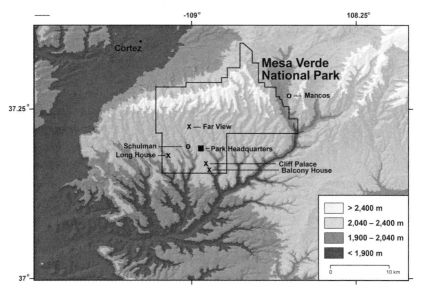

FIGURE 8.1 A digital representation of the terrain at Mesa Verde is mapped, along with the location of National Park boundary, major archaeological sites (x), selected tree-ring collection sites (o), and the headquarters weather station (black square).

century (for example, Euler et al. 1979, Dean & Funkhouser 1995, and Grissino-Mayer 1995). These are further supported by the gridded reconstructions of the Palmer drought severity index (PDSI) available in the North American Drought Atlas (Cook et al. 1999, 2004, 2007, 2010; Stahle & Dean 2011). However, the degree to which the drought may have promoted Ancestral Pueblo migrations remains unclear. Archaeological and dendrochronological evidence suggest that late thirteenth-century abandonment proceeded in an orderly fashion in the Kayenta region (Dean 2002), but appears to have been more chaotic in the Mesa Verde sector (Kohler et al. 2010). Agent-based models operating in a framework of reconstructed climate, soils, potential crop production, and wild food availability suggest that the carrying capacity of the environment was not completely exhausted during the late thirteenth century drought (see Dean et al. 2000, Kohler et al. 2005). According to these models, a reduced human population could have persisted in the region despite the extended dry conditions and, thus, drought alone cannot explain abandonment of the region.

Tree-ring based reconstructions of the maize niche in southwestern Colorado support conclusions derived from this agent-based model regarding the region's carrying capacity during the worst of the thirteenth-century droughts (Bocinsky & Kohler 2014, Schwindt et al. 2016). However, the archaeological evidence indicates that the Mesa Verde region was abandoned by the late thirteenth century, which leads Schwindt et al. (2016) to posit that earlier population increases in the region and expansion of the subsistence base into areas only marginally suitable for maize agriculture led to social stresses during subsequent dry periods that contributed to instability and eventual societal collapse. The maize niche reconstructions reported by Bocinsky and Kohler (2014) are potentially testable in the archaeological record, but several factors in the study model render it inconclusive: the use of tree-ring chronology predictors located far from southwestern Colorado, the reconstruction of a discrete temperature record (that is, growing degree days) from moisture-sensitive tree-ring chronologies that integrate positive precipitation and negative temperature signals into a single ring-width response, and the lack of seasonal estimations of climate influence on the maize growing niche.

Because most tree-ring chronologies from the Colorado Plateau are sensitive to soil moisture variations determined by precipitation amounts during the winter-spring prior to the onset of the growing season (Fritts, Smith, & Stokes 1965), the degree to which the late thirteenth-century drought may have impacted the Ancestral Pueblo subsistence system, and especially its summer crop production, has been called into question (Gladwin 1947, Burns 1983, Adams & Petersen 1999, Kohler et al. 2005). Cool-season moisture conditions currently dominate the paleoclimate component of agent-based models (Dean et al. 2000) and maize-niche reconstructions (Bocinsky & Kohler 2014) for the region, all of which are based on annual ring-width chronologies. Discrete proxies of seasonal moisture conditions are needed to estimate prehistoric agricultural potential more realistically (Wright 2010). To that end, separate tree-ring chronologies based on the springwood and summerwood components of the annual ring have recently been developed for discrete reconstructions of cool- and warm-season moisture conditions (also referred to as earlywood and latewood; Meko & Baisan 2001, Stahle et al. 2009, Griffin et al. 2011, Stahle et al. 2015).

Mesa Verde Douglas fir (*Pseudotsuga menziesii*) trees have one of the strongest climate signals ever detected in tree-ring data. The physiology underlying their sensitive climate response is examined in a now-classic paper by Hal Fritts, David Smith, and Marvin Stokes (1965). The superb dating properties of the Mesa Verde Douglas fir have often been used to demonstrate the exact cross-synchronization of ring-width time series, which can be achieved with meticulous microscopic comparisons using the crossdating technique that Douglass developed (1939, 1941). Schulman (1946) first documented the strong correlation between annual growth rings of Mesa Verde Douglas fir and winter-spring precipitation.

The annual growth rings of Douglas fir and other conifers reflect the climates of the previous autumn, winter, and spring—mostly prior to the summer season of cambial activity and radial growth at Mesa Verde and elsewhere on the Colorado Plateau (Fritts, Smith, & Stokes 1965). Consequently, these annual ring-width chronologies do not provide an unambiguous proxy of growing season climate conditions. Maize cultivation on the arid and relatively high-elevation Colorado Plateau can be constrained by low soil moisture, cool temperatures, and the length of the growing season (Hack 1942, Burns 1983, Petersen 1994). Winter-spring moisture is needed for germination and growth, but summer precipitation can also be important for maturation and yield (Adams & Petersen 1999, Benson et al. 2013). In fact, ongoing field experiments with the Pueblo Farming Project at the Crow Canyon Archaeological Center have highlighted the importance of winter-spring and July precipitation totals to selected varieties of maize (Mark Varien 2014, personal communication). If substantiated by further research, these findings would be favorable for improved tree-ring reconstructions of maize yields and maize niche space, because the earlywood and latewood components of the annual rings in such Southwestern conifers as Douglas fir and ponderosa pine (*Pinus ponderosae*) are especially sensitive to October-May and June-July precipitation totals, respectively (see, for example, Stahle et al. 2009, 2015; Griffin et al. 2012).

The annual rings of Douglas fir include anatomically distinctive earlywood and latewood cells. Because the boundary between these components of the species is usually abrupt, earlywood and latewood width within each ring can be measured precisely, and separate subannual ring-width chronologies can be computed to reconstruct past seasonal moisture levels (Schulman 1942, 1952, Meko & Baisan 2001, Stahle et al. 2009, Griffin et al. 2011). Here we report new chronologies of earlywood, latewood, and total ring width for Mesa Verde and use them to reconstruct cool and early warm season moisture levels and water-year total precipitation for the weather recording station on Chapin Mesa near the headquarters of Mesa Verde National Park. These new reconstructions are based on measurements of living trees and archaeological wood and charcoal from the national park, and they are restricted to the strongest seasonal climate signals detectable in these Douglas-fir proxies, namely September though June effective moisture for earlywood, just May and June effective moisture for latewood (that is, adjusted latewood, see below), and water-year total precipitation for annual ring width.

To a large degree we are repeating the analyses of Fritts, Smith, and Stokes (1965), who used annual ring-width data from archaeological and living Douglas-fir specimens to infer past climate for Mesa Verde, and Smith and Nichols (1967), who developed a 1529-year composite chronology of annual ring widths from living and archaeological Douglas fir. However, we add many new collections from living trees, partial ring chronologies of earlywood and latewood width, new reconstructions of winter-spring and early summer effective moisture, and a reconstruction of water-year precipitation totals. We compare these new reconstructions from Mesa Verde with other seasonal reconstructions based on earlywood and latewood width from the region and use them to identify episodes of simultaneous cool- and warm-season moisture deficits during the two centuries prior to the depopulation of southwestern Colorado by the Ancestral Pueblo. Two decadal regimes of cool and warm season drought are identified at Mesa Verde and in the nearby Mancos River region during the early and late thirteenth century. These highly unfavorable multi-season moisture conditions may have contributed to the environmental and social stresses that culminated in the final abandonment of the region.

Data and Methods

Instrumental Climate Data

Data used in these analyses were obtained from two sources: (1) the weather station located on Chapin Mesa near the National Park headquarters where daily observations were started in 1922 (Figure 1; NOAA station ID 055531, Western Regional Climate Center, Desert Research Institute); and (2) the PRISM grid point located at 37.2°N, 108.5°W, and 2139-meter elevation, which is the point closest to the headquarters instrument site. The PRISM acronym refers to "Parameter-elevation Regressions on Independent Slopes Model" (Daly et al. 2002). The PRISM model essentially distributes the irregular and often sparsely located instrumental measurements of precipitation and temperature across a digital terrain surface, accounting for the effects of elevation, rain shadows, temperature inversions, and other factors on the local gridded climate. The database provides grid point estimates of monthly total precipitation, maximum temperature, minimum temperature, and other climate variables with a 4 kilometer resolution. The PRISM monthly precipitation data nearest to headquarters are nearly identical to the headquarters station record on a seasonalized basis. (For example, the two series are correlated for October-June total precipitation at r = 0.99, for May-June totals of the highest quality subperiod, AD 1948–2011, at r = 0.98, and for the AD 1924–1948 subperiod at r = 0.94 and 0.93.). The number of missing daily observations used to compute the monthly data for the headquarters record increases prior to 1949, and the PRISM model estimates these compromised monthly values from other stations in the Four Corners region. Both records are representative of the mesa top environment of Mesa Verde itself, which is close to many of the major archaeological sites.

Tree-Ring Data

Over 160 tree-ring specimens obtained from living trees and archaeological wood and charcoal were re-measured to develop separate earlywood (EW), latewood (LW), and total ring (TR) width chronologies. The tree-ring specimens were all Douglas fir from Mesa Verde and were originally dated at the Laboratory of Tree-Ring Research in Tucson (Table 8.1). The collection includes archaeological samples dated by A. E. Douglass and cores from living trees dated by E. Schulman. The specimens were remeasured for EW, LW, and TR width to a precision of 0.001 millimeters using the procedures outlined in Schulman (1952) and Stahle et al. (2009). This level of measurement precision is necessary to quantify the often-minute latewood variance on microscopic rings.

The measurements of EW, LW, and TR width were screened for dating and measurement accuracy with correlation analyses among all dated specimens using the computer program COFECHA (Holmes 1983. The program ARSTAN (Cook 1985, Cook & Krusic 2005) was used to compute the detrended robust

TABLE 8.1 High-quality Measured and Dated Tree-ring Specimens

The individual tree-ring specimens used to develop the earlywood, latewood, and total ring-width chronologies for Mesa Verde, Colorado, are tabulated (identification number, inner and outer measured ring). Many additional dated tree-ring specimens from Mesa Verde are on archive at the University of Arizona Laboratory of Tree-Ring Research, of course, but here we selected all high-quality Douglas fir specimens that were relatively long, exhibited ring-width sensitivity, and covered important time periods over the past 1500 years. A few long series could not be located at the time of re-measurement. [Column 1 = sequence number; column 2 = series identification number; column 3 = first year dated and measured, column 4 = last year dated and measured] Schulman's old tree is identified as SOT01a-k (sequence numbers 60 to 70).

#	SPECIMEN	FIRST	LAST	#	SPECIMEN	FIRST	LAST
1	GP3165A	942	1118	32	MV I 029	960	1196
2	GP3 1 65B	942	1090	33	MV1029B	961	1202
3	GP3 1 66	815	1060	34	MV1050B	930	1010
4	GP3166B	816	1022	35	MV1158A	888	1023
5	GP3166C	818	979	36	MV1158B	887	1010
6	GP3780	990	1145	37	MV1159A	858	1013
7	oP378013	990	1131	38	MV1159B	857	1030
8	GP3780C	990	1130	39	MVI I 59C	857	1002
9	GP3780D	990	1130	40	MV1159D	857	1010
10	GP3782	826	1120	41	MV1172	1060	1230
11	GP4477A	623	1000	42	MV 1172E	1090	1231
12	GP4477B	623	903	43	MVI 1 72C	1060	1222
13	GP4479	734	1141	44	MV1177A	846	1040
14	GP4480	850	1150	45	MV1177B	848	1040
15	GP4481	962	1142	46	MV1337	960	1106
16	GP4482	978	1130	47	MV1337B	939	1105
17	GP4483	989	1145	48	MV1337C	935	1090
18	GP5437	947	1010	49	MVI 390	695	801
19	GP5442	969	1023	50	MV2213A	554	690
20	GP6364	860	960	51	MV2213B	554	636
21	GP6502	880	1164	52	MV2258	690	803
22	GP6505	955	1130	53	MV2320	480	599
23	GP6524	1040	1127	54	MV2320B	480	545
24	MV169A	794	880	55	RFN1001	1283	1580
25	MV169B	850	950	56	MVD1	1405	1938
26	MV267	808	1010	57	MVD2	1462	1932
27	MV286	1024	1133	58	MVD8	1550	1937
28	MV286B	1024	1133	59	MVD1O	1680	1890
29	MV525	1181	1270	60	SOTO1A	1270	1820
30	MV525B	1181	1268	61	SOTO1B	1275	1850
31	MV525C	1210	1269	62	SOTO1C	1242	1950

(continued)

TABLE 8.1 Continued

#	SPECIMEN	FIRST	LAST	#	SPECIMEN	FIRST	LAST
63	SOTO1D	1220	1953	105	MVB08A	1619	1971
64	SOTO1E	1311	1962	106	MVB08B	1536	1971
65	SOTOIF	1250	1962	107	MVB09A	1832	1971
66	SOTO1G	1227	1962	108	MVBO9B	1830	1971
67	SOTO1H	1237	1520	109	MVBIOA	1800	1971
68	SOTO1 1	1260	1957	110	MVB1OB	1789	1971
69	SOTO1J	1252	1663	111	MVB11A	1648	1971
70	SOTOIK	1234	1961	112	MVBI 1 B	1600	1971
71	SOTO2A	1420	1750	113	MVB12A	1420	1971
72	SOTO2B	1480	1800	114	MVB12B	1390	1950
73	SOTO2D	1470	1861	115	BBDO11	1638	1988
74	SOTO3A	1591	1962	116	BBDO12	1697	1989
75	SOTO3B	1622	1962	117	BBDO21	1762	1989
76	SOTO3C	1591	1954	118	13BDO22	1812	1989
77	SOTO3D	1470	1962	119	BBDO31	1600	1989
78	SOTO3E	1470	1962	120	BBDO32	1622	1989
79	SOTO3F	1420	1960	121	BBDO51	1607	1989
80	SOTO4A	1337	1960	122	BBDO52	1629	1989
81	SOTO4B	1325	1962	123	BBDO61	1600	1989
82	SOTO4C	1375	1962	124	BBDO71	1572	1989
83	SOTO4D	1321	1962	125	BBDO72	1635	1989
84	SOTO4E	1326	1920	126	BBDO8 1	1600	1989
85	SOTO5A	1472	1930	127	BBDO82	1704	1989
86	SOTO5B	1490	1880	128	BBDO91	1763	1989
87	SOTO5C	1470	1950	129	BBDO92	1810	1988
88	SOTO5D	1528	1930	130	BBD101	1390	1989
89	SOTO6A	1469	1750	131	BBD102	1399	1930
90	SOTO6B	1670	1800	132	BBDII 1	1785	1988
91	SOTO6C	1527	1950	133	BBD112	1798	1989
92	SOTO6D	1490	1960	134	SCMo 1 A	1792	2008
93	MVBO 1 A	1695	1972	135	SCMO 1 B	1797	1955
94	MVBOIB	1642	1950	136	SCMO2A	1726	2008
95	MVB03A	1631	1971	137	SCMO2B	1720	2008
96	MVB03B	1619	1971	138	SCMO4A	1798	2008
97	MVB04A	1820	1971	139	SCMO4B	1823	2008
98	MVB04B	1690	1971	140	SCMO5A	1784	2008
99	MVB05A	1825	1971	141	SCMO5B	1775	1948
100	MVB05B	1746	1971	142	SCMo6A	1537	2008
101	MVB06A	1630	1920	143	SCMO6B	1520	2008
102	M VE106B	1700	1972	144	SHMO2A	1434	2008
103	MVBO7A	1644	1972	145	SHMIOB	1585	2008
104	MVBo7B	1597	1971	146	SHMO3A	1693	2008

(continued)

TABLE 8.1 Continued

#	SPECIMEN	FIRST	LAST	#	SPECIMEN	FIRST	LAST
147	SHMO3B	1701	2008	157	SHM13A	1479	2008
148	SHMO6A	1812	2008	158	SHM15A	1587	2008
149	SHMO6B	1623	1834	159	SHM15B	1515	2008
150	SHMO7A	1531	2008	160	SHM16A	1890	2008
151	SHMO7C	1740	2008	161	SHM16B	1890	2008
152	SHMO8A	1630	2008	162	SHM17A	1818	2008
153	SHMO9A	1854	2008	163	SHM17B	1818	2008
154	SHMO9B	1854	2008	164	SHM20A	1544	2008
155	SHM10A	1609	2008	165	SHM2 1A	1839	2008
156	SHMIOB	1609	2008	166	SHM 21B	1587	2008

mean index standard chronologies of EW, LW, and TR width. No dating or measurement errors were discovered in this exceptionally sensitive tree-ring collection, although the LW width series are not as strongly cross-correlated as the EW or TR width series. For the EW, LW, and TR series, the mean correlation (RBAR) statistics from ARSTAN are 0.71, 0.42, and 0.70, respectively. The coherence among these Douglas-fir series is exceptionally strong for EW and TR width. The mean correlation among the LW series is not as strong as might be typical of arid site TR width collections, but that is a very high standard indeed and their agreement is still highly significant and useful for paleoclimatic reconstruction.

For this analysis, we detrended each measured time series twice, first fitting a negative exponential curve to remove the non-climatic growth trend due to the increasing age and size of the trees, and then a cubic smoothing spline with a 50 percent frequency response equal to 67 percent of the series length to minimize poor curve fitting effects especially near the beginning and end of the time series. This detrending prescription was used for all three data types (EW, LW, TR width) and preserves more low frequency variance in the longer series. The numerical ring-width measurements and derived chronologies, along with the seasonal climate reconstructions, are all provided at http://www.uark.edu/dendro/MVdata.xisx.

To remove the strong correlation between the derived EW and LW standard chronologies (r = 0.81; AD 480–2008), we regressed the LW width chronology on the EW width chronology using robust regression to discount outliers (using standard chronologies in both cases). The residuals from the regression of LW on EW were then used to represent the early warm season climate independent of the physiological persistence of growth within the annual ring [that is, the adjusted LW (LW_{adj}); Meko & Baisan 2001; Stahle et al. 2009]. We used correlation analysis to model the EW and LW_{adj} response to monthly moisture variables and to identify the optimal seasons for climate reconstruction.

The absolute tree-ring dating of Mesa Verde ruins was not achieved using long-lived trees local to the mesa. Rather the ruins were initially dated with reference to the Central Pueblo Chronology developed from old living trees, colonial era building timbers, and prehistoric wood and charcoal which were concentrated in the vicinity of northeastern Arizona and northwestern New Mexico (Douglass 1935). The famous temporal gap between the chronology based on the living trees and the floating chronology compiled from the archaeological specimens was finally closed with certainty with wood recovered near Showlow, Arizona (Douglass 1935, Haury 1962). Old living trees that actually date back over 700 years to the era of heavy prehistoric occupation are very rare at Mesa Verde. Only one Douglas-fir tree has been found that actually dates to the thirteenth century and crossdates with the archaeological wood on the mesa: an 800-year old individual discovered by Schulman (1947) several years after the initial dating of the ruins. Thus, the period when severe drought appears to have persisted over the Colorado Plateau and the mesa was being depopulated by Ancestral Pueblo is represented locally by tree-ring data available from only one tree. (Descriptive data on Schulman's old tree are listed in Table 8.1 as SOT01a-k).

There is no doubt regarding the absolute dating of Schulman's old tree or of the archaeological sites at Mesa Verde, of course, which all correlate with the regional master chronology first developed by Douglass and subsequently by others. But the magnitude and persistence of late thirteenth century drought at Mesa Verde itself is less certain because the sample size of the tree-ring data is so low. We have not been able to add any additional tree-ring sequences to this crucial time period, but we do provide statistical evidence that lends some confidence to the portion of the Mesa Verde Douglas-fir chronology that is represented only by Schulman's old tree (below). We also include analyses of cool- and warm-season precipitation reconstructions developed from recent remeasurements of EW and LW_{adj} at El Malpais, New Mexico (Grissino-Mayer 1995, Stahle at al. 2009), for a broader perspective on seasonal drought during the thirteenth century at Mesa Verde.

Reconstruction Methods

The seasonal climate response of the EW, LW_{adj}, and TR width chronologies was investigated using correlation analysis between the chronologies and monthly precipitation, maximum temperature, and an effective moisture index during the year prior to and concurrent with tree growth (year t-1 and t, respectively). The effective moisture index was computed for the PRISM grid point as the difference between total monthly precipitation and potential evapotranspiration, or P-PE. Potential evapotranspiration was computed from the monthly mean temperature of minimum and maximum temperatures available at the PRISM grid point using the Thornthwaite method (1948, Burnette & Stahle 2013). The monthly P-PE estimates were transformed into anomalies by subtracting the mean P-PE for the 1895–2011 period of record. Positive P-PE values represent wet and cool conditions, and negative values are dry and warm. Because one goal of this research has been to reconstruct restricted seasonal moisture variables from EW and LW$_{adj}$ chronologies, this P-PE index does not include any prescribed month-to-month persistence to model soil moisture storage.

To develop the reconstructions, the tree-ring chronologies and three lead and lagged versions were entered into forward stepwise regression as potential predictors of October-June P-PE for EW, May-June P-PE for LW$_{adj}$, and October-September precipitation for TR width. Autoregression in the predictor and predictand time series during the calibration period was identified using the corrected Aikaike Information Criteria (AICc; Cook et al. 1999), and any persistence in the time series was removed prior to calibration. The derived reconstructions were tested against independent climate data withheld from the calibration, and the degree of fit between observed and estimated data was evaluated graphically and with the Pearson correlation (r), reduction of error (RE), and coefficient of efficiency (CE) statistics (Cook et al. 1999). To estimate the amount of variance in effective moisture that might be represented during the poorly replicated portion of the chronology in the late thirteenth and early fourteenth century, we also calibrated EW and LW$_{adj}$ chronologies based only on the eleven cores from Schulman's old tree.

Results

Seasonal Moisture Response

The correlation coefficients computed between the tree-ring chronologies and monthly precipitation, maximum temperature, and P-PE are plotted in figure 8.2. EW width is significantly and positively correlated with precipitation during the winter-spring season preceding tree growth from September through May (fig. 8.2a) and is negatively correlated with monthly mean maximum temperature from September through July (fig. 8.2b). This is the same basic precipitation and temperature response previously documented for Douglas-fir EW width on the Colorado Plateau (Cleaveland 1983, 1986; Stahle et al. 2009) and for TR width at Mesa Verde (Fritts et al. 1965) and can be represented by a single variable that measures effective moisture, the P-PE index. Because the EW chronology is positively correlated with monthly P-PE continuously from September through June (fig. 8.2c), the effective moisture signal averages the late spring-early summer precipitation and temperature responses (ending in May and July, respectively).

The cool season moisture response of EW width at Mesa Verde is nearly identical to the response of total ring width (fig. 8.2), simply because EW represents the major fraction of annual ring width (that is, mean EW width = 0.348 mm, mean LW width = 0.111 mm, mean TR width = 0.459 mm, or 76 and 24 percent on average for EW and LW, respectively). Chronologies of EW and TR width can both be used to develop verifiable reconstructions of water-year precipitation or P-PE totals at Mesa Verde. But the strongest climate signal for EW is confined to moisture variation during the winter-spring season (fig. 8.2c), largely prior to the onset of radial growth that may begin in March, but 90 percent of which occurs only in May for Mesa Verde Douglas-fir (Fritts et al. 1965). Thus the conclusion that "climate previous to the growing season has a greater effect on ring width than climate during the growing season" (Fritts, Smith, & Stokes 1965) applies to both EW and TR width. The LW_{adj} chronology, on the other hand, is significantly correlated with climate early in the growing season concurrent with the formation of LW cells (fig. 8.2def).

The response of the raw LW chronology, prior to the adjustment of LW width for dependency on EW width, is also dominated by climate during the winter-spring, well before the formation of LW cells (not shown). The regression of the LW chronology on the EW chronology is necessary to obtain a direct estimate of early growing season climate conditions not heavily influenced by preceding climate and tree growth (EW) variables. The correlation between the LW_{adj} chronology and monthly precipitation, maximum temperature, and P-PE values indicates that the adjustment has worked reasonably well. Adjusted LW is significantly and positively correlated with precipitation and P-PE only in the months of May and June concurrent with tree growth, and it is negatively correlated with maximum temperature only in June and July (fig. 8.2def).

The monthly moisture response of EW does overlap with the response of LW_{adj} in May and June, depending on which climate variable is being considered (fig. 8.2). However, the two series are not strongly correlated (r = 0.13, AD 480–2008); the EW chronology is dominated by climate prior to the onset of growth, and the LW_{adj} is dominated by climate concurrent with growth in the May-June period. The two partial ring chronologies therefore provide reasonably discrete estimates of cool and early warm season moisture for Mesa Verde.

Calibration and Verification of the Seasonal and Water-Year Moisture Reconstructions

The EW chronology for Mesa Verde Douglas-fir and six lead and lagged versions (the robust mean standard EW width chronology in years *t-3, t-2, t-1, t, t+1, t+2, t+3*) were submitted to forward stepwise regression to estimate Sept-June P-PE for the PRISM grid point nearest to headquarters at Mesa Verde National Park. The potential predictors were screened for correlation with Sept-June P-PE (*p* < 0.10) and only the EW chronology in year *t* was selected—that is, the chronology representing EW growth beginning in May near the end of the Sept-June P-PE season). The resulting bivariate regression model used to estimate Sept-June effective moisture was:

$$\hat{Y}_t = -9.059 + 10.937X_t \qquad (1)$$

where $\hat{y}_t$ is reconstructed P-PE in year *t* for the Sept-June season and X_t is the EW width chronology also in year *t*. Prior to the regression, autoregressive modeling detected no persistence in either variable based on the AICc during

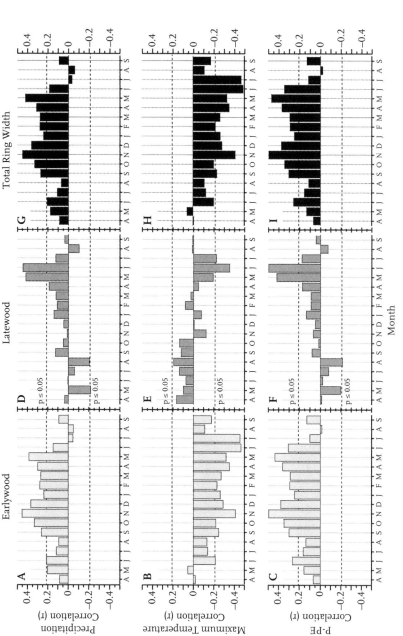

FIGURE 8.2 Correlations of the earlywood width chronology with monthly precipitation (*a*), maximum temperature (*b*), and P-PE (*c*) for the year prior and concurrent with tree growth (prior April through current September), using the PRISM grid point closest to the headquarters rain gage at Mesa Verde, National Park (1923–2008). Adjusted LW width is also correlated with monthly precipitation (*d*), maximum temperature (*e*), and P-PE (*f*) for the same PRISM grid point. Total ring width (TRW) is correlated with these same variables for the water year (October–September; *g*, *h*, and *i*).

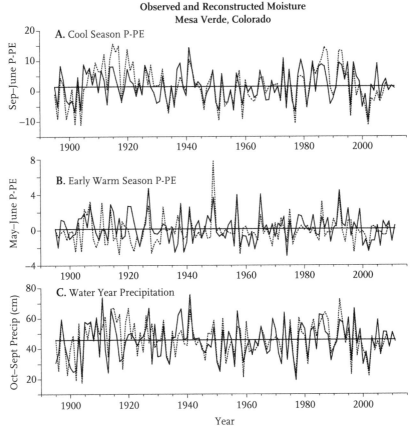

Observed and Reconstructed Moisture
Mesa Verde, Colorado

A. Cool Season P-PE

B. Early Warm Season P-PE

C. Water Year Precipitation

Year

FIGURE 8.3 (*a*) The observed (solid) and reconstructed (dashed) cool-season moisture balance values (P-PE) are plotted for the Sept-June season at the PRISM grid location closest to the headquarters weather station at Mesa Verde National Park (based on EW width). The calibration period was AD 1923–1966 and the verification period was 1967–2008. The EW chronology is also correlated with the PRISM P-PE at the Mesa Verde grid location from 1895–1922, though not as well as after 1922, when local weather observations began at the headquarters station. (*b*) Same as (*a*) for the early warm season (May-June, based on adjusted LW width). The LW$_{adj}$ is not correlated with the PRISM-based P-PE before 1923, the year observations began at the headquarters weather station at Mesa Verde. (*c*) Same as (*a*) for water-year precipitation (October-September) reconstructed from TRW.

the short calibration interval, 1923–1966 (Cook et al. 1999). The EW chronology explains nearly two-thirds of the variance in the Sept-June effective moisture index during the 1923–66 calibration period (coefficient of determination, adjusted for the loss of one degree of freedom, $R^2_{adj} = 0.648$) and the residuals from regression do not exhibit serial correlation or other serious deficiencies (see Durbin-Watson statistic, Table 8.2). The observed and reconstructed P-PE

TABLE 8.2 Calibration and Verification Statistics
Calibration and verification statistics computed for the reconstruction of cool (September-June) and early-summer (May-June) moisture balance (P-PE), and water year precipitation totals (October-September) at Mesa Verde, Colorado. The calibration interval is listed first (e.g., 1923–1966), followed by the verification interval (e.g., 1967–2008) for each reconstruction. The coefficients of the regression models, the variance explained (R^2_{adj} = coefficient of determination adjusted downward for the loss of degrees of freedom), the standard error of the estimates (SE), and the Durbin-Watson statistic (DW) are listed for each reconstruction. The Pearson correlation coefficient, comparing reconstructed with instrumental P-PE data during the statistically independent verification periods, is shown for the reconstructions, along with the reduction of error (RE) and coefficient of efficiency (CE) statistics calculated on observed and reconstructed data in the verification period. All tests indicate successful verification, although the early-summer reconstruction based on LW_{adj} is considerably weaker than the EW reconstruction of the September-June moisture balance or the TRW estimate of water year precipitation.

TIME PERIOD	COEFFICIENTS		CALIBRATION			VERIFICATION		
Sept–June	b_0	b_1	R^2_{adj}	SE	DW	r	RE	CE
1923–1966	−9.06	10.94	0.64	2.79	1.93			
1967–2008						0.79	0.63	0.62
May–June	b_0	b_1	R^2_{adj}	SE	DW	r	RE	CE
1923–1966	−4.83	2.48	0.37	1.40	2.12			
1967–2008						0.48	0.21	0.20
Oct–Sept	b_0	b_1	R^2_{adj}	SE	DW	r	RE	CE
1949–2008	0.00	9.40	0.65	2.69	2.03			
1896–1948						0.57	0.26	0.26

indices for the September-June season are plotted in figure 8.3a from 1923–2008, and observed and reconstructed values are significantly correlated during the 1967–2008 validation period at r = 0.79 (p <0.0001, Table 8.2). The RE and CE statistics indicate that the reconstruction is reproducing approximately 61 percent of the instrumental variance when compared with independent P-PE indices from 1967–2008.

The Douglas-fir adjusted LW chronology was submitted to a forward stepwise regression to estimate the PRISM May-June effective moisture (with leads and lags) and once again only the LW_{adj} chronology in year t passed the screening. Thus, a bivariate regression model was also used to estimate May-June P-PE:

$$\hat{Y}_t = -4.831 + 2.479X_t \tag{2}$$

where $\hat{Y}_t$ is reconstructed P-PE in year t for the May-June season and X_t is the adjusted LW width chronology also in year t. No autoregressive structure was detected in either regression variable, but in this case the LW_{adj} only explains 37 percent of the variance during the calibration period ($R^2_{adj} = 0.372$, 1923–1966; Table 8.2). The relationship between LW_{adj} and May-June P-PE is also subject to some temporal instability (figure 8.3b), and this is evident in the validation statistics where the observed and reconstructed values are correlated at r = 0.483 from 1967–2008, and the RE and CE indicate only modest validation skill (0.21 and 0.20, respectively; Table 8.2). The LW_{adj} chronology calibrates and verifies somewhat better against May-June precipitation totals (for example, $R^2_{adj} = 0.43$ and RE = 0.28 for 1922–1966), but there is still considerable temporal variability in the relationship, and the model estimates some negative May-June precipitation totals. The P-PE reconstruction from LW_{adj} was therefore deemed better in overall performance.

May and June are two of the driest months on average at Mesa Verde and precipitation during this season may be spatially discontinuous. For this reason it can be more difficult to calibrate LW_{adj} chronologies against station or regional average precipitation data (for example, Stahle et al. 2009). But some of the noise in the relationship between May-June moisture levels and the LW_{adj} chronology at Mesa Verde must arise from the strong correlation between EW and raw unadjusted LW width (r = 0.81, AD 480–2008). When this much shared variance is removed in regression, in this case presumably due to the common winter-spring climate signal and physiological persistence in growth, some of the residual variance must be random noise unrelated to climate. That some 20 to 37 percent of the variance in May-June moisture levels can be recovered from the adjusted LW chronology at Mesa Verde is interesting in light of the very strong common signal between EW and raw LW width.

The Douglas-fir TRW chronology was also submitted to a forward stepwise regression to estimate water-year precipitation totals (October-September, with leads and lags) based on the calibration period of 1949–2008. With only the TRW chronology in year t passing the screening, the following regression model was used to estimate water-year precipitation:

$$\hat{Y}_t = -0.004 + 9.40X_t \tag{3}$$

where $\hat{Y}_t$ is reconstructed water-year precipitation in year t and X_t is the TR width chronology also in year t. No autoregressive structure was detected in either regression variable during the calibration period. The TRW chronology explains 65 percent of the variance in water-year precipitation during the calibration period, but the verification statistics indicate that observed and reconstructed precipitation are not as tightly coupled during the period 1896–1948 (for example, r = 0.57; Table 8.2).

These calibration and verification results are based on well-replicated twentieth century tree-ring chronologies. However, several cores from only one tree represent the chronologies for the period in the late thirteenth century when Ancestral Pueblo left Mesa Verde (that is, 1271–1282). It is therefore reasonable to question the strength of the seasonal and water-year response of this single old Douglas-fir tree. Our analyses indicate that the climate signals in Schulman's single old tree are quite good. Seasonal precipitation totals instead of P-PE and different calibration and verification periods had to be used to derive the best models during the 1895–1962 period in common to both the Schulman chronologies and the PRISM climate data. Nevertheless, the EW chronology from Schulman's old tree explains 56 percent of the October-May precipitation, the LW_{adj} chronology explains 25 percent of the May–June precipitation, and the TRW chronology explains 58 percent of the October-September precipitation (all calibrated from 1940 to 1962). Verification of the reconstructions during the 1895–1940 period is only marginal, but it is still extraordinary that EW, LW_{adj}, and TR width data from only one tree can successfully estimate cool, water-year, and—to a lesser extent—even early warm-season moisture levels. In fact, the two reconstructions of cool season moisture are correlated at r = 0.93 from 1220 to 1962 (that is, September-June P-PE versus October-May total precipitation during the period covered by our EW data derived from Schulman's old tree) and the two early warm-season reconstructions, at r = 0.77 (that is, May-June P-PE versus May-June total precipitation, 1220–1962). Even during the 162-year period from 1801–1962 when Schulman's old tree was really old (over 600 years), the two reconstructions are still well correlated at r = 0.93 and 0.66 for the cool and early warm seasons, respectively.

Analysis and Discussion

Seasonal Moisture Reconstructions for Mesa Verde

The EW reconstruction of September-June effective moisture is plotted along with a smoothed version emphasizing decadal regimes over the past 1529 years (480–2008, fig. 8.4a). The driest single year in the entire reconstruction was estimated for 2002, which was also the most negative P-PE value measured for the September-June period in the instrumental PRISM data from 1923–2011 (the period of operation for the nearby headquarters weather station). The worst multiyear drought episode in the September-June reconstruction is estimated to have occurred in the late sixteenth century (fig. 8.4a), when severe sustained drought conditions were present over a large fraction of western North America (see, for example, Meko et al. 1995, Stahle et al. 2000, Herweijer, Seager, & Cook 2006). The medieval era was marked by several prolonged cool season droughts at Mesa Verde, especially during the twelfth and thirteenth centuries. This 200-year period of dryness was occasionally interrupted by above average conditions, of course, but prolonged drought nevertheless prevailed and was finally terminated by a major multidecadal pluvial in the early fourteenth century (fig. 8.4a).

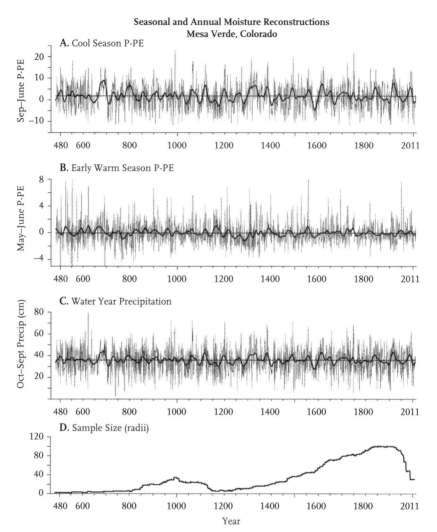

Seasonal and Annual Moisture Reconstructions
Mesa Verde, Colorado

A. Cool Season P-PE

B. Early Warm Season P-PE

C. Water Year Precipitation

D. Sample Size (radii)

FIGURE 8.4 *(a)* The cool season moisture balance reconstruction for Mesa Verde, Colorado, based on EW width is plotted from AD 480 to 2008 (September-June total P-PE in gray, 30-year smoothing in black). Note the generally below average cool season moisture conditions estimated for the twelfth, late thirteenth, late fourteenth, and late sixteenth centuries. *(b)* Same as *(a)* for the early warm season (May-June total P-PE in gray, 30-year smoothing in black). Prolonged below average early warm-season moisture conditions are reconstructed during the late-thirteenth century, but from 1271 to 1282 the cool and warm season reconstructions are both based on just nine radii from Schulman's single old tree *(d)*. Sample size is also low before 700 and may be partly responsible for the heightened variance during the first two centuries of this reconstruction. *(c)* Same as *(a)* for water-year precipitation totals based on TRW (October-September).

The annual and decadal values of the May-June effective moisture reconstruction are plotted in figure 8.4b, and the number of dated radii each year is plotted for both EW and LW in Figure 4c. The variance structure of instrumental and reconstructed May-June P-PE is substantially more skewed than Sept-June P-PE (figs. 8.3–8.4), and early summer P-PE is dominated by the episodic occurrence of very high values. The decadal variability of reconstructed May-June P-PE is also much lower than the decadal variance in the Sept-June reconstruction, in part because so much common variance between EW and LW was removed in order to adjust the LW. The most severe and prolonged May-June drought is reconstructed during the late thirteenth century, when sample size is restricted to Schulman's Old Tree, even though several individual years are estimated to have been well above average. Prolonged May-June drought is also estimated during the early thirteenth century, and during the eighth and fifteenth centuries (fig. 8.4b).

Simultaneous Cool and Early Growing Season Moisture Estimates

The cool and early warm season reconstructions were normalized, zero-centered, and smoothed to compare ca. 10-year regimes of seasonal moisture at Mesa Verde, Mancos River, Colorado (from Stahle et al. 2015), and El Malpais National Monument, New Mexico (from Stahle et al. 2009). The z-scores were computed for each reconstructed value by subtracting the median and then dividing by the inter-quartile range. These smoothed cool- and warm-season reconstructions are plotted together from 1051 to the present in figures 8.5 and 8.6, and just from 1101 to 1350 during the heavy Ancestral Pueblo occupations of southwestern Colorado in figure 8.7. The reconstructions indicate that the droughts during the early- and late-thirteenth century were among the most severe and sustained dual-season droughts of the last millennium (that is, during the 1051-2002 period in common to the three site recons; figs. 8.5 and 8.6). A multi-season megadrought during the sixteenth century likely exceeded all droughts of the past millennium, but the prolonged deficits in cool-season moisture in the late thirteenth century at all three locations were unusual in the context of the past millennium, and they were likely made worse in terms of environmental and socioeconomic impacts by the co-occurrence of shorter drought periods during the growing season.

Dual-season drought also occurred earlier in the thirteenth century over southwestern Colorado and west-central New Mexico (fig. 8.7). Dual-season drought began in the 1210s at Mancos River and El Malpais, developed in the 1220s, and persisted into the 1240s at Mesa Verde. (The sample size is low at Mesa Verde during this period.) Note especially the large magnitude warm-season droughts reconstructed at Mancos River in the 1210s, 1220s, and 1250s (fig. 8.7). These warm season extremes at Mancos River were matched by decadal excursions at El Malpais, but with somewhat different timing and intensity. Prolonged dual-season drought is reconstructed from 1170 to 1190 at

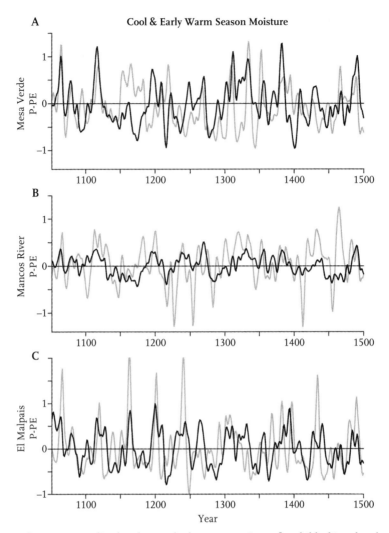

FIGURE 8.5 *(a)* Normalized and smoothed reconstructions of cool (black) and early warm season (gray) moisture conditions are plotted from AD 1051 to 1500 for Mesa Verde *(a)*, Mancos River *(b)*, and El Malpais *(c)* [i.e., Sept-June and May-June for Mesa Verde, Sept-May and June-July for Mancos, and Nov-May and July for El Malpais; spline smoothing to highlight 10-year variability (Cook and Peters 1981)].

El Malpais during a period of contrasting seasonal moisture regimes in southwestern Colorado (fig. 8.7). Finally, as figure 8.7 shows, the early fourteenth century was general wetter in both seasons, with the exception of the 1320s at Mesa Verde, which is represented only by Schulman's old tree.

The smoothed moisture reconstructions from Mesa Verde, Mancos River, and El Malpais are more consistent during the cool season (figs. 8.5–8.7). The early warm-season estimates are less coherent due in part to low sample

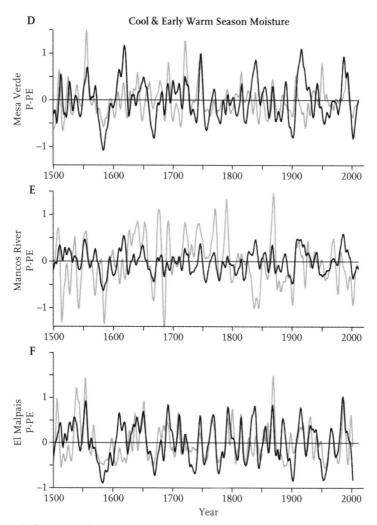

FIGURE 8.6 Same as fig. 8.5 for the period 1501-present for Mesa Verde *(d)*, Mancos River *(e)*, and El Malpais *(f)*.

size at Mesa Verde, slight differences among the sites in the monthly climate response of adjusted LW width, and the fact that a large proportion of shared variance was removed from the LW_{adj} chronologies when they were regressed against EW width. Additional warm-season moisture proxies will be needed from the region to improve the estimates of growing season moisture. But the reconstructions all indicate that the thirteenth century was marked by at least two episodes of simultaneous drought during the cool- and early-warm season over southwestern Colorado and northwestern New Mexico.

The new reconstructions generally do not indicate that the late thirteenth century experienced drought in the cool season, but wetness in the summer,

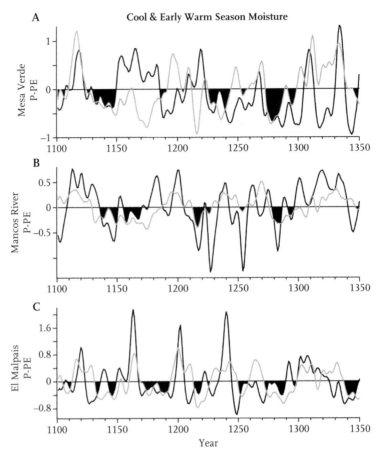

FIGURE 8.7 Similar to Figure 5, but here the normalized and 10-year smoothed reconstructions of cool and early warm-season moisture conditions are plotted only from AD 1101 to 1350 for Mesa Verde *(a)*, Mancos River *(b)*, and El Malpais *(c)*. Note that the cool season is plotted here in gray, warm season in black, and the episodes of simultaneous cool- and warm-season drought are shaded black (i.e., normalized P-PE or PPT for both the cool and warm season are ≤ 0.00).

a hypothesized climate scenario that may have ameliorated to some extent the impacts of prolonged cool season drought on native subsistence systems (Gladwin 1947). Fritts, Smith, & Stokes (1965) also did not find support for a strongly out-of-phase pattern of cool and warm season moisture during the Great Drought of the late thirteenth century. But their inferences and ours do not entirely rule out the wet summer hypothesis for the late thirteenth century, because full summer climate conditions (JJAS) are not represented in our reconstructions based on adjusted LW width, and the summer signal detected by Fritts, Smith, and Stokes cannot be isolated from the suite of previous summer, autumn, winter, and spring climate factors that explain annual ring width in their model. Figure 8.7 does indicate that most of the thirteenth century was relatively dry during May-June at Mesa Verde, June-July at Mancos River, and July at El Malpais. Thus, to the extent that growing season moisture was important to the subsistence base of the Ancestral Pueblo at Mesa Verde, reconstructed warm-season conditions were unlikely to have mitigated winter-spring drought effects during the late thirteenth century. In fact, some of the most extremely dry warm-season conditions are estimated during the thirteenth century in the Mesa Verde-Mancos River area, including one of the worst episodes of simultaneous cool- and warm-season drought reconstructed during the 250-year period from 1101 to 1350 or during the entire period covered by these dual season reconstructions (see also Stahle et al. 2015). These unfavorable seasonal moisture conditions likely contributed to the growing instability in Ancestral Pueblo societies in southwestern Colorado. The decline in regional populations had apparently begun well before the late thirteenth-century dual season drought, perhaps in partial response to simultaneous cool- and warm-season droughts over the region in the early thirteenth century (fig. 8.7).

Finally, the reconstructions in figures 8.4a, b, and c do not indicate that the thirteenth-century drought was exceeded by at least ten other droughts over the past 1500 years, as was posited by Fritts, Smith, and Stokes, based on five-year running means of the total ring-width index of selected archaeological and modern Douglas fir. This is particularly true when we consider simultaneous cool- and warm-season drought at Mesa Verde, Mancos River, and El Malpais (figs. 8.5–6). By these measures, the thirteenth century appears to have been one of the most persistently dry warm-season regimes, with some of the most intense episodes of simultaneous cool and early summer drought of the last 1000 years. Dual season drought conditions over the Mesa Verde region appear to have been equaled or exceeded by simultaneous droughts only in the mid-seventeenth, late nineteenth, and early twenty-first centuries, and especially during the late sixteenth-century megadrought.

Conclusions

The new seasonal moisture reconstructions indicate that the "Great Drought" of the late thirteenth century included severe and sustained moisture deficits during both the cool and early warm season. They also indicate that the Great Drought was preceded by dual-season drought conditions in the early- to mid-thirteenth century and that dry conditions in the early warm season prevailed over a large portion of the Four Corners during the entire century. The archaeological record indicates that immigration and depopulation were multidecadal processes at Mesa Verde and culminated in complete abandonment by the close of the thirteenth century (Varien 2010, Schwindt et al. 2016). The co-occurrence of cool and early warm-season drought during the thirteenth century may have contributed to the environmental and social stresses that stimulated these migrations among Ancestral Pueblo.

The new cool and early warm-season moisture reconstructions could yield improved estimates of crop yields and the maize niche over the Mesa Verde region (Kohler 2010, Bocinsky & Kohler 2014, Schwindt et al. 2016). Douglas-fir LW chronologies at Mesa Verde, Mancos River, and El Malpais are not correlated with precipitation or temperature during August or September (fig. 8.2), but selected LW chronologies of ponderosa pine (*Pinus ponderosa*) do respond to late summer precipitation elsewhere in the Southwest (Griffin 2013). Development of ponderosa pine LW chronologies spanning the period of Puebloan occupations is a possibility (for example, Guiterman et al. 2016) and could improve reconstructions of regional summer moisture considerably.

Finally, the chronological gap in the late thirteenth and early fourteenth century that was so problematic during the development of Southwestern dendrochronology (Douglass 1935, Haury 1962) remains a serious replication issue impacting the paleoclimatic inferences on occupation and migration that can be derived from these outstanding tree-ring proxies. The long tree-ring chronologies developed from living trees and subfossil wood at El Malpais and Mancos River are proof that replication for this important episode in environmental and Ancestral Pueblo history can be improved. Given the importance of environmental variability to archaeological analyses of Ancestral Pueblo migrations and the very sparse tree-ring record presently available during key time periods at Mesa Verde, we still need to develop exactly dated, well-replicated 800-year-long tree-ring chronologies from multiple species better to understand the environmental conditions during the occupation and depopulation of the region.

Acknowledgements

Funding for this research was provided by the National Oceanic and Atmospheric Administration CCDD grant number NA08OAR4310727, NSF award number 0823090, EPA STAR Fellowship #FP917185 (Griffin). Steve Nash, Mr. Blake Mitchell, Grant West, Chris Baisan, Jesse Edmondson, Daniel Stahle, George San Miguel, Marilyn Colyer, and Tom Wolfe provided us assistance. We thank the University of Arizona Laboratory of Tree-Ring Research, Jeff Dean, and Tom Swetnam for access to modern and archaeological tree-ring collections from Mesa Verde.

References

Adams, K. R., and K. L. Petersen. 1999. Environment. In *Colorado Prehistory: A Context for the Southern Colorado River Basin,* ed. W. D. Lipe, M. D. Varien, and R. H. Wilshusen. Denver: Colorado Council of Professional Archaeologists, 4–50.

Barry, M. S., and L. V. Benson. 2010. Tree-ring dates and demographic change in the southern Colorado Plateau and Rio Grande regions. In *Leaving Mesa Verde: Peril and Change in the Thirteenth-Century Southwest*, ed. T.A. Kohler, M.D. Varien, and A.M. Wright. Amerind Studies in Archaeology, Vol. 5. Tucson: University of Arizona Press, 53–74.

Benson, L. V., D. K. Ramsey, D. W. Stahle, and K. L. Petersen. 2013. Some thoughts on the factors that controlled prehistoric maize productivity in the American Southwest with application to southwestern Colorado. *Journal of Archaeological Science* 40: 2869–2880.

Bocinsky, R. K., and T. A. Kohler. 2014. A 2000-year reconstruction of the rain-fed maize agricultural niche in the US Southwest. *Nature Communications* 5: 5618.

Burnette, D. J., and D. W. Stahle, 2013. Historical perspective on the Dust Bowl drought in the central United States. *Climatic Change* 116:479–494.

Burns, B. T. 1983. "Simulated Anasazi storage behavior using crop yields reconstructed from tree rings: AD 652–1968." PhD diss. Tucson: University of Arizona.

Cleaveland, M. K. 1983. "X-Ray Densitometric Measurement of Climatic Influence on the Intra-Annual Characteristics of Southwestern Semiarid Conifer Tree Rings." PhD diss. Tucson: University of Arizona.

Cleaveland, M. K. 1986. Climatic response of densitometric properties in semiarid site tree rings. *Tree-Ring Bulletin* 46:13–29.

Cook, E. R. 1985. "A Time Series Analysis Approach to Tree Ring Standardization." PhD diss. Tucson: University of Arizona. https://arizona.openrepository.com.

Cook, E. R., and K. Peters. 1981. The smoothing spline: a new approach to standardizing forest interior ring-width series for dendroclimatic studies. *Tree-Ring Bulletin* 41:4–53.

Cook, E. R., D. M. Meko, D. W. Stahle, and M. K. Cleaveland. 1999. Drought reconstructions for the continental United States. *Journal of Climate* 12:1145–1162.

Cook, E. R., C. A. Woodhouse, C. M. Eakin, D. M. Meko, and D. W. Stahle. 2004. Long-term aridity changes in the western United States. *Science* 306:1015–1018.

Cook, E. R., and P. J. Krusic. 2005. "Program ARSTAN: A Tree-Ring Standardization Program Based on Detrending and Autoregressive Time Series Modeling, with Interactive Graphics." Manuscript on file, Tree-Ring Lab, Lamont Doherty Earth Observatory of Columbia University, Palisades, NY.

Cook, E. R., R. Seager, M. A. Cane, and D. W. Stahle. 2007. North American drought: reconstructions, causes, and consequences. *Earth Science Reviews* 81:93–134.

Cook, E. R., R. Seager, R. R. Heim, R. S. Vose, C. Herweijer, and C. W. Woodhouse. 2010. Megadroughts in North America: placing IPCC projections of hydroclimatic change in a long-term paleoclimate context. *Journal of Quaternary Science* 25(1): 48–61.

Daly, C., W. P. Gibson. G. H. Taylor, G. L. Johnson, and P. Pasteris. 2002. A knowledge-based approach to the statistical mapping of climate. *Climate Research* 22:99–113.

Dean, J. S., and G. S. Funkhouser. 1995. Dendroclimatic reconstructions for the southern Colorado Plateau. In *Climate Change in the Four Corners and Adjacent Regions: Implications for Environmental Restoration and Land-Use Planning*, ed. W. J. Waugh. Grand Junction, CO: U.S. Department of Energy, Grand Junction Projects Office.

Dean, J. S., G. J. Gumerman, J. M. Epstein, R. L. Axtell, A. C. Swedlund, M. T. Parker, and S. McCarroll. 2000. Understanding Anasazi culture change through agent-based modeling. In *Dynamics in Human and Primate Societies,* ed. T. Kohler and G. Gumerman. Oxford, UK: Oxford University Press.

Dean, J. S. 2002. Late Pueblo II-Pueblo III in Kayenta-Branch Prehistory. In *Prehistoric Culture Change on the Colorado Plateau,* ed. S. Powell and F. E. Smiley. Tucson: University of Arizona Press.

Douglass, A. E. 1929. The secret of the Southwest solved by talkative tree-rings. *National Geographic Magazine* 56(6): 736–770.

Douglass, A. E. 1935. *Dating Pueblo Bonito and Other Ruins of the Southwest.* Contributed Technical Papers, Pueblo Bonito Series No. 1. Washington, DC: National Geographic Society.

Douglass, A. E. 1936. The central Pueblo chronology. *Tree-Ring Bulletin* 2(4): 29–34.

Douglass, A. E. 1939. Typical site of trees producing the best crossdating. *Tree-Ring Bulletin* 6(2): 10–11.

Douglass, A. E. 1941. Crossdating in dendrochronology. *Journal of Forestry* 39: 825–831.

Euler, R. C., G. J. Gumerman, T. N. V. Karlstrom, J. S. Dean, and R. H. Hevly. 1979. The Colorado Plateaus: Cultural dynamics and paleoenvironment. *Science* 205: 1089–1101.

Fritts, H. C., D. G. Smith, and M. A. Stokes. 1965. The biological model for paleo-climatic interpretation of Mesa Verde tree-ring series. *American Antiquity* 31(2): 101–121.

Gladwin, H. S. 1947. Tree-ring analysis, tree rings and drought. *Medallion Papers* No. 37. Gila Pueblo, AZ: Globe.

Griffin, R. D., D. M. Meko, R. Touchan, S. W. Leavitt, and C. A. Woodhouse. 2011. Latewood chronology development for summer-moisture reconstruction in the US Southwest. *Tree-Ring Research* 67(2): 87–102.

Griffin, R. D., C. A. Woodhouse, D. M. Meko, D. W. Stahle, H. L. Faulstich, C. Carrillo, R. Touchan, C. Castro, and S. W. Leavitt. 2012. North American monsoon precipitation reconstructed from tree-ring latewood. *Geophysical Research Letters* 40:1–5. doi:10.1002/grl.50184.

Griffin, R. D. 2013. "North American monsoon paleoclimatology from tree rings." PhD diss. Tucson: University of Arizona. https://arizona.openrepository.com/arizona/handle/10150/301558

Grissino-Mayer, H. D. 1995. "Tree-ring reconstructions of climate and fire history at El Malpais National Monument, New Mexico." PhD diss. Tucson: University of Arizona.

Guiterman, C. H., T. W. Swetnam, and J. S. Dean, 2016. Eleventh-century shift in timber procurement areas for the great houses of Chaco Canyon. *Proceedings of the National Academy of Science* 113: 1186–1190.

Hack, J. T. 1942. *The Changing Physical Environment of the Hopi Indians of Arizona.* Papers of the Peabody Museum of American Archaeology and Ethnology 35(1). Cambridge, MA: Harvard University Press.

Haury, E. W. 1962. HH-39: recollections of a dramatic moment in Southwestern archaeology. *Tree-Ring Bulletin* 24(3–4):11–14.

Herweijer, C., R. Seager, and E. R. Cook. 2006. North American droughts of the mid to late nineteenth century: a history, simulation, and implication for mediaeval drought. *Holocene* 16:159–171.

Holmes, R.L. 1983. Computer-assisted quality control in tree-ring dating and measurement. *Tree-Ring Bulletin* 43: 69–78.

Kohler, T. A. 2010. A new paleoproductivity reconstruction for Southwestern Colorado, and its implications for understanding thirteenth-century depopulation. In *Leaving Mesa Verde: Peril and Change in the Thirteenth-Century Southwest*, ed. T. A. Kohler, M. D. Varien, and A. M. Wright. Amerind Studies in Archaeology, Vol. 5. Tucson: University of Arizona Press, 102–127.

Kohler, T. A., G. Gumerman, and R. Reynolds. 2005. Simulating ancient societies: computer modeling is helping unravel the archaeological mysteries of the American Southwest. *Scientific American* 293: 76–83.

Kohler, T. A., M. D. Varien, and A. M. Wright, eds. 2010. *Leaving Mesa Verde: Peril and Change in the Thirteenth-Century Southwest*. Amerind Studies in Archaeology, Vol. 5. Tucson: University of Arizona Press.

Meko, D. M., and C. H. Baisan. 2001. Summer rainfall from latewood width. *International Journal of Climatology* 21: 697–708.

Meko, D. M., C. W. Stockton, and W. R. Boggess. 1995. The tree-ring record of severe sustained drought. *Water Resources Bulletin* 31: 789–801.

Petersen, K. L. 1994. A warm and wet Little Climatic Optimum and a cold and dry Little Ice Age in the southern Rocky Mountains. *Climatic Change* 26: 243–269.

Schulman, E. 1942. Dendrochronology in pines of Arkansas. *Ecology* 23: 309–318.

Schulman, E. 1946. Dendrochronology at Mesa Verde National Park. *Tree-Ring Bulletin* 12(3): 2–8.

Schulman, E. 1947. An 800-year old Douglas Fir at Mesa Verde. *Tree-Ring Bulletin* 14(1): 18–24.

Schulman, E. 1952. Dendrochronology in Big Bend National Park. *Tree-Ring Bulletin* 18: 18–27.

Schwindt, D. M., R. K. Bocinsky, S. G. Ortman, D. M. Glowacki, M. D. Varien, and T. A. Kohler. 2016. The social consequences of climate change in the central Mesa Verde region. *American Antiquity* 81: 74–96.

Smith, D. G., and R. F. Nichols. 1967. A tree-ring chronology for climatic analysis: the dendrochronology of the Wetherill Mesa Archaeological Project. *Tree-Ring Bulletin* 28: 7–12.

Stahle, D. W., E. R. Cook, M. K. Cleaveland, M. D. Therrell, D. M. Meko, H. D. Grissino-Mayer, E. Watson, and B. H. Luckman. 2000. Tree-ring data document 16th century megadrought over North America. *Eos, Transactions of the American Geophysical Union* 81(12): 212, 125.

Stahle, D. W., M. K. Cleaveland, H. Grissino-Mayer, R. D. Griffin, F. K. Fye, M. D. Therrell, D. J. Burnette, D. M. Meko, and J. V. Diaz. 2009. Cool and warm season precipitation reconstructions over western New Mexico. *Journal of Climate* 22: 3729–3750.

Stahle, D. W., J. R. Edmondson, J. N. Burns, D. K. Stahle, D. J. Burnette, E. Kvamme, C. LeQuesne, and M. D. Therrell. 2015. Bridging the gap with subfossil Douglas-fir at Mesa Verde, Colorado. *Tree-Ring Research* 71: 53–66.

Stahle, D. W., and J. S. Dean. 2011. North American tree rings, climatic extremes, and social disasters. In *Dendroclimatology: Progress and Prospects*, ed. M. K. Hughes, T. W. Swetnam, and H. F. Diaz. Developments in Paleoenvironmental Research, Vol. 11. Dordrecht: Springer, 297–327.

Thornthwaite, C. W. 1948. An approach toward a rational classification of climate. *Geographical Review* 38: 55–94.

Varien, M. D. 2010. Depopulation of the northern San Juan Region. In *Leaving Mesa Verde: Peril and Change in the Thirteenth-Century Southwest*, ed. T. A. Kohler, M. D. Varien, and A. M. Wright. Amerind Studies in Archaeology, Vol. 5. Tucson: University of Arizona Press, 1–33.

Wright, A. M., 2010. The climate of the depopulation of the northern Southwest. In *Leaving Mesa Verde: Peril and Change in the Thirteenth-Century Southwest*, ed. T. A. Kohler, M. D. Varien, and A. M. Wright. Amerind Studies in Archaeology, Vol. 5. Tucson: University of Arizona Press, 75–101.

| # Fourteenth to Sixteenth Centuries AD

The Case of Angkor and Monsoon Extremes in Mainland Southeast Asia

ROLAND FLETCHER, BRENDAN M. BUCKLEY,
CHRISTOPHE POTTIER, AND SHI-YU SIMON WANG

Angkor, the capital of the Khmer Empire in Southeast Asia, was the most extensive low-density agrarian-based urban complex in the world. The demise of this great city between the late 13th and the start of the 17th centuries AD has been a topic of ongoing debate, with explanations that range from the burden of excessive construction work to disease, geo-political change, and the development of new trade routes. In the 1970s Bernard-Phillipe Groslier argued for the adverse effects of land clearance and deteriorating rice yields. What can now be added to this ensemble of explanations is the role of the massive inertia of Angkor's immense water management system, political dependence on a meticulously organized risk management system for ensuring rice production, and the impact of extreme climate anomalies from the 14th to the 16th centuries that brought intense, high-magnitude monsoons interspersed with decades-long drought. Evidence of this severe climatic instability is found in a seven-and-a-half century tree-ring record from tropical southern Vietnam. The climatic instability at the time of Angkor's demise coincides with the abrupt transition from wetter, La Niña-like conditions over Indochina during the Medieval Warm Period to the more drought-dominated climate of the Little Ice Age, when El Niño appears to have dominated and the ITCZ migrated nearly five degrees southward. As this transition neared, Angkor was hit by the double impact of high-magnitude rains and crippling droughts, the former causing damage to water management infrastructure and the latter decreasing agricultural productivity. The Khmer state at Angkor was built on a human-engineered, artificial wetland fed by small rivers. The management of water was a massive undertaking, and the state potentially possessed the capacity to ride out drought, as it had done for the first half of the 13th century. Indeed, Angkor demonstrated just how powerful a water management system would be required and, conversely, how formidable a threat drought can be. The irony, then, is that extreme flooding destroyed Angkor's water management capacity and removed a system that was designed to protect its population from climate anomalies.

Introduction

The global context of the rise of the Khmer Empire and its capital at Angkor
was the increase in planetary temperature during the Medieval Warm Period
of the ninth to thirteenth centuries AD (Lamb 1965; Fagan 2009). Enclosed
form urban centers, such as Angkor Borei (Stark 2006) and Wat Phu (Santoni
2008), had been developing for over six hundred years in the Mekong delta
and the interior of Mainland SE Asia. In the eighth and ninth centuries these
compact urban centers gave way to an open, dispersed urban form, the ultimate expression of which was Greater Angkor. At its peak in the thirteenth
century, the Khmer capital was the most extensive low-density agrarian-based
urban complex in the world, covering about 1000 square kilometers and
inhabited by around 750,000 people (Pottier 1998; Coe 2003; Fletcher et al.
2003; Evans et al. 2007). Angkor was strategically located between the Tonle
Sap—a massive lake that expands in monsoon season to more than 10,000
sq km—and the Kulen Hills, where it dominated a water catchment basin of
about 3000 sq km (fig. 9.1). Also available to the state were several small rivers
that could be diverted and controlled for its urban-rural needs. Angkor's location on this almost level plain gave it the added advantage of having no physical obstacles to suburban expansion except distance. As warm temperatures
favored the greater crop yields needed to sustain a robust and growing state,
Angkor could expand in all directions, clearing more land for rice production.
In a region where returns from rice production are poor and intensification of

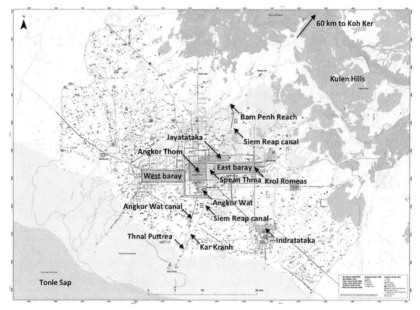

FIGURE 9.1 Greater Angkor, with sites discussed in the text. (Map courtesy of Damian
Evans and Christophe Pottier.)

yield not feasible, the only option was to clear more forest for rice fields as the population grew (Fletcher 2003).

The population of Angkor lived on house mounds around small water tanks. In the central area bounded by two massive reservoirs, West Baray and East Baray, the great temple complex of Preah Khan to the north, and the city center of Angkor Wat to the south (Evans et al. 2013: 125986–7) the mounds were located in a street grid. Overlapping with that central pattern was a vast landscape extending across the suburbs of the rest of Greater Angkor marked by larger mounds of grouped housing, with shared, larger water tanks and adjacent small shrines. Many people also lived along the canal and road embankments, where their occupation debris can be seen today. Small numbers of people—up to a few thousand in Ta Prohm and Angkor Wat, for example—lived within the temple enclosures (Evans et al. 2013: 12597). The extensive urban world in which they resided, with an aggregate density below 10 persons per hectare and an occupation density gradient decrease from the core to the periphery, can be seen in the distribution of water tanks, shrines, and other installations. Surrounding the nine square kilometers of Angkor Thom (literally: Great City), with its population of between 30,000 and 60,000 people, was an adjacent central area population of 250,000 inhabitants The outer suburbs, extending 10 to 15 kilometers from the urban core, held half a million people, with densities dropping towards rural levels at the periphery. All dwellings, from transient shacks and houses to the splendid residences of the elite, were built of timber. Domestic houses were surrounded by economic plants, bushes and trees that provided resources and shade, as they do today. What would have been a heavily forested tropical landscape was comprehensively cleared for the production of rice in bunded fields (fig. 9.2). Binding it all together was a network of canals and great road embankments.

The Development of Greater Angkor: Ninth to Twelfth Centuries AD

For several centuries before the warm climate commenced, the Khmer had been developing a unique tool of water management centered on the use of *baray*, which served as a simulation of the backswamps of a natural delta either to stockpile water or to hold surplus water so that it could be drained into the lake. Sometime after the fifth and sixth centuries AD, a large reservoir, Thnal Mrek, was built atop the Kulen Hills (Tang 2011: 45–47, 58) in what we now know would become the early ninth century, dispersed urban center of *Mahendraparvata* (Chevance 2011). A possible proto *baray* is also observable under the West Baray in association with the Ak Yum temple that, if of the same date, was created in the seventh to eighth centuries. In the ninth century a *baray* was being built at the other early Khmer capital, Hariharalaya (Pottier 1996).

FIGURE 9.2 Extent of the bunded rice fields identified in the Angkor plain. (Data
from Hawken 2011, based on Angkor archaeological maps in Pottier 1999 and
Evans 2007.)

Dynasties came and went, but the development of Angkor's physical structure was characterized from its earliest days in the ninth century less by political
history than by a relentless focus on its water management system. Continual
efforts to integrate the system with the ritual nodes of the urban complex
coincided with great dynastic shifts, but its "hydraulic system" remains the
constant through Angkor's four main phases of development (Fletcher et al.
2008a): (1) towards the east in a process initiated by Yasovarman I in the late
ninth century, when the East Baray was established; (2) towards the west with
the development of the West Baray in the eleventh century; (3) towards the
south with the building of the Angkor Wat temple complex during the reign
of Suryavarman II, an expansion that also potentially connected the complex
via waterways to the West and the East Baray; and (4) towards the north with
the construction of the last great *baray*, Jayatataka, in the late twelfth century,

followed by the extension through the middle of Angkor of a canal carrying the flow of the Siem Reap River.

By the twelfth century the Khmer had created a vast artificial delta with intake channels in the north and a fan of western, central, and eastern channels in the south flanking the East and the West Baray (fig. 9.1). This versatile system allowed water to be shunted across almost level gradients in numerous directions and to be conserved or discarded into the lake. The elements of the system, recognized by Groslier (1979), do not, however, fit the conventional Western definition of "irrigation," which presumes strong central control and intensification. Neither presumption is appropriate in Angkor; nor is the multi-crop myth (Van Liere 1980, Pottier 2001, Evans 2007). Although state control was required for the construction of the massive canals and reservoirs, there is no indication that the state managed the water supply. Rather, the system is more simply understood as a risk management mechanism that worked to mitigate the uncertainties of monsoon variation in a landscape of largely rainfed, bunded rice fields (fig. 9.2). What the Khmer engineers and farmers had created was a paradox—a versatile system capable of coping with unpredictable situations, yet necessarily built on such a huge scale that it introduced massive inertia into the capacity of local communities within Greater Angkor to make sustainable adjustments to changing circumstances.

The water network had two primary functions that could mitigate dry or wet climate conditions: to pass water through the the bunded fields in the northern half of Greater Angkor and in to the southern fields; and to dispose of water as it moved southwards for flood management (Pottier 2001; Fletcher et al. 2003; Fletcher et al, 2008a; Kummu 2008). In addition, the E-W banks just north of central Angkor and the West Baray served to shunt water around the western end of the *baray*. From there it could have been disposed of down the SW canal and also put into the SW grid and moved to the SE along the upslope side of the SE road/canal embankment, to be distributed to the fields. What the distribution of water served to achieve was to ensure rice production in the high yield, bunded fields of south Angkor. Water from the *baray* served a comparatively small area of the whole rice production landscape of the Angkor plain and would have helped to ensure supply for about 200,000 people, an adequate protection against a normal poor monsoon (Fletcher et al. 2003: 116). Since it did not function to intensify yields, it did not require continual central management.

In the late twelfth century, the Khmer were increasingly focused on the urban center, notably with a concentration of water channels running through the middle of Angkor and draining into the lake (Pottier 2000). This involution, which began just as the empire was reaching its apex, had profound consequences. Significantly, it also involved a substantial increase in the commitment of the population of Greater Angkor to temple maintenance and management. A key issue in the sustainability of the urban complex is the evidence that the stock of major temples the city was supporting effectively doubled in the late twelfth and early thirteenth centuries, under Jayavarman

VII's administration. Even with a decrease in the staff of the older temples, a huge additional commitment would have been added to the city. One staff person in a later twelfth-century temple, such as Ta Prohm, required the support of approximately five providers (farmers) delivering rice on an as-needed basis (Cœdès 1906.). The inscriptions indicate that Ta Prohm and Preah Khan alone had a total support population of more than 150,000 people (see Cœdès 1941) who also had to live nearby and hence be part of the population of Greater Angkor. In the late twelfth century, just two medium-sized temples were absorbing the labor of one-fifth of the population of Greater Angkor.

On top of these demands must be added a much larger support population for the central temple of the capital, the Bayon, and support populations for earlier big state temples, such as the Baphuon and Angkor Wat, which are mentioned by Zhou Daguan—a Chinese diplomat whose account of his late thirteenth-century visit constitutes our sole written source on Angkor for a period of three centuries. He gave no indication that the older temples were derelict or non-operational (Pelliot 1951, Harris 2007). The serious conclusion, therefore, is that the population of Greater Angkor was substantially committed to supporting or working for the great temples in a largely closed and self-referential economic cycle into the thirteenth century—even as political and social practices changed. For example, the Bayon—Javayarman VII's state temple—was the last monumental ritual construction, even though he reigned in the middle phase of the history of the state. After the twelfth century, Angkor Thom, which was built in the 1180s, remained the permanent core of the city, while the Bayon was repeatedly modified over the following centuries (Jacques 1999; Dagens 2003; Polkinghorne, Pottier, & Fisher 2013).

At the level of ordinary daily life, changes were occurring in Angkor from the twelfth century onwards. There is some indication that the number of kiln sites within Greater Angkor declined (Chhay et al. 2013). Even in the port area near Phnom Krom, the known ceramic production site at Tuol Kok Pey ceased in the thirteenth century (Brotherson et al. 2012). By then a smaller number of larger kiln sites further away from Angkor, such as Cheung Ek, continued to supply the city and other markets (Desbat 2011).

As far as we can judge, the state required some form of tax yield, whether indirectly through the ritual establishments or (from the late twelfth century) through production, as well as a land tax of some kind (see Dagens 2003: 146). Local communities could be left to their own devices to work out how to produce the tax and to make corvee labor and other services available (Sahai 1970, Sedov 1967). The agricultural system was predicated on extensification, rather than intensification, as a means to obtain more yield, and on stability and risk management to ensure a stable crop supply. It is parsimoniously understandable as a supply stabilizer system, not as a mechanism to enhance rice yield.

The material configuration of the water management system was sophisticated and complex, using minimal gradients and simple if massive "constructions" on an enormous scale. However, these massive constructions were not easily modified, and they both redefined and dominated the topography of central and southern Angkor (as they still do today). How such a complex, inertial, stabilizing system interacted with a cultural, regional, and climatic milieu that was changing is at the heart of our discussion.

The Demise of Angkor

In the 300 years from the early thirteenth to the mid-sixteenth centuries, Angkor declined from a vast megacity at its peak to a mere relic of its former self (Evans et al. 2007, Fletcher & Evans 2012). How and why the city failed has been the subject of much debate, in large part because of the scarcity of original texts and dated archaeological remains for this period. An important historical event, the Iconoclasm—the apparent defacement of the Buddhist narrative bas-reliefs on Angkor's temples—is known from the archaeological record, but until recently we have relied on an educated guess to date the event to some time after the death ca. AD 1220 of the Khmer king Jayarvarman VII; Jacques 2007: 280–3). A single seed found in 1996 in foundation deposits from Jayarvarman's Terrace of the Elephants yields a radiocarbon conventional age of 717 ± 20 years BP (Pottier 1997a)—that is, around the mid-14th century AD. Between 1220 and the end of the fifteenth century, only 26 inscriptions are known, of which ten are one or two lines (Vickery 2004a). There are no securely dated inscriptions between 1328 (K. 470) and 1546 (K. 296). Moreover, Angkor is scarcely referred to in fourteenth and fifteenth century texts, even by the Chinese, who fail to mention its takeover in 1431 by the kingdom of Ayutthaya.

Indeed, the Khmer state would have experienced profound social and political change with the progressive movement of the Thai and Vietnamese southwards on its eastern and western flanks. Thai tribes had begun to move out of the mountains of southern China down into the Khmer controlled lowlands of the Chao Praya basin in the eleventh and twelfth centuries and reached Sukhothai on the state's northwestern edge late in the thirteenth century (Jacq-Hergoualc'h 2004: 48). As the new populations began to gain autonomy and break away from the Khmer empire, land routes would have been severed, the empire's regional tax base of staple crops lost, and new and unpredictable alliances and threats of attack introduced. To the east, the Dai Viet, previously confined to the Red River basin and its vicinity, finally began to prevail in a centuries old war with the Chams to the south. Although the process was slow and erratic, the Vietnamese had reached the city-state of Vijaya to the east (on the present-day South China Sea) by 1471 (Vickery 1985), where they could have cut off Khmer access to the coast through the former Cham territories. The movement of the Thai and the Vietnamese was a gigantic pincer cutting off the old imperial territories and maritime access of the Khmer on the west, east, and southeast.

Maritime developments would transform the region's economies further in the sixteenth century, when the wealth that could be gained from Arab, Indian,

SE Asian, and Chinese ship-borne trade began to supersede the old wealth of stable rice production in interior river basins (Reid 1993). European traders had also started to infiltrate this interconnected and competing mercantile system as they circumvented the core Arab states from the fifteenth century onwards (Reid 1993). As has been argued, the "globalization" of commerce in East and Southeast Asia beginning in the early fifteenth century, and the rise of coastal trading cities surely played a role in undermining the Khmer empire and its Angkor urban complex, and may have had a role in the extensive depopulation of the surrounding region (Lombard 1970, Vickery 1977, Reid 1993).

Also coming along the long-established trade network from India to Southeast Asia was the other great transformer of society: Therevada Buddhism (Harris 2008). The Thai kingdom of Sukhothaya had embraced Therevada by the early fourteenth century. Through connections between the communities of Chao Praya basin and the western side of Cambodia, Therevada Buddhism seems to have been introduced to Cambodia by the thirteenth and fourteenth centuries (Giteau 1975: 11-12; see Coe 2003: 201). Zhou Daguan describes its saffron-robed practitioners in the streets (Dagens 2003: 181), which they shared with Brahmins and Shaivite priests. We should not suppose that Therevada principles made the Cambodian state any less willing to fight. But Therevada Buddhism could have allowed state power and tax to be decoupled from the maintenance of huge ritual facilities. The logic of the great shrines with their thousands of staff and their immense support populations (see Cœdès 1906, 1941) is part of Brahmanism and Mahayana Buddhism but is not inherent to Therevada Buddhism (see, for example, Groslier 1974). With its advent the rationale for the management and economic arrangement of the population of Greater Angkor in relation to the major shrines (Dagens 2003: 140-150) begins to vanish, quickly or slowly, from the thirteenth century.

Angkor Thom was certainly in use in the second half of the sixteenth century (Groslier 2006 edition: 1958: 21 & sqq.), but sixteenth and seventeenth century Khmer textual sources refer to Angkor Wat primarily as a pilgrimage center (Thompson 2004). By the late sixteenth century, Greater Angkor had experienced a profound urban decline, though scattered occupation occurred throughout its former extent, as indicated by Chinese trade wares of the sixteenth to nineteenth centuries found in the Greater Angkor Project ceramics surveys. The seat of state power is usually considered to have moved to the Quatre Bras region around present-day Phnom Penh (Cœdès 1913, Cœdès 1968, Vickery 1977, Vickery 2004a).

While many social, political, and economic problems might have been within Angkor's capacity to manage or rectify, one class of change was not so tractable or predictable. Angkor was subject to severe and unstable climate change, from flood-inducing monsoons to drought. The resulting damage could well have discredited the relevance of the titanic rituals in Angkor's major shrines to a viable and productive water economy. If the central system was destabilized, the management indicated in the Preah Khan inscriptions (Cœdès 1941) would also have weakened the provincial economies.

The decline would have been a substantial and varied process, affecting different sectors of society differently. Political and social factors affecting the people living in Angkor Thom may have been of little direct consequence in the suburbs beyond the West and the East Baray, where subsistence-level farming might have survived as it had before, without the state's massive infrastructure. We should not suppose that what the elite did, and when, was the same as the dispersal of the rest of the population. No primary Khmer historical record tells us when the elite left. Secondary sources ascribe the event to the 1430s, when Ayutthaya took control of Angkor for a short time (Vickery 1977 and 2004a) and sent two individuals, *bañā kaev* and *bañā daiy*, to the Siamese capital (Vickery 2004b). Although we know that the population did decline dramatically, no text tells us how. Nor do we have evidence that the population died abruptly—at least not in Angkor, where there are no skeletal traces of a demographic disaster.

Population Decline of Angkor in the Late Sixteenth Century AD

The overall population growth of Angkor can be read from the cumulative surface area of water in the *baray* between the seventh century and the end of the urban complex in the late sixteenth century (figs 9.3 and 9.4). Angkor's population growth can be represented schematically by the cumulative amount of water in the *baray*. The old assumption that each *baray* replaced its predecessor is invalid. Evidently, the *baray* did not silt up, and textual references to the East Baray occur through to the late Angkorian period (the lake *lac Ma-sseu-lou* described by Wang Ta-yuan published in 1349; Pelliot 1951: 141). Instead, the *baray* are a cumulative stock of water for the urban area of Greater Angkor within which they are located. Because their exit channels serviced the southern half of the urban landscape (see Sonnemann 2011: 187–98), they provide a kind of index for the amount of water that was needed to serve the population. The cumulative water stock is an index of the rate of demand for water as it was gauged by the engineers. The broadly ogival curve of the cumulation matches almost exactly the usual growth pattern of a great imperial capital and can therefore be used to represent population growth, because the peak water supply corresponds with the peak of the city in the twelfth century and, hence, with the peak population that its rice fields could support (Fletcher et al. 2003: 116, Lustig 2001, Hawken 2011). In consequence, it is possible to graph the population decline between its peak in the twelfth century and final demise in the late sixteenth century (fig. 9.3). The average growth rate from the ninth to the twelfth century was about 2000 persons per annum, with a short period of much steeper growth between AD 900 and 1000. The average decline was about 1800 persons per century, with either a steady decline through the thirteenth century or a precipitous drop in the fourteenth century. If the drop was delayed until the fifteenth century, it was catastrophic, but with no textual, demographic, or skeletal evidence to support such a scenario, we suggest a more gradual decline.

If we accept the overall rate of population decline through the thirteenth and fourteenth centuries, the population of Greater Angkor would have been reduced by half sometime between 1250 and 1350—that is, before the climatic instability of the fourteenth century, but during the drought of the thirteenth century (fig. 9.4). By these estimates, Angkor was already a shadow of its former self when the Ayutthayan incursion occurred in the 1430s. The significance of such a population decline is that the labor available for corvee work to repair infrastructure, let alone build new monuments, would have been drastically reduced. Assuming that hard manual labor was performed mainly by men and that each family of five contained at least one child, there would have been a maximum of about 300,000 adults in Angkor, per the average decline, when the climate crisis began in the fourteenth century. The authorities would have had access to the labor of about 150,000 men to cope with repairs and crises across the entire urban complex. Of these a significant number would no longer be available if a water crisis was badly affecting their own farmlands and houses.

The Asian Monsoon and Abrupt Climate Change

Angkor would reach its peak and experience its demise as capital of the Khmer kingdom between the fourteenth and sixteenth centuries AD, a period of transition from the Medieval Warm Period to the Little Ice Age when the climate was changing intensely and abruptly on a vast, planetary scale. While there had been occasional monsoon peaks and episodes of drought between the tenth and twelfth centuries, what now began was rapid fluctuation between extreme monsoons and extreme droughts (Buckley et al. 2010, 2014). This vast scale of change was a further imposition among the many other stresses to the Khmer state.

Given its outsized role in the tropics in general and in the demise of Angkor in particular, the phenomenon of monsoon deserves closer study. The Asian monsoon, an annually occurring, hemispheric-scale wind reversal caused by the differential heating of the Eurasian continent and its adjacent oceans between summer and winter. It dominates the Earth's climate, bringing reliable seasonal rains and/or periods of annual drought to half of the globe and more than 60 percent of our planet's human population (Wang 2006; Neelin 2007; Clift and Plumb 2008). Believed to have developed about ten million years ago (Molnar 2005), the monsoon exhibited greater intensity by the early Holocene than it does today (Kutzbach 1981; Wang et al. 2005, Maher 2008). Nevertheless, in the past millennium alone it has, on occasion, deviated from its mostly predictable strength and timing, with substantial consequences for the societies of Southeast Asia.

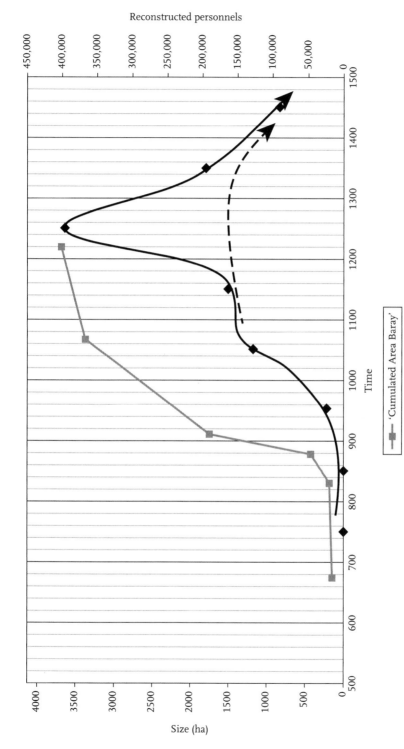

FIGURE 9.3 Cumulative volume of *baray* water storage capacity (gray line) over reconstructed curve (in black).

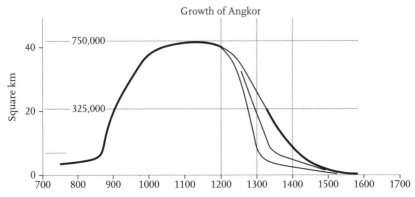

FIGURE 9.4 Reconstructed population growth and decline.

Over land, changes in the position and expansion of the Inter-Tropical Convergence Zone (ITCZ)—an equatorial belt of converging trade winds and rising air—coupled with the Asian monsoon's strength can be of great consequence to those who depend upon the reliability and predictability of monsoon rains (Fleitmann et al. 2007; Yancheva et al. 2007; Sachs et al. 2009). Three regional monsoon systems also contribute to the complex climate dynamics that determine rainfall distribution and intensity in the Indochina Peninsula (Chen & Yoon 2000). In summer, moisture from the Indian Summer Monsoon (ISM) is carried across the Indian subcontinent and the Arabian Sea to the west and the Bay of Bengal to the east. During the warm season, the Western North Pacific Summer Monsoon (WNPSM) is integrated into the western portion of the North Pacific anticyclonic gyre (a gigantic subtropical high-pressure system formed by the descending branch of the Hadley circulation) and interacts with the ISM to form a broad region of convergence between two opposite direction flows—that is, the westerly monsoon flow and the easterly Trade Winds that meet around the Philippine Sea. This convergence zone pumps moisture into the atmosphere, generating large volumes of rainfall and releasing latent heat. Its position and intensity also change year to year depending on the strength of the subtropical high-pressure system and/or the ISM. Moreover, long-term changes in the monsoon systems can and have had great societal and ecosystems impacts. (For a wind climatology map of Cambodia that outlines these monsoon subregions, see Buckley et al. 2014: 3.)

Several identifiable dynamical factors related to atmosphere-oceanic coupling can perturb the monsoon. Foremost among these is the El Niño-Southern Oscillation (ENSO) phenomenon (Kumar et al. 1999; Krishnamurthy and Goswami 2000), which is second only to the monsoon in global reach. The ENSO's multi-decadal expression, the Interdecadal Pacific Oscillation (or IPO; Meehl and Hu 2006) has been linked to decadal-scale climate anomalies that have delivered epic droughts over much of Asia, including Southeast Asia, during the past millennium—events that provide a context for the magnitude of what occurred in Angkor in the mid-second millennium AD (Buckley et al. 2010, 2014; Cook et al. 2010). Other global climate features that have been shown to interact with the Asian monsoon are North Atlantic sea surface temperature (NASST - Goswami et al. 2006), the Indian Ocean Dipole (IOD—Ashok et al. 2001) and Eurasian snowpack anomalies (Hahn & Shukla 1976; Bamzai & Shukla 1999).

Variability of the Monsoon: Dynamical Explanations

Seasonal Evolution of the Monsoon

Significant debate exists about how to define this seasonal wind and determine its onset and withdrawal with any degree of accuracy (Wang & Linho 2002; Zhang et al. 2004; Cook & Buckley 2009; Fosu & Wang 2014). Zhang et al. describe the Asian Summer Monsoon's evolution and mechanics through analyses of atmospheric dynamics that encompass such large-scale circulation parameters as the meridional temperature gradient of the troposphere and the establishment of an easterly vertical wind shear, both caused by upper-level warming due to convective transport of latent heat. According to their study, the earliest onset of the broad monsoon begins over the Indochina Peninsula, following developments over the Indian Ocean and the Bay of Bengal.

The monsoonal southwesterly winds first develop during the month of May and trigger the monsoon flow-terrain interaction in western Myanmar and western Thailand, thus signifying the monsoon onset in that particular region (fig. 9.5a) (Wang et al. 2013). Meanwhile India remains rather dry under the predominant northwesterly flows. This situation changes in June (fig. 9.5b), when the cross-equatorial monsoon flow quickly intensifies and expands northward, bringing moisture over the southern part of India and initiating the Indian Summer Monsoon (ISM) onset from south in early June to north in late June. The intensified monsoon westerlies penetrate the Indochina Peninsula into the South China Sea, staging the monsoon onset there (see, for example, Wang 2006; Saha 2010). During July and August (fig. 9.5c), after the ISM reaches the mature stage with widespread monsoon rains over India, the westerlies further extend into the western tropical Pacific with the northward displacement of the easterly Trade Wind. At this stage the Western North Pacific Summer Monsoon (WNPSM) reaches its highest intensity while tropical cyclonic activity increases dramatically (see, for example, Chen et al. 2004).

Beginning in mid-September, the intensity of the westerly and southwesterly monsoon flow decreases rapidly and has completely withdrawn by October (fig. 9.5d). What replaces the southwesterly monsoon in East Asia is the development of northeasterly flow associated with the cooling and high-pressure system developing over mainland East Asia, centered in Siberia. The cool, relatively dry northwesterly winds out of the continent then travel a great

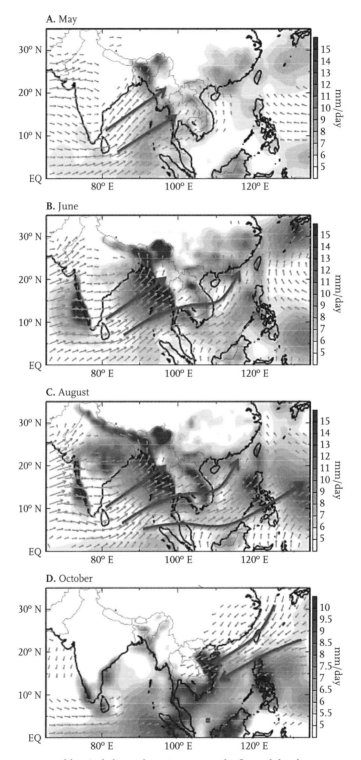

FIGURE 9.5 Monthly wind climatology (vectors) at the 850-mb level, or ca. 1500-meter altitude, with superimposed precipitation over land and oceans. Bold arrows indicate the main direction and extension of the major monsoonal flows. Precipitation data sources: for land (Global Precipitation Climatology Centre—GPCC); for sea (Global Precipitation Climatology Project—GPCP).

distance over the ocean before encountering Vietnam (fig. 9.5d), and consequently produce copious rainfall over coastal/central Vietnam that persists until December. Therefore, while the region surrounding Greater Angkor benefits from summer rains, central Vietnam receives the most rainfall during the summer-winter transition of the monsoons. The latter might have shaped the development of the kingdoms that held dominion over the regions to the east of the Annamite Range, as discussed below.

Climate Oscillations

As noted, the Asian Summer Monsoon is affected by a number of "remote forcings" through the atmospheric teleconnection mechanism (ENSO, IPO). Such remote modulations occur through either the tropical east-west circulation (Walker Circulation) in response to tropical sea surface temperature (SST) and convective activity, or through a dispersion of atmospheric energy from the west and from higher latitudes, forming the so-called Eurasian wave train (Ding & Wang 2007). In some years, the two processes may either be in phase or out of phase, leading to constructive or destructive influences on the monsoon rains, respectively. During the autumn season, ENSO affects western tropical Pacific circulation and modifies the intensity and moisture content of the northeasterly monsoon flows toward Indochina and central Vietnam (Tsay 2004). These climate oscillations are, therefore, of great concern to the human populations living within the monsoon's influence.

Weather Extremes

Summer rainfall on the Indochina Peninsula has two main origins: (a) steady monsoon rain caused by terrain-flow interaction and enhanced diurnal convection, and (b) torrential rain brought in by tropical disturbances (Chen & Yoon 2000). Neighboring the western tropical Pacific, the world's largest breeding area for tropical cyclones, Indochina receives its share of tropical cyclones. The Western Pacific tropical cyclones almost always propagate towards the west once they form, following the tropical easterly winds that steer them. These storms can then turn northward if they encounter mid-latitude frontal systems, or they may continue to move westward across the South China Sea, into Indochina or southern China. Tropical cyclones can and do penetrate far into Indochina. They also form in the South China Sea (SCS), where they have a greater tendency to propagate westward into Indochina than their Western Pacific counterpart (Liu & Chan 2003).

When tropical cyclones reach mainland Southeast Asia, their intensity in terms of wind speed decreases and they are downgraded from cyclone status and become what are known as residual lows (Yoon & Chen 2005). Although not fierce in wind power, these residual lows contain as much moisture as a tropical cyclone and continue to produce large amounts of rainfall as a slower-moving storm system, dumping copious amounts of rain for an average duration of three to four days. Yoon and Huang (2012) have noted that residual lows coming from tropical cyclones and equatorial easterly waves (a type of tropical weather system not strong enough to be declared a tropical cyclone but associated with organized convective clouds) contribute to about 50 percent of the warm season rainfall over Indochina. Climate oscillations like ENSO have a direct modulating effect on the residual low activity as well (Chen & Yoon 2000). Yet, even though an El Niño (warm phase of ENSO) reduces the seasonal number of tropical cyclones, it does not eliminate them. Tropical cyclones and easterly waves still may form and propagate into Indochina. Perhaps more importantly, any changes in the residual low activity may impose a stronger effect on the wet/dry climate anomalies over Indochina as such storms are the only source in the region that can overwhelm any hydrologic system in a short period of time, regardless of the ENSO-monsoon status. It must therefore be considered that the source of large deluge events as recorded at Angkor, Hanoi, and Hué, with their impacts on infrastructure, described below, may be derived from an increase in tropical cyclonic activity, and are not derived from an increase in monsoon strength per se.

An Explanation for Persistent Drought over Indochina

The impact of ENSO on the climate of Indochina is well known (Buckley et al. 2007, 2010, 2014; Sano et al. 2009). Anomalous conditions in response to El Niño/La Niña occur mostly at the interannual (year-to-year) timescale. However, ENSO also exhibits a broad, lower-frequency energy spectrum at the decadal scale (Zhang et al. 1997; Zhang & Levitus 1997). An example is the Interdecadal Pacific Oscillation (IPO), which imposes a basin-scale interdecadal variability (Folland et al. 2002). Buckley et al. (2010) found that southern Vietnamese tree-ring records respond with remarkable fidelity to the IPO index, acquiring a correlation coefficient of 0.673. In fact, the relationship was so stable that it allowed for a successfully calibrated and verified reconstruction following standard dendroclimatological procedures (Buckley et al. 2010, see Supplementary Online Material for details). This strong relationship indicates that climate over Indochina not only responds to the interannual ENSO variation, but also reflects the decadal-scale variability of the IPO, belying the region's susceptibility to persistent phases of either El Niño or La Niña over an extensive period of time. The two Victorian droughts depicted in figure 9.6 present a great example of what might be termed "normal" El Nino conditions. The Strange Parallel Drought of the eighteenth century, however, was a persistent multi-decadal drought that is more indicative of the lower frequency expressions of El Niño. The possibility of this type of extended drought occurring with current population density is a real cause for concern.

Drought over Indochina may also be the result of climate forcing other than ENSO. For instance, Buckley et al. (2014) demonstrate through an empirical orthogonal function (EOF) analysis applied to the combined MADA and NADA (North American Drought Atlas of Cook et al. 2004) over the 730 years of the common period that the first EOF responds closely to ENSO and the

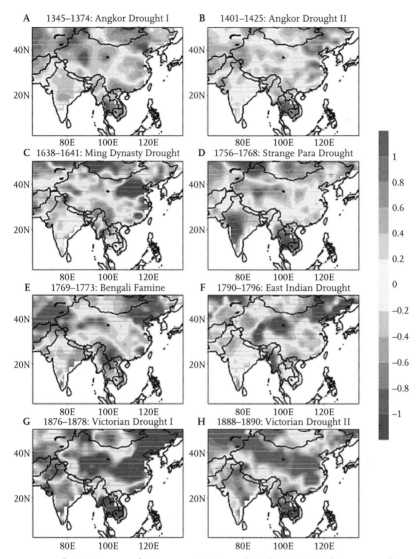

A   1345–1374: Angkor Drought I   B   1401–1425: Angkor Drought II

C   1638–1641: Ming Dynasty Drought   D   1756–1768: Strange Para Drought

E   1769–1773: Bengali Famine   F   1790–1796: East Indian Drought

G   1876–1878: Victorian Drought I   H   1888–1890: Victorian Drought II

FIGURE 9.6 PDSI maps of Monsoon Asia Drought Atlas (MADA) data averaged for the period of years indicated and the corresponding historical drought across Monsoon Asia. An understanding of the spatial footprint of these historically significant droughts is important for determining the primary causes for droughts of a particular magnitude.

IPO that commonly results in opposite climate anomalies between South Asia and North America. The second mode (EOF2) reveals the same phase in South Asia and North America and explains about 11 percent of the total variance; and it responds to an overall tropical warming/cooling of SST, in contrast to the eastern Pacific warming or cooling. For the two historical Angkor droughts, shown in figure 9.7, the spatial pattern of drought over the two main continents

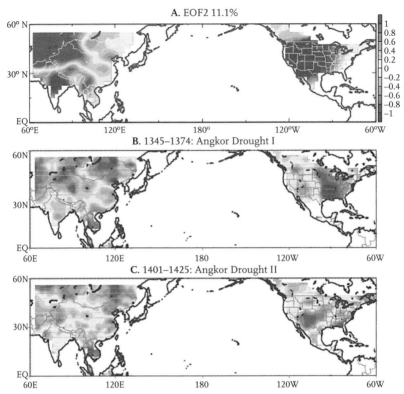

A. EOF2 11.1%

B. 1345–1374: Angkor Drought I

C. 1401–1425: Angkor Drought II

FIGURE 9.7 North American and Asian drought conditions (derived from the second EOF mode of the combined MADA and NADA data for AD 1300–2005) and the two corresponding historical Angkor droughts.

appears similar—that is, both show the same phase of drought over much of North America—suggesting that the climatic conditions represented by EOF2 may have contributed to the formation of the two Angkor droughts. This pattern is not common over the period of record and requires future research on the dynamic process leading to the EOF2 of drought.

Monsoon Variability over the Past Millennium

Speleothem records from China and India, the regions with the highest population density and the greatest vulnerability to variations in the strength of the monsoon, offer some of the most reliable high-resolution proxies for monsoon strength. Sinha et al. (2010) show that the speleothem record from Dandak Cave, located in the core monsoon region of India, agrees well with the Wanxiang Cave record from China (Zhang et al. 2008) and the Vietnam tree-ring record of Buckley et al. (2010). (For a map of the Monsoon Asia Drought Atlas with underlying tree-ring locations and speleothem sites, see Buckley et al. 2014, p. 5.) The agreement among these paleoclimate records is particularly remarkable during the fourteenth and fifteenth centuries, one of the most tumultuous periods in pre-modern Southeast Asia, as well as in China and northern Eurasia (see, for example, Kahan 1985; Lieberman 2003). It was at this time of drastic climatic instability with repeated fluctuations between mega-monsoons and mega-droughts that the Khmer Empire fell into disarray and its capital was largely abandoned. This was an equally tumultuous time in China, as Zhang et al. (2008) note, where drought is considered to have contributed to the demise of the ethnic Mongolian Yuan Dynasty in 1368, and in the kingdom of Tonkin (in present-day Vietnam), which faced epic drought and famine.

Figure 9.8 shows maps of the regional expression of several of the main droughts discussed in this paper, as well as the grid spacing and geographical extent of the Monsoon Asian Drought Atlas (MADA) drought indices produced by Cook et al. (2010). There are two primary periods of drought that occurred during the late fourteenth and early fifteenth centuries (figs.9.7a and 9.7b) referred to by Buckley et al. (2010 and 2014) as Angkor Droughts I and II, respectively. In the absence of historical records for Angkor, we can look to descriptions of this period of anomalous climate in the chronicles of Hanoi and Hué, as well as to texts from Lanna (present-day north Thailand) that told of monks from Chiang Mai being forced to return from Sri Lanka due to the severity of the drought there (Thera 1967; Wyatt & Wichienkeeo 1998).

Analogous highly variable dry and wet periods also occurred in the 16th to 18th centuries, according to the MADA, and for these the historical record is rich with information that can provide some context for the societal impact that extreme climate change might have had on Angkor. The Ming Dynasty ended during the drought of 1638–41 (fig. 9.6c). However, it was the latter half of the eighteenth century, when decades-long droughts wracked South and Southeast Asia, as well as northern Eurasia, that we see some of the greatest societal turmoil in the face of monsoon extremes (Lieberman 2003; Kahan 1985). The Hanoi chronicles refer to drought with increasing frequency in the eighteenth century, separately or cumulatively citing eighteen years when it occurred. Tonkin missionaries describe a 1703 famine that caused the death of almost half the population of some provinces, a 1712 drought, a1713 flood due to dyke breaches, and famine and an epidemic in 1714–15. In 1776, three years after Tonkin's occupation of Hué, Hanoi's rulers are described as conducting a ceremony to bring rain. Floods are also reported regularly, with 21 years mentioned cumulatively or separately, including in 1787 (the eve of Tonkin's fall to the Tay Son).

The Hué chronicle, in stark contrast, does not mention drought until after 1788, when Nguyen Anh established Saigon as a rival capital in Cochinchina. However, floods are noted in Saigon in 1713, 1730, 1736 and 1754, and in Cambodia in 1732. In 1789, the new Saigon regime sent rice to Siam to alleviate the effects of a drought that would drive up the price of rice in 1791. A Bangkok Royal chronicle describes high water, a failed rice crop, and high prices in 1785 (Flood & Flood 1978). The Hué chronicle describes drought as a significant

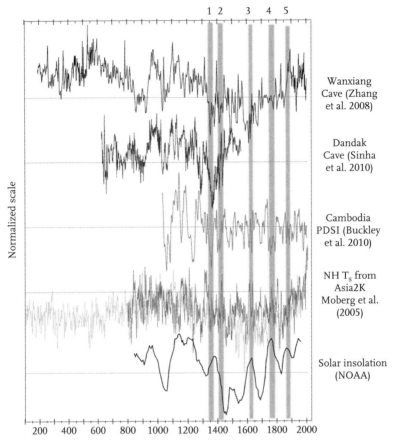

FIGURE 9.8 Climate reconstructions from Speleothem [18]O records of Wanxiang Cave and Dandak Cave, PDSI reconstruction for the area surrounding Angkor, two North Hemispheric surface air temperature reconstructions (darker: PAGES Asia2K; lighter: Moberg et al. 2005 multi-proxy reconstruction), and Solar insolation. Wide gray lines indicate key periods of drought: (1) Angkor Drought 1, (2) Angkor Drought II, (3) Ming Dynasty Drought); (4) Strange Parallels Drought, and (5) Victorian Holocaust Drought.

factor in the military campaigns of Nguyen Anh, noting that the Tay Son-held Binh Thuan Province was abandoned upon his arrival there in 1794 due to years of drought; the Province would face drought again in 1798. On the eve of Nguyen Anh's defeat of the Tay Son in 1801, other parts of the realm are said in the Hué chronicle to have faced years of drought and lost harvest. (For a MADA reconstruction of PDSI for Hanoi and Hué superimposed with flood, drought, famine, and epidemic records, see Buckley et al. 2014, p. 12.)

The "Strange Parallels Drought," so named by Cook et al. (2010) after Victor Lieberman's (2003) book by that name, spanned from India to Vietnam for more than a decade from 1756–1768, and corresponded to a period when all of Southeast Asia's main polities dissolved (fig. 9.6d; Lieberman 2003). The Bengali Famine drought of 1769–1773 (fig. 9.6e) and the East Indian Drought of 1790–1796 (fig. 9.6f) each resulted in millions of deaths due to starvation, marking the eighteenth century, with its successive decades of weakened monsoons, as one of the most tumultuous of the past millennium.

Two events of the recent past stand out for their severity: the Victorian Holocaust Droughts of the late nineteenth century, triggered by the El Niño events of 1877–79 and 1888–90, were responsible for millions of deaths globally and hit South Asia particularly hard (figs. 9.6g–9.6h). The first Victorian Drought had a near-global impact, but was particularly severe across a wide swath of Asia, with famine and deaths in India and northern China (Davis 2001). These historical accounts of drought and associated famine, faithfully recorded by the MADA data, give us insight into the general causes of monsoon failure for this region and the social and economic damage caused by climate extremes.

Angkor's Vulnerability to Monsoon Instability

Even absent written records from Angkor in the thirteenth to sixteenth centuries—and before we examine the archaeological record—we can reasonably point to two environmental vulnerabilities that would have been exposed by extreme positive anomalies in monsoon rainfall: erosion and an aging infrastructure. From the seventh century onwards, mainland Southeast Asia had been profoundly altered by centuries of forest clearance to extend rice production. By the twelfth century much of the Angkor plain and the low-lying slope below the Kulen-Khror hills had been converted to bunded rice fields (fig. 9.2, Groslier 1979, Fletcher et al. 2008a; Hawken 2011). Instead of a continuous forest canopy, patches and strips of anthropogenic woodland were clustered on the slightly raised mounds around the houses and scattered across the landscape. The clearing of shallow slopes would have created the risk of increased run-off and erosion during heavy monsoon rains (Fletcher et al. 2003). LiDAR surveys undertaken by the Khmer Archaeology LiDAR Consortium (KALC) have revealed rice fields under today's dense tree cover even on the Kulen Hills, as well as additional patches of fields further north at Koh Ker (fig. 9.1; Evans et al. 2013). Thung (2002) has argued that much of central Cambodia between the Dangrek Mountains and the hills north of the Tonle Sap had been cleared for farming, though how completely is not as yet resolved, resulting in the sparse tree cover and minimal soils persisting in many places to the present day. Upland deforestation was also in progress (Penny et al. 2014:1), with potentially very serious ecological consequences. Groslier (1979) and Van Liere (1980) both noted millions of abandoned rice fields on aerial photos, suggesting far more extensive earlier agriculture than practiced in the nineteenth and early twentieth centuries.

Meanwhile, Angkor's water network had served the city as a risk management system against the instability of the monsoon for more than half a millennium. The last *baray* had been built more than a century earlier. The oldest extant *baray*, the Indratataka, was about 500 years old. By the thirteenth century the immense infrastructure of Angkor was aging, intractable, and profoundly convoluted, with numerous redundancies and old, superseded features—in short, it was vulnerable to disruption (Fletcher et al. 2008a). The discovery of earlier major breaches indicates that the system was vulnerable to damage. At Bam Penh Reach, where the Siem Reap River is diverted to bring water to the East Baray, a large, early dam had failed and been rebuilt sometime in the tenth to eleventh centuries (Fletcher et al. 2008b, Lustig et al. 2008). When the system was new, the city population burgeoning, and the Khmer empire becoming more expansive and powerful, damage could apparently be rectified effectively.

Archaeological Evidence of a Water Network Crisis in Angkor

Fieldwork between 2004 and 2009 revealed what appeared to be contradictory evidence about the water network after the twelfth century, suggesting both that large quantities of water had been moving rapidly through the system and that there had been severe water shortages. The *baray* sites indicated water shortage, while the canals indicated water overload. The initial tree-ring data from Vietnam (Buckley et al. 2010) resolved the interpretative dilemma by showing a period of persistent monsoon deterioration through the first half of the thirteenth century, followed by a period of marked improvement in rainfall, and then severe fluctuations between mega-monsoons and megadroughts from the early fourteenth century onwards (fig. 9.9) While its risk management capacity meant that the system could cope with droughts, as it had since at least the ninth century, there is some evidence, as at Bam Penh Reach, that its dams were vulnerable to heavy monsoons. Extreme fluctuations between repeatedly large monsoons, which caused structural damage, and extreme droughts, which reduced crop yields, were perhaps beyond the capacity of the network.

The extant *baray* originally had several inlets and outlets. The major eastern exit channels, which are most clearly identified in the East Baray and have also been found by excavation at the Indratataka and the West Baray, are all either buried or blocked today (Sonnemann 2011: 187–91, 191-5). Blockage began as early as the late twelfth century at the West Baray, which was also deteriorating by the thirteenth century (Penny et al. 2005). The date of the blocking of the Indratataka exit is unknown. A possible eastern exit from the Jayatataka located by GPR (Sonnemann 2011: 195-8) appears not to have been extended further east and would have been superseded when the outer enclosure of Ta Som temple was built. The East Baray was blocked by a masonry wall of twelfth

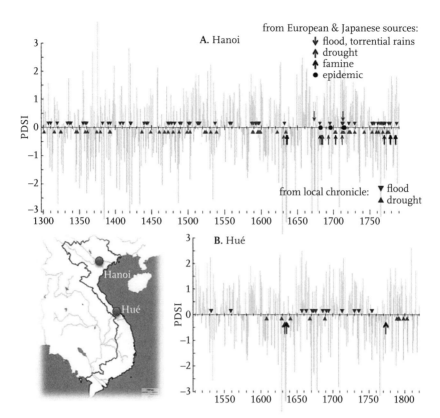

FIGURE 9.9 Major events in Angkor, AD 850 to 1450 (Buckley et al. 2010), juxtaposed with the Buckley et al. (2010) PDSI reconstruction. Gray arrows indicate major conflicts or dynastic changes. Rectangles indicate the estimated timespans for *baray* construction. Triangles represent dedication dates of major temples. ICO and ZD are abbreviations for the Iconoclasm and Zhou Daguan's visit to Angkor in 1295–6. The black circle marks the date of the last identified Angkor inscription and the black arrow indicates the Ayutthaya takeover. (Adapted from Buckley et al. 2014.)

to thirteenth-century style from the masonry-lined clearing Krol Romeas (fig. 9.10), and the eastern end of the large masonry-built, north wall was demolished and replaced by a set of narrow sluices, each less than one meter wide. A canal was dug down the outer, eastern side of the northern half of the east bank of the *baray* bringing water into the *baray* through what had been the exit channel (fig. 9.11). The dispersal of water through the old, massive eastern channel of the East Baray ceased and the *baray* was converted into a holding tank for water with dispersal through an opening in the southern or eastern bank. The new canal brought water from the same point at the NE corner of the *baray* where the old entry channel is located.

The source of the water for the East Baray had not changed. What had changed, presumably, was the available quantity of water, since this effort to bring water in half way down the east bank indicates that, at least temporarily,

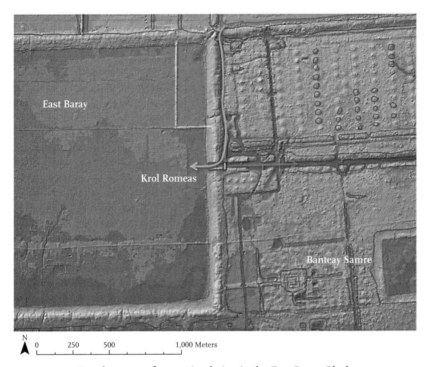

N

| 0 | 250 | 500 | | 1,000 Meters |

FIGURE 9.10 Development of water circulation in the East Baray. Black arrows show the original intake and exit channels passing between the laterite walls of Krol Romeas. Once blocked with a perpendicular wall (gray rectangle), Kroll Romeas is used as an intake fed by a new canal starting outside of the NE corner of the *baray* (gray arrow). (Archaeological details from Pottier 1999; background from KALC LiDAR survey, 2012.)

only half the *baray* could be filled. Serious water shortages would have commenced in the cumulatively drier conditions of the first half of the thirteenth century, then interspersed with heavy monsoons from the later thirteenth century into the early sixteenth century. The monsoon stabilized in the sixteenth century, but the *baray* were never again used as a means to shunt large packets of water back out to the east against the slope and then down into the southern canals, whether because of a lack of necessity and/or the lack of labor to remodel them.

Erosion Damage within the Central Urban Area

Between the Jayatataka and the East Baray, the Siem Reap canal turned westwards into a huge channel where it was highly constrained by the massive *baray* banks. At the western end of the East Baray is a N-S dam between the south bank of the Jayatataka and the north bank of the East Baray through which two channels passed, each with a stone bridge with culverts to manage

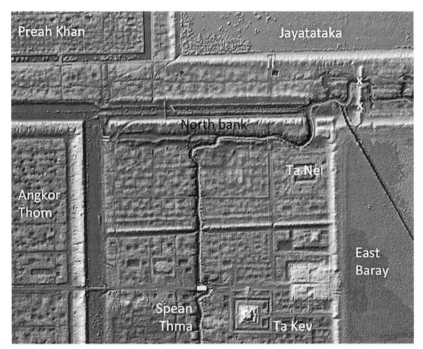

FIGURE 9.11 Area between the East Baray and the northeast corner of the moat of Angkor Thom. Masonry remains, bridge, and culvert are indicated with gray =, x, and rectangular marks. (Background from the KALC LiDAR survey, 2012)

the water flow (fig. 9.11; Trouvé 1933: 1120–8). What can now be observed on LiDAR images is that the water bypassed the southern bridge and destroyed the northern bridge, breaking through the dam at a very high level, and then gouged into the foot of the southern face of the Jayatataka south bank. The implication is that water had accumulated behind the dam until it over-flowed—something that would have occurred if great amounts of water and /or debris and vegetation were blocking the bridge culverts through the dam. A forceful flow of water produced multiple channels to the dam's west that then bent sharply to the south through the wide gap in the eastern end of the huge North bank between the East Baray and the NE corner of the moat of Angkor Thom. This gap now contains a dam and sluice built by the French in the 1940s. To the south of the North bank, the Siem Reap now turns abruptly westwards at a low elevation. During the period of flooding the water took several paths at a higher level than the current river, tearing away the north-ern edge of the residential area west of Ta Nei and flowed through the hous-ing area, causing severe damage before debouching back into the main Siem Reap canal. Given that these channels were elevated, we can assume that they emanated from the Siem Reap canal before it eroded down to its pres-ent level.

The channel immediately south of the western half of the North bank is also very deep, and to its south the northern edge of the western residential area was apparently torn away. Over a kilometer south along the canal, the water tore a path around the eastern end of the Spean Thma Bridge (fig. 9.11), bypassing the old elevated canal that flowed through it and ruining the bridge. The water surface of the dry season Siem Reap River in the twentieth century was more than four meters below the Angkorian flow level (Parmentier 1933), which suggest that, once this downcutting had happened and all the inlet and outlet connections between the surrounding water network and the old Siem Reap canal had been severed, water could never again be moved between the eastern and the western portions of the network across the line of that canal.

Deposition in and around the Southern Canals

The large southern canals that run directly to the Tonle Sap are filled with cross-bedded sand of medium to coarse grain size, indicating that large quantities of rapidly moving water was slowing down abruptly as it reached the flood plain near the lake (Fig. 9.12). Sand also fills the channel of the southwestern disposal canal from the West Baray and the Angkor Wat canal near the junction with the SE Canal (pers. comm. Sam Player, PhD student, University of Sydney)) and further south to the west of Kar Kranh. In the former Siem Reap canal at Kar Kranh and at Thnal Puttrea, the sand postdates the early fourteenth century, according to radiocarbon dating of vegetation preserved by waterlogging in the sand and of a wooden pirogue (Buckley et al. 2014, p. 10), also retrieved from the thick layer of sand in the canal. The sand in the southern part of the Siem Reap canal is the downstream signature of a process whose effects can be seen far upstream near the Kulen Hills, where the Siem Reap River is deeply incised as much as eight meters below the surrounding land surface, producing the eroded sediment that was then deposited downstream in the old canals. The river that now flows in the old Siem Reap canal channel breaks away from the line of the former canal just where that canal crosses the old SE canal from the West Baray's SW corner grid. To the south, the Siem Reap River now meanders off to the SE within a broad low ridge formed of levee banks. The former line of the canal continued southwards through Kar Kranh and Wat Atvear and then on south to Thnal Puttrea near Phnom Krom (fig. 9.12).

Once the Siem Reap canal and the Angkor Wat canal to the west were filled by sand, continual overflow from the former junction of the Siem Reap canal and the SE canal fanned out across the landscape and over the old canals, forming a bird's foot configuration of outflow channels between flat-topped ridges that spreads out from the junction that is clearly visible on the satellite remote sensing images. The flat-topped ridges are raised about one to two meters above the surrounding landscape, and meandering between them is a delta-like feature of channels that have cut into the former canals and are filled by a fine white deposit, which is stratified in the sloping layers of the shallow

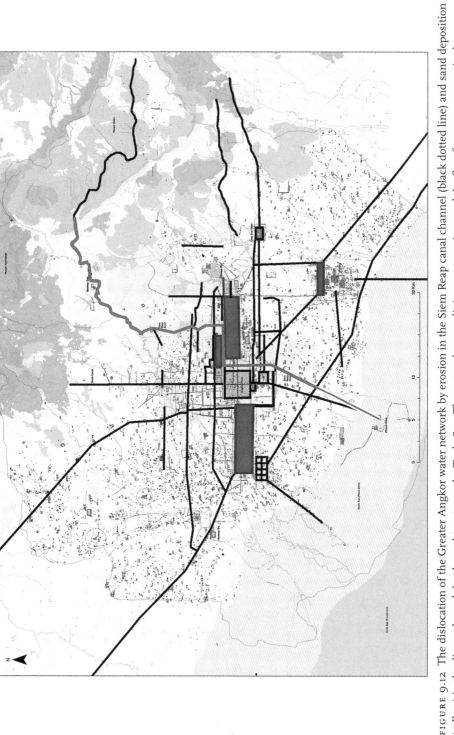

FIGURE 9.12 The dislocation of the Greater Angkor water network by erosion in the Siem Reap canal channel (black dotted line) and sand deposition (yellow) in the disposal canal that brought water to the Tonle Sap. The network was split into two portions and the flow from east to west in the northern half and from west to east in the southern half could no longer occur, permanently altering the functioning of Angkor's massive hydro-infrastructure. (Data used for this image courtesy of the Khmer Archaeology LiDAR Consortium.)

ridge west of Kar Kranh. Elsewhere the deposit forms layers more than a meter thick, and in places just west of the southern half of Siem Reap town it can be seen in extensive shallow quarries near the main roads. Further, surrounding the ridges and meandering channel is a lobe of very fine white sediment that either represents a process concurrent with or later than the ridges extending all the way down to the lake. What we appear to be seeing is severe erosion in the northern part of the network, leading to massive re-deposition of sand in the southern canals during a period of the mega-monsoons in the fourteenth century, followed by the formation of the delta-like feature and the meandering channels, and then by the deposition of the lobe of fine white sediment.

Discussion of the Physical Evidence

The weakening of the water network by the extreme mega-monsoons of the early fourteenth century could have been progressive, but the physical evidence suggests the impact of a few abrupt events. In particular, south of the North Cross Wall, the gouged channels do not display multiple overlaps. The eastern damage is at the former ground level, while the western erosion goes directly down to the present lower level of the river. Nevertheless, these different effects could be the result of one major storm season, since the water in the Siem Reap canal would have diminished extremely rapidly once the erosion began. The late sand deposition in the southern canals was also rapid. There are four deposition cycles from coarse to fine-grained in Thnal Puttrea, but these could simply be successive storms in one monsoon. There is no trace of trampling by animals in the fine-grained layers that would have lain on the surface through an annual dry season.

The abrupt, double impact of extreme monsoons followed by severe drought over about 150 years had the capacity to disrupt the urban agrarian economy of Angkor. An old, massive infrastructure was hit by extreme quantities of water that it could not manage and, presumably, by the effects of modifications undertaken to cope with immediate problems in the network. As the network was progressively broken and the risk management system that had supported Angkor for more than half a century ceased to operate, the urban population lost its self-sufficient means of food procurement. With the network degraded and crop yields reduced, the relationship between the institutional staff of the major temples and their farmer support population could not have been maintained. Nor could the elite have kept their large households functioning. The operational logic of the urban complex began to vanish. The fact that there was no reversal of the downgrading of the *baray* into predominantly holding tanks suggests that Angkor's elite had lost their power to marshal labor from the population for major engineering work.

A complication for the people of Angkor was that cross-dependency on features with multiple roles had become a feature of community life. For example, when a canal failed a road on its embankment was disrupted. Because rice fields that provided a carbohydrate crop were also ponds providing fish, loss of the bunded fields adversely affected both the protein supply and the staple crop. In addition, if parts of the highly elaborate and hyper stable water system failed, then disruption and damage would proliferate downstream, causing flooding, water shortages, and food crises in close proximity to adequate water and stable food supplies on either side of the great barrier banks and canals.

The water network was also interlocked with the road system of Angkor, a large part of which fed water into and out of the center along the upslope sides of the road embankments. The integrated transport network combining boats and roads facilitated interregional connections to a vast hinterland (Hendrickson 2010). Many different resources, such as salt, forest animal products, fish, charcoal, timber, iron, and stone in varying quantities came in on the same routes, connecting different resource chains and interlinking the economic prospects of the transporters, the people who serviced them along the routes, and the markets to which they were delivered. The market system also appears to have been highly integrated, with a very consistent repertoire of products reaching every part of Greater Angkor, as illustrated by the occurrence of general Chinese trade wares throughout the general domestic areas of the urban complex at about 6 percent of the total assemblage (Cremin 2006). A failure in part of the market supply chain could have been highly disruptive in more distant locations. The market network also facilitated the transmission of useful trading information, but could as easily deliver and magnify rumor, apprehension, fear, and intercommunity antagonism.

The drought of the 1360s was extremely severe and surely crippled the food economy if the water network was no longer working. We should therefore envisage that parts of Angkor became unusable and the temple economy could not function by the second half of the fourteenth century. The farmers of Angkor had to cope as best they could, but the elites had the capacity to move away to their upcountry estates, stay with related elite families elsewhere, or relocate with their rulers to other urban centers. Once the elite were gone, the shopkeepers and traders who were dependent on their trade would also have begun to leave, causing a further decrease in population and an irreversible cascade of decline.

With the water network ruptured, life amidst the wreckage must have become extremely difficult. A mega-monsoon, far larger than the great monsoons of the 19th and 20th centuries, surely scattered a great deal of debris across the landscape. The disrupted landscape would have become dangerous and unpredictable as further elements of the network failed episodically and erratically. Some areas might have retained working embankments and good yields, while adjacent failed embankments would have led to flooding. Natural water flow would have reasserted its path from NE to SW. Canals would have become clogged with debris and sediment, making them difficult to navigate. Flooding in 2000 that led to breached roadways would have brought poverty to the villages cut off from market places (Devine & Van Rouen 2001). Until the water flow of the Angkor plain returned to a new equilibrium, the damaged structures would have caused disruption and risk and prompted the population to move away. The eventual resurrection of natural water flow amidst the wreckage would have taken many years—how many we do not know. Only after about two hundred years, in the mid-sixteenth century, was there again a visible royal presence in Angkor. Work must have been done on the water system for Angkor Thom in the interim, however, since Portuguese visitors a few decades earlier referred to its moats and canals (Groslier 1958: 71, 77).

Implications

The climatic crisis in Angkor has implications on many levels: for our understanding of the local and regional history from the 13th to 16th centuries, for the capacity of massive infrastructure systems to withstand abrupt climate change affected by climatic impact, and for the relationship between climate instability and the capacity of societies to make adjustments to extreme changing circumstances. Given the paucity of primary texts that would tell us what happened in Angkor, we rely on the material evidence and palaeo-climatic proxy data to begin to assemble a case study for the Khmer state.

As mentioned earlier, a phenomenon we can now place in context is the Iconoclasm—the period in which the images of Buddha were excised by the thousands from great temple walls of the late twelfth and early thirteenth centuries. While the climate crisis does not resolve what this event was or why it occurred, it provides a circumstance of which we had no previous knowledge. The Iconoclasm can now be placed between the period of continual decline in monsoon rainfall from the 1220s to the 1260s and the arrival in 1295 of Zhou Daguan, who makes no mention of such a profound cultural event, even though he does refer to the politics of the ascension of Indravarman III in the same year (Pottier 1997a, 1997b, Harris 2007: 82). In a reappraisal of the Iconoclasm, it will be worth considering that, following the establishment of Mahayana Buddhism as the state religion in the late twelfth century and the death of Jayavarman VII at some time in the second decade of the thirteenth century, the monsoon—the weather condition critical to Angkor's food economy—became weaker decade by decade. Both the population and the elite had to confront how to cope and how to explain what was happening. A reaction against Mahayana Buddhism and a belief that other faiths might better respond to the cumulative drought would not have been surprising.

Another event that looks different in light of the documented climate crisis—and that requires further study—is the Ayutthayan incursion and period of occupation after 1431, which would have been in the midst of severe climatic instability. The severe monsoon damage to Angkor is dated to the second half of the fourteenth century, raising the possibility that the population and perhaps even the elite had already left a damaged city before the fifteenth century. By the 1430s the water management system of Angkor would no longer have been functioning, crop production would have been considerably reduced, and the urban area may have been substantially unusable. The climate crisis introduces the possibility that the attack on Angkor by the rulers of Ayutthaya, a possibly related royal house (Vickery 2004b) from the former Khmer imperial area of the Chao Praya basin, will need reappraisal. Still to be determined is when the elite left Angkor. If the elite had begun to leave in the second half of the fourteenth century because Angkor had become uninhabitable and taken the remaining wealth of the city with them, then little would have been left for Ayutthaya to sack. Nor does Ayutthaya mention very much. What needs to be identified is whether a corresponding increase in occupation occurred around Srei Santhor, Pursat, Lovek, Udong and Phnom Penh to the south a half-century before the Ayutthayan event? (See C. Jacques 1999, 2008, for the possibility of a separate royal house in the south before the 1430s.)

Additionally, if the move from Angkor to the Quatre Bras region can no longer be viewed solely as a response to a political-military crisis, but also or rather as a response to a climate crisis, then the imputed sense of defeat and retreat disappears. Instead, the narrative becomes an affirmation that the Khmer survived and that Angkor transmuted to a new settlement pattern, even as the Khmer strove to retain it while concurrently moving the center of the state to the southeast. Khmer society displayed flexibility and an ability to remake its world. The move to the southeast could be regarded as a transition from the restrictive, damaged, complex, and massively engineered rice production system of Angkor to an area favored by natural annual flooding. The southern floodplain would have survived even a major drought, since the Mekong River carries vastly more water than can be retained within its channel bed; the expansion of the Tonle Sap by about 7000 sq km in a normal monsoon is testament to that.

On a global scale, a comparison of what happened to the very different societies and economies of the great low-density agrarian urban settlements of the tropical world before the sixteenth century AD is salutary. What is disturbing is that despite distinct, specific regional differences in sociopolitical history, when low-density urbanism ceased to function in Mesoamerica in the ninth to twelfth centuries, Sri Lanka in the twelfth to thirteenth centuries, and in the Khmer world between the fourteenth and sixteenth centuries, the great urban heartlands were substantially depopulated and urbanism was reconstituted in small, compact settlements around the periphery of the former heartland. Severe climate instability, coinciding with dependence on substantial and inflexible infrastructure, tightly engaged with local staple crop production, led to an urban diaspora in all three cases. (Fletcher 2012: 309-10; Lucero et al. 2015).

Conclusions

Angkor was unlikely to have survived the numerous changes taking place between then fourteenth and the sixteenth centuries AD and would have been extremely vulnerable to the double impact of mega monsoons and severe droughts in the period of climatic instability. The complex and massive water management network was altered, hit, and then irretrievably damaged. Yet the Khmer strove to retain and retrieve Angkor with such determination that for a while in the mid-sixteenth century Angkor Thom had water and the rulers were again in residence. While old Angkor, the great, dispersed urban complex, did not recover, the south central area of the former capital began to transmute into the later cluster of Buddhist wats on the southern channel of the Siem Reap River from Angkor Wat to the lake. Meanwhile, sixteenth and seventeenth-century inscriptions in Angkor Wat attest to its transformation into a Buddhist pilgrimage site. Like Rome's move to Constantinople, the Khmer state's shift sideways from Angkor to the Phnom Penh region was a display of flexibility and versatility.

Rather than a collapse of the Khmer civilization, the demise of Angkor triggered the formation of a new urban network out towards the periphery of the old state, where smaller towns reconstituted a new mercantile world around royal power. Angkor had faced change on every scale, from the vast, slow processes of climate to the rapid decisions of local communities and individuals, without the capacity to make sustainable adjustments in situ. The massive and intractable water network and the urban structure of Greater Angkor fractured and could not adjust. The Khmer political system and the population did so quite readily, even if, most likely, with substantial hardship.

The degree to which climate change impacts a society is a function both of the intensity and variability of the process and of its relationship to the society's material infrastructure and social conditions. The specifics have to be understood. Climate alone does not determine the outcome (see, for example, Annamalei et al. 2007). Nor is the operation of a society simply a means of adjustment; what a community has been doing can rebound on it. As we have seen, old, massive infrastructure in particular contains serious risks, and a society's resilience and flexibility are critical. The case of Angkor, finally, raises questions about what we might begin to face in similar circumstances.

Acknowledgements

This research was funded by the US National Science Foundation Grants GEO 09-08971 and AGS 130–3976. The Greater Angkor Project at the University of Sydney is funded by grants from the Australian Research Council and National Geographic Committee for Research Exploration. The MAFKATA mission is funded by the Ecole française d'Extrême-Orient and the Archaeological Commission of the French Ministry of Foreign Affaires. The LiDAR survey was conducted for the eight institutions of the Khmer Archaeology LiDAR Consortium (KALC) and partially funded by the Simone and Cino Del Duca Foundation and the National Geographic Committee for Research Exploration. The authors would like to thank Damian Evans, Chhay Rachna, Wayne Johnson, and the many members of the University of Sydney's Greater Angkor Project team, our international collaborators at the Ecole française d'Extrême-Orient, the APSARA National Authority of Cambodia, Professor Miriam Stark of the University of Hawaii, and Professor Rob Gillies of the Climate Center at Utah State University in Logan, Utah. Thanks are also due to the Robert Christie Research Centre and to the Faculty of Arts and Social Sciences, both of the University of Sydney.

References

Ashok, K., T. Yamagata, and Z. Guan. 2001. Impact of the Indian Ocean dipole on the relationship between the Indian monsoon rainfall and ENSO. *Geo. Res. Lett.* 28: 4499–4502.

Bamzai, A. S., and J. Shukla. 1999. Relation between Eurasian snow cover, snow depth, and the Indian summer monsoon: an observational study. *J Climate* 12: 3117–3132.

Brotherson, D., R. Chhay, and P. Suy. 2012. Greater Angkor Project: Preliminary Report. Phnom Krom–Tuol Kok Pey: June 2012.

Buckley, B. M., K. J. Anchukaitis, D. Penny, R. Fletcher, E. R. Cook, M. Sano, C. Nam le, A. Wichienkeeo, T. M. Ton, and M. H. Truong. 2010. Climate as a contributing factor in the demise of Angkor, Cambodia. *Proc. Nat. Acad. Sci.* 107: 6748–6752.

Buckley, B. M., R. Fletcher, S.-Y. Wang, B. Zottoli, and C. Pottier. 2014. Monsoon extremes and society over the past millennium on mainland Southeast Asia. *Quaternary Science Reviews* 95: 1–19. doi 10.1016/j.quascirev.2014.04.022.

Buckley, B.M., K. Palakit, K. Duangsathaporn, P. Sanguantham, and P. Prasomsin. 2007. Decadal scale droughts over northwestern Thailand over the past 450 years and links to the tropical Pacific. *Climate Dynamics* 29: 63–71. doi 10.1007/s00382-007-0225-1.

Chhay, R., R. Fletcher, W. Johnson, T. Lustig, L. Benbow, C. Khieu, A. Wilson, T. Heng, C. Pottier, B. Chourn, P. So, S. Mackay, and A. Biggs. 2010. Greater Angkor Project: Preliminary Report. Bam Penh Reach: 2005–09.

Chen,T.-C., and J.-h. Yoon. 2000. Interannual Variation in Indochina summer monsoon rainfall: possible mechanism. *J Climate* 13: 1979–1986.

Chen, T.-C., S.-Y. Wang, M.-C. Yen, and W. A. Gallus. 2004. Role of the monsoon gyre in the interannual variation of tropical cyclone formation over the western North Pacific. *Wea. Forecasting* 19: 776–785.

Chevance, J. B. 2011. Le phnom Kulen à la source d'Angkor, nouvelles données archéologiques. Ph.D thesis. Paris: Université Sorbonne Nouvelle.

Clift, P. D., and R. A. Plumb. 2008. *The Asian monsoon: causes, history and effects.* Cambridge: Cambridge University Press.

Coe, M. E. 2003. Angkor and the Khmer Civilization. New York: Thames and Hudson.

Cœdès, G. 1906. La stèle de Ta-Prohm. *Bull. de l'Ecole française d'Extrême-Orient* 6: 44–86.

Cœdès, G. 1913. Études cambodgiennes VIII. La fondation de Phnom Peñ au XVe siècle d'après la chronique cambodgienne. *Bull. de L'École Française d'Extrême-Orient.* 13(6): 6–11.

Cœdès, G. 1941. Études cambodgiennes IV. La stèle du Práh Khằn d'Ankor, *Bull. de l'Ecole française d'Extrême-Orient* 41: 255–302.

Cœdès, G. 1968. *The Indianized States of South-East Asia.* . ed. W. F. Vella. trans. S. B. Cowing. Honolulu: East-West Center Press.

Cook, B. I., and B. M. Buckley. 2009. Objective determination of monsoon season onset, withdrawal and length. *JGR-Atmospheres* 114: D23109. doi:10.1029/2009JD012795.

Cook, E. R., C. A. Woodhouse, C. M. Eakin, D. M. Meko, D. W. Stahle. 2004. Long-term aridity changes in the western United States. *Science* 306(5968): 1015–1018. doi: 10.1126/science.1102586.

Cook, E. R., K. J. Anchukaitis, B. M. Buckley, R. D. D'Arrigo, W. E. Wright, and G. C. Jacoby. 2010. Asian monsoon failure and megadrought during the last millennium. *Science* 328(5977): 486–489. doi: 10.1126/science.1185188.

Cremin, A. 2006. Chinese ceramics at Angkor. *IPPA Bulletin* 26:121–123 and online at www.ippa...

Dagens, B. 2003. *Les Khmers.* Paris: Belles lettres.

Davis, M. 2001. *Late Victorian Holocausts: El Niño Famines and the Making of the Third World.* Brooklyn, NY: Verso Books.

Desbat, A. 2011. Pour une révision de la chronologie des grès Khmers. *Aséanie* 27: 11–34.

Devine, A. and V. Rouen. 2001. Hunger Gap. *The Cambodian Daily Weekend* 152: 6–7.

Ding, Q., and B. Wang. 2007. Intraseasonal teleconnection between the summer Eurasian wave train and the Indian monsoon. *J. Climate* 20: 3751–3767.

Evans, D. 2007. "Putting Angkor on the Map: A New Survey of a 'Khmer Hydraulic City,' in Historical and Theoretical Context. " PhD diss., University of Sydney.

Evans, D., R. Fletcher, C. Pottier, J.-B. Chevance, D. Soutif, B. S. Tan, S. Im, D. Ea, T. Tin, S. Kim, C. Cromarty, et al. 2013. Uncovering archaeological landscapes at Angkor using lidar. *Proc Natl Acad Sci* 110: 12595–12600.

Evans, D., C. Pottier, R. Fletcher, S. Hensley, I. Tapley, A. Milne, and M. Barbetti. 2007. A comprehensive archaeological map of the world's largest pre-industrial settlement complex at Angkor, Cambodia. *Proc Natl Acad Sci*, 104: 14277–14282.

Fagan, B. 2009. *The Great Warming: Climate Change and the Rise and Fall of Civilisations.* London: Bloomsbury Press.

Fleitmann, D., S. J. Burns, A. Mangini, M. Mudelsee, J. Kramers, I. Villa, U Neff, A. A. Al-Subbarye, A. Buettner, D. Hippler, and A. Matter. 2007. Holocene ITCZ and Indian monsoon dynamics recorded in stalagmites from Oman and Yemen (Socotra). *Quaternary Science Reviews* 26(2007): 170–188.

Fletcher, R. J., M. Barbetti, D. Evans, H. Than, I Sorithy, K. Chan, D. Penny, C. Pottier, and T. Somaneath. 2003. Redefining Angkor: structure and environment in the largest, low-density urban complex of the pre-industrial world. *UDAYA* 4: 107–125.

Fletcher, R., and D. Evans. 2012. The dynamics of Angkor and its landscape. In *Old Myths and New Approaches: Interpreting Ancient Religious Sites in Southeast Asia,* ed. A. Haendel. Melbourne: Monash University Publishing, 42–62.

Fletcher, R., D. Penny, D. Evans, C. Pottier, M. Barbetti, M. Kummu, T. Lustig, and APSARA. 2008a. The water management network of Angkor, Cambodia. *Antiquity* 82: 658–670.

Fletcher, R., C. Pottier, D. Evans, and M. Kummu. 2008b. The development of the water management system of Angkor: a provisional model. *IPPA Bulletin* 28: 57–66.

Flood, E. T., and C. Flood. 1978. *The dynastic chronicles, Bangkok era, the first reign.* Translation of Thiphakonwongmahakosathibodi Chaophraya. Tokyo: Center for East Asian Cultural Studies.

Folland, C. K., J. A. Renwick, M. J. Salinger, and A. B. Mullan, 2002: Relative influences of the Interdecadal Pacific Oscillation and ENSO on the South Pacific Convergence Zone. *Geophys. Res. Lett.* 29: 1643.

Fosu, B. O., and S.-Y. Wang. 2014. Bay of Bengal: Coupling of pre-monsoon tropical cyclones with the monsoon onset in Myanmar. *Climate Dynamics.* doi:10.1007/s00382-014-2289-z.

Giteau, M. 1975. *Iconograpie du Cambodge post-angkorien.* Paris: EFEO.

Goswami, B. N., M. Madhusoodanan, C. Neema, and D. Sengupta, 2006: A physical mechanism for North Atlantic SST influence on the Indian summer monsoon. *Geophy. Res. Lett.* 33: L02706. doi 10.1029/2005GL024803.

Groslier, B. P. 1958. Angkor et le Cambodge au XVIème siècle d'après les sources portugaises et espagnoles. Paris: Annales du Musée Guimet, PUF.

Groslier, B. P. 1974. Agriculture et religion dans l'empire angkorien. *Études Rurales,* 53–56: 95–117.

Groslier, B. P. 1979. La cité hydraulique angkorienne: exploitation ou surexploitation du sol?. *Bull. de l'Ecole française d'Extrême-Orient* 66: 161–202.

Groslier, B.P. 2006. *Angkor and Cambodia in the Sixteenth Century.* Trans. M. Smithies. Bangkok: Orchid Press.

Hahn, D. G. and J. Shukla. 1976. An apparent relationship between Eurasian snow cover and Indian monsoon rainfall. *Journal of the Atmospheric Sciences* 33: 2461–2462.

Harris, P. 2007. *A Record of Cambodia: The Land and the People.* Thailand: Silkworm Books.

Harris, I. 2008. *Buddhism in Cambodia.* Honolulu: University of Hawaii Press.

Hawken, S. 2011. "Metropolis of Ricefields: A Topographic Classification of a Dispersed Urban Complex." PhD dissertation, Department of Archaeology, University of Sydney, Australia.

Hendrickson, M. 2010. Historic routes to Angkor: development of the Khmer Road system (9th to 13th centuries CE) in mainland Southeast Asia. *Antiquity* 84(324): 480–496.

Jacq-Hergoualc'h. M. 2004. *Le Siam.* Paris: Les Belles Lettres.

Jacques, C. 1999. Les derniers siècles d'Angkor. Académie des Inscriptions & Belles-Lettres. *Comptes-Rendus des Séances.* Jan.-Mar.: 367–390.

Jacques, C. 2007. Angkor: Residences des Dieux. Geneva: EditionOlizane.

Jacques, C. 2008. L'histoire enfouie: Angkor et le Cambodge du XIII au XIX siècle. In *Angkor VIIIe-XXIe Siècle, Mémoire et Identité Khmères,* dir. H. Tertrais. Paris: Autrement Ed., 73–91.

Kahan, A. 1985. *The Plow, the Hammer, and the Knout: an Economic History of Eighteenth-Century Russia*. University of Chicago Press.

Krishnamurthy, V., and B. N. Goswami. 2000. Indian monsoon-ENSO relationship on interdecadal timescale. *J. Clim.* 13: 579–595.

Kumar, K. K., B. Rajagopalan, and M. A. Cane. 1999. On the weakening relationship between the Indian monsoon and ENSO. *Science* 284: 2156–2159.

Kummu, M. 2008. Water management in Angkor: Human impacts on hydrology and sediment transportation. *Journal of Environmental Management* 90: 1413–1421. doi:10.1016/j.jenvman.2008.08.007.

Kutzbach, J. E. 1981. Monsoon climate of the early Holocene: climate experiment with the Earth's orbital parameters for 9000 years ago. *Science* 214(4516): 59–61.

Lamb, H. H. 1965. The early medieval warm epoch and its sequel. *Palaeogeography, Palaeoclimatology, Palaeoecology* 1: 3. doi: 10.1016/0031-0182(65)90004-0

Lieberman, V., 2003. *Strange Parallels: Southeast Asia in Global Context, c. 800-1830*. Cambridge: Cambridge University Press.

Lieberman, V., and B. M. Buckley. 2012. The impact of climate on Southeast Asia c. 950–1820: new findings. *Modern Asian Studies* 1–48. doi: 10.1017/S0026749X12000091.

Liu, K., and J. C. Chan. 2003. Climatological characteristics and seasonal forecasting of tropical cyclones making landfall along the South China coast. *Mon. Wea. Rev.* 131: 1650–1662.

Lombard, D. 1970. Pour une histoire des villes du Sud-Est asiatique. *Annales* 25: 842–856.

Lucero, L. J., R. Fletcher, and R. Coningham. 2015. From "collapse" to urban diaspora: the transformation of low-density, dispersed agrarian urbanism. *Antiquity* 89 (347): 1139–1154.

Lustig, E. 2001. "Water and the Transformation of Power at Angkor, 10th to 13th Centuries A.D." Unpublished BA (Hons) thesis. University of Sydney, Department of Archaeology.

Lustig, T., R. Fletcher, M. Kummu, C. Pottier, and D. Penny. 2008. Did traditional cultures live in harmony with nature? In *Modern Myths of the Mekong*, ed. M. Kummu, M. Keskinen, and O. Varis. Helsinki: University of Technology, 81–94.

Maher, B. A. 2008. Holocene variability of the East Asian summer monsoon from Chinese cave records: a re-assessment. *The Holocene* 18(6): 861–66.

Meehl, G. A., and A. Hu. 2006. Megadroughts in the Indian monsoon region and southwest North America and a mechanism for associated multidecadal Pacific sea surface temperature anomalies. *J. Climate* 19: 1605–1623.

Molnar, P. 2005. Mio-Pliocene growth of the Tibetan Plateau and evolution of East Asian climate. *Palaeontol. Electron.* 8(1): A–23.

Neelin, J. D. 2007. Moist dynamics of tropical convection zones in monsoons, teleconnections and global warming. In *The Global Circulation of the Atmosphere*, ed. T. Schneider and A. Sobel. Princeton, NJ: Princeton University Press, 267–301.

PAGES/Ocean2k Working Group. 2012. Synthesis of marine sediment-derived SST records for the past 2 millennia: First-order results from the PAGES/Ocean2k project. AGU Fall Meeting, abstr. PP11F-07.

Parmentier, H. 1933. Examen du nivellement d'Ankor. *Bull. de l'Ecole française d'Extrême-Orient* 33: 310–318.

Pelliot, P. 1951. Mémoires sur les coutumes du Cambodge de Tcheou Ta-kouan. Version nouvelle suivie d'un commentaire inachevé. Paris: Adrien-Maisonneuve.

Penny, D., C. Pottier, M. Kummu, R. Fletcher, U. Zoppi, M. Barbetti, and S. Tous. 2005. Hydrological history of the West Baray, Angkor, revealed through palynological analysis of sediments from the West Mebon. *Bull. de l'École Française d'Extrême-Orient* 92: 497–521.

Penny, D., J.-B. Chevance, D. Tang, and S. De Greef. 2014. The environmental impact of Cambodia's ancient city of Mahendraparvata (Phnom Kulen). *PLoS ONE* 9(1): e84252. doi:10.1371/journal.pone.0084252.

Polkinghorne, M., C. Pottier, and C. Fisher. 2013. One Buddha can hide another. *Journal Asiatique* 301(2): 575–624. doi: 10.2143/JA.301.2.3001711.

Pottier, C. 1996. Bakong et son implantation. *Bull. de l'École Française d'Extrême-Orient* 83: 318–326.

Pottier, C. 1997a. Embryons et tortues: les dépôts de fondation découverts au perron nord de la terrasse des Éléphants. *Bull. de l'École Française d'Extrême-Orient* 84: 402–407.

Pottier, C. 1997b. La restauration du perron nord de la terrasse des Éléphants à Angkor Thom. Rapport sur la première année de travaux (avril 1996-avril 1997). *Bull. de l'École Française d'Extrême-Orient* 84: 376–401.

Pottier, C. 1998. Elaboration d'une carte archéologique de la région d'Angkor, *Southeast Asian Archaeology, International Conference of the European Association of Southeast Asian Archaeologists* 1994, Vol. I, Leiden: 179–194.

Pottier, C. 2000. A la recherche de Goloupura, *Bull. de l'École Française d'Extrême-Orient*, 87: 79–107.

Pottier, C. 2001. Some evidence of an inter-relationship between hydraulic features and rice field patterns at Angkor during ancient times. *The Journal of Sophia Asian Studies* 18: 99–119.

Reid, A. 1993. *Southeast Asia in the Age of Commerce 1450–1680, Volume Two: Expansion and Crisis*. Bangkok: Silkworm Books.

Sachs, J. P., D. E. Sachs, R. H. Smittenberg, Z. Zhang, D. S. Battisti, and S. Golubic. 2009. Southward movement of the Pacific intertropical convergence zone AD 1400–1850. *Nature Geoscience* 2: 519–525.

Saha, S. S. Moorthi, H. L. Pan, X. Wu, J. Wang, S. Nadiga, et al. 2010. The NCEP climate forecast system reanalysis. *Bulletin of the American Meteorological Society* 91: 1015–1057.

Sahai, S. 1970. Les institutions politiques et l'organisation administrative du Cambodge ancient. Paris: EEFEO.

Sano, M., B. M. Buckley, and T. Sweda. 2009. Tree-ring based hydroclimate reconstruction over northen Vietnam from Fokienia hodginsii: eighteenth century mega-drought and tropical Pacific influence. *Climate dynamics* 33: 331.

Santoni, M. 2008. La mission archéologique française et le Vat Phu: recherches sur un site historique exceptionnel du Laos. In *Recherches nouvelles sur le Laos*, ed. Y. Goudineau and M. Lorillard. Paris: EFEO, 81–111.

Sano, M., B. M. Buckley, and T. Sweda. 2009. Tree-ring based hydroclimate recon-struction over northern Vietnam from *Fokienia hodginsii*: eighteenth century mega-drought and tropical Pacific influence. *Climate Dynamics* 33: 331–340.

Sedov, L. A. 1967. Angkorskaiia Imperiia (The Angkor Empire). Moscow: Izdatel'sto "Nauka."

Sinha, A., L. Stott, M. Berkelhammer, H Cheng, R. L. Edwards, B. M. Buckley, M. Aldenderfer, and M. Mudelsee. 2010. A global context for megadroughts in monsoon Asia during the past millennium. *Quaternary Science Reviews* 30: 47-62. doi:10.1016/j.quascirev.2010.10.005

Sodhi, N. S., L. P. Koh, B. W. Brook, and P. K. L. Ng. 2004. Southeast Asian biodiversity: an impending disaster. *TRENDS in Ecology and Evolution.* 19(12): 654–60. doi:10.1016/j.tree.2004.09.006

Sonnemann, T. 2011. "Angkor Underground: Applying GPR to Analyse the Diachronic Structure of a Great Urban Complex." PhD diss. University of Sydney.

Stark, M. 2006. Early mainland Southeast Asian landscapes in the first millennium A.D. *Annu. Rev. Anthropol* 35: 407–32.

Tang, A. 2011. "Water Management on the Kulen Plateau 5th–12th century CE." Honors thesis. University of Sydney.

Thera, R. P. 1967. *Jinakalamali Prakorn*. Trans. into Thai by S. M. Mitra. Bangkok: Nara Publishing.

Thompson, A. 2004. Pilgrims to Angkor: A Buddhist "Cosmopolis" in Southeast Asia? In *Indochina: Trends in Development*, ed. N. Bektimirova and N. V. Dolnikova. Moscow: Institute of Asian-African Studies.

Thung, H. L. 2002. Angkor revisited: lessons to learn? *SPAFA Journal* 12(2): 5–17.

Trouvé, G. 1933. Chaussées et canaux autour d'Angkor Thom. *Bull. de l'École française d'Extrême-Orient* 33(2): 1120–1128.

Tsay, J.-D. 2004. "Water Vapor Budget of Cold Surge Vortices." MS, Iowa State University, 120 pp.

Van Liere, W. J. 1980. Traditional water management in the lower Mekong Basin. *World Archaeology* 11(3): 265–280.

Vickery, M. 1977. "Cambodia After Angkor, the Chronicular Evidence for the Fourteenth to Sixteenth Centuries. PhD diss. New Haven, CT: Yale University.

Vickery, M. 1985. The reign of Sūryavarman I and royal factionalism at Angkor." *Journal of Southeast Asian Studies* 16(2): 226–244.

Vickery, M. 2004a. "A Beginner's Guide to Early Cambodian History" (Originally prepared at the Royal University of Fine Arts; copied and distributed.)

Vickery, M. 2004b. "Cambodia and Its Neighbors in the 15th Century." Asia Research Institute Working Paper, Series No. 27.

Wang, B. 2006. *The Asian Monsoon*. Chichester: Praxis Publishing Limited.

Wang, B., and LinHo. 2002. Rainy season of the Asian-Pacific summer monsoon. *J. Climate* 15: 386–398.

Wang, S.-Y., B. M. Buckley, J.-H. Yoon, and B. Fosu. 2013. Intensification of pre-monsoon tropical cyclones in the Bay of Bengal and its impacts on Myanmar. *Journal of Geophysical Research* 118: 1–12.

Wang, Y. J., H. Cheng, R. L. Edwards, Y. Q, He, X. G. Kong, Z. S. An, J. Y. Wu, M. J. Kelly, C. A. Dykoski, and X. D. Li. 2005. The Holocene Asian monsoon: links to solar changes and North Atlantic climate. *Science* 308: 854–857.

Wyatt, D. K., and A. Wichienkeeo. 1998. *The Chiang Mai Chronicle*. Chiang Mai, Thailand: Silkworm, 1998.

Yancheva, G., N. R. Nowaczyk, J. Mingram, P. Dulski, G. Schettler, J. F. W. Negendank, J. Liu, D. M. Sigman, L. C. Peterson, and G. H. Haug. 2007. Influence of the intertropical convergence zone on the East Asian monsoon. *Nature* 445: 74–77. doi:10.1038/nature05431.

Yoon, J.-H., and T. C. Chen. 2005. Water vapor budget of the Indian monsoon depression. *Tellus A* 57: 770–782.

Yoon, J.-H., and W.-R. Huang, 2012. 02 Indian monsoon depression: climatology and variability. In *Modern Climatology*, ed. S.-Y. Wang, and R. R. Gillies. Rijeka, Croatia: InTech.

Zhang, Z., J. C. L. Chan, and Y. Ding. 2004. Characteristics, evolution and mechanisms of the monsoon onset over Southeast Asia. *Int. J. Clim.* 24: 1461–1482.

Zhang, P., H. Cheng, R. L. Edwards, F. Chen, Y. Wang, X. Yang, J. Liu, M. Tan, X. Wang,and J. Liu. 2008. A test of climate, sun, and culture relationships from an 1810-year Chinese cave record. *Science* 322: 940–942.

Zhang, R.-H., and S. Levitus. 1997. Structure and cycle of decadal variability of upper-ocean temperature in the North Pacific. *J. Climate* 10: 710–727.

Zhang, Y., J. M. Wallace, and D. S. Battisti. 1997. ENSO-like interdecadal variability: 1900–93. *J. Climate* 10: 1004–1020.

INDEX

Note: Page references followed by a "*t*" indicate table; "*f*" indicate figure.

dendroclimatology, 18, 248
Diamond, Jared, 4
Digital Elevation Model (DEM), 185*f*
domestication syndrome, 34, 44
Dorians, 164
Douglas fir, 250, 251
    seasonal moisture response, 257
Douglass, A. E., 248
drought. *See also* megadroughts
    categories of, 216–18
    persistent, over Indochina, 290–92
drought-induced food shortages and
        famine, 166–67
dry-farming agriculture, 96, 97, 100,
        110–11, 248
    adaptation by, 112
    Akkadian collapse and, 117
    4.2 ka BP megadrought and, 105
    Oaxaca Valley piedmont and, 197
dual-season drought, 265, 269
dynamic starch analysis, 44
Džuljunica, 83–84, 83*f*

Early Bronze Age III Anatolian, 13
Early Bronze Age IV Levantine, 13
Early Classic Maya, 16
Early Dynastic period, 9
Early Intermediate Period, 236
    on *altiplano*, 237
Early Iron Age, 14, 162
early irrigation agriculture, 94, 96
Early Mushabian, 51
Early Natufian culture, 51, 52, 52*f*, 53
Early Teotihuacan dry period, 195
East African monsoon, 169
East Asian monsoon, instabilities
        in, 102
East Baray reservoir, 277, 278*f*, 278, 283,
        296, 299*f*
    erosion damage, 298–300
    exit channel blockages, 296–97
    water circulation evolution, 298*f*
Eastern Mediterranean
    climatic setting, 169–70
    collapses in, 162, 164–66
East Indian Drought of 1790-1796, 295
Ebla, 97, 111, 173

ecological systems, stress factor
        responses of, 5–6
Ecuador, 234
Egypt
    bas-reliefs, 14
    First Intermediate Period in, 13
    Old Kingdom collapse, 6, 13, 100, 162
    Sea People invasions and, 164–66
8.2 ka cal BP climate event, 11–12,
        73, 74, 76
    Mesopotamia impact of, 94
    neolithization and, 94
Ein Gedi shore, 173
el-Kowm, 49
El Malpais National Monument,
        19, 265–67
El Niño, 4, 13
El Niño-Southern Oscillation (ENSO), 1,
        233, 234
    Indochina climate impact of, 290–92
    monsoon changes from, 289
    tropical cyclones and, 289
    variability in, 200
el-Wad Terrace, 53
Emar year-names, 166
ENSO. *See* El Niño-Southern Oscillation
environmental determinism, 116
environmental stressors, 7
Epiclassic megadrought, 196–97, 200
Epi-Paleolithic era, 49, 54
    hunting in, 44
    Levant entities in, 47, 48*f*
    social entities in, 46
Eski Açigöl core, 100, 103, 175
es-Sifiya, 77
Euphrates River, 93–94, 118
    flow reduction in, 105, 111–112
    4.2 ka BP megadrought and, 112,
        114, 116
    Late Bronze Age food shortages
        along, 166
    as part of Mesopotamia, 93
    Younger Dryas migrations to,
        10, 56, 57
Europe, spread of farming to, 11–12
evapotranspiration, 217, 256
Eynan (Ain Mallaha) site, 53

lake bottom sediments, 2
Lake Chichancanab, 209, 212, 215f
Lake Kinneret, 48, 49
Lake Lisan, 36, 41, 42
  levels of, 42f, 43, 47
Lake Ojibway, 69, 73
Lake Punta Laguna, 212
lake sediment cores, paleoclimate data
    from, 186
Lake Tiberias, 14
Lake Titicaca, 4, 17–18, 232, 233f, 234,
    236, 243
  levels of, 236f
  pampas regions around, 242
  sediment records, 240
Lake Van, 10, 175
Lake Van core, 103
Lake Yammoûneh, 38, 41
Lake Yoa, 100
Languedoc Late Neolithic, 13
Lanna, 292
Late Bronze Age, Dead Sea in, 173
Late Bronze Age collapse, 14–15, 162,
    164–65, 170
  radiocarbon chronology for, 166,
    167, 168f
Late Bronze Age megadroughts,
    spatio-temporal patterns, 162
Late Cypriot IIC ceramics, 169
Late Glacial Maximum (LGM), 9, 34, 36
  Lake Lisan levels in, 47
  Levant climate proxies, 38, 41
Late Helladic IIB vessels, 169
Late Helladic IIIB-IIIC vessels, 169
Late Horizon, 236
Late Intermediate Period, 18
  droughts in, 240
  South American cultures in, 236
Late Mushabian, 51
Late Natufian, 51, 52, 53–54
  distribution of, 55f
Late Neolithic, 85
Late Pleistocene, cultivation
    experiments in, 44
Late Postclassic Period Maya, 207
Late Pre-Pottery Neolithic B
    (LPPNB), 76

Late Uruk period, 94
  megadrought in, 96
Late Yarmoukian Crisis, 77
Lattimore, Owen, 5
LC21 core, 72, 172
Leilan Acropolis, 108, 115f
Leilan IIc Akkadian imperial collapse, 13
Leilan IIc period, 110
Leilan radiocarbon chronology, 99
Leilan Region Survey, 110, 111f, 117
Levant, 34
  archaeological record from LGM to
    mid-Holocene, 45
  climate and paleoclimatic
    sources, 34–38
  climate-change models for, 38, 172
  climate proxies for, 38, 41–44, 169–175
  climatic and archaeological record
    interpretation for, 15, 46–55
  drought events in, 11, 14, 15
  dry farming in, 110–111
  8.6 ka cal BP event in, 76
  4.2 ka BP abandonments/riverine
    refugia in, 112
  origins of agriculture in, 9, 34, 54, 57
  current phytogeographic belts, 35f
  dry-farming in, 110–11
  Late Paleolithic/Early Epi-Paleolithic
    entities, 48, 48f
  Late Pleistocene foragers in, 44, 53–54
  mobility increases in, 78
  Neolithic distribution of farming/
    foraging, 56f
  regional climate variation in, 36, 106
  Sea Peoples in, 165–166, 167
  3.2 ka BC event in, 167, 173
  Yarmoukian Rubble Layers, 76–77
  Younger Dryas event in, 10
LGM. *See* Late Glacial Maximum
LIA. *See* Little Ice Age
Libya, 165
LiDAR (Light Detection and Ranging),
    20, 295, 299
Lieberman, Victor, 294–95
Little Ice Age (LIA), 9, 70, 73, 284
  QIC record and, 232
Long Count calendar system, 207

Pannonian Basin, 84
PDSI. *See* Palmer Drought
    Severity Index
Peleshet, 14
period IIId, 112
permanent relocation, 8
Peru, 234
    Inca Empire in, 236
Peru Current, 234
Peten region, 209
Phnom Krom, 280
Phnom Penh, 282, 305, 306
phytolith analysis, 44
piquant trièdre techniques, 51
Pisidia, 74, 80–82
planktonic foraminifera proxy records
    3.2 ka BP and, 172
    8.2 ka cal BP and, 83*f*
    oxygen isotope measurement
        and, 104*f*
    Younger Dryas and, 37
Pleistocene era, paleoclimate proxies for,
    2–3, 9, 34, 70, 175
    Heinrich 1 Stadial (H1), 41–42, 57, 194
    Heinrich Event 1, 39*f*, 44, 52
    Heinrich Event 2, 44
    Late Glacial Maximum, 9, 38–41, 44,
        47, 175
    Post-Late Glacial Maximum, 41–42
    Terminal, 57
    wild animal species
        manipulation in, 44
pollen cores, 38. *See also specific cores*
    Dead Sea reconstructions and, 172
pollen records
    arboreal, 43
    for coastal Syria, 170
    marine, 41, 42
ponderosa pine, 251
Popocatéptl eruption, 195
post-Late Glacial Maximum, Levant
    climate proxies in, 41–42
potassium, in GISP2 records, 70
Pottery Neolithic, 76, 76*f*, 77, 78*f*, 85
PPN. *See* Pre-Pottery Neolithic
Preah Khan, 277, 280, 282
Pre-Pottery Neolithic (PPN), 85

PRISM model, 252, 256, 258, 263
Proto-Halaf culture, 78
proto weeds, 44
Puduhepa (Queen), 166
Puebla Valley, monumental architecture
    in, 195
Pueblo Farming Project, 251
Pueblo societies, 248. *See also* Ancestral
    Pueblo peoples
Puuc Hills, 209
Puuc Maya, 207, 221, 223

Qalkan, 47
QIC. *See* Quelccaya ice cap
QND. *See* North Dome QIC core
QSD. *See* Summit Dome QIC core
quartz spikes, 103
Quatre Bras, 282, 305
Quelccaya ice cap (QIC), 232, 233*f*,
    234, 243
    climate record from, 4, 18, 234, 235*f*,
        238*t*, 237, 239*f*
    radiocarbon dating, 241*f*
Quercus calliprinos, 43
Questioning Collapse (McAnany and
    Yoffee), 5
Qunf Cave, 100

radiocarbon paleoclimate proxy dating
    Angkor demise, 14th–16th C. AD, and,
        281, 300
    Classic Maya Collapse and, 16–17,
        212, 215
    first millennium Holocene
        predomestication and, 34, 38
    4.2 ka BP megadrought and, 13, 99,
        101–5, 108, 110, 115, 117
    lake and ocean sediment in, 2, 7
    Late Bronze Age aridification
        event and, 14–15, 165–166,
        167–69, 172–76
    Neolithic dispersal (migrations) and,
        73, 85, 94
    Teotihuacan, AD 550–600 collapse
        and, 186–189, 196, 200
    Tiwanaku state decline and, 242
    rainfall reconstruction, 175, 191, 194*f*